Climate Change and Carbon Markets: Late or Too Late?

John Ure

Climate Change and Carbon Markets: Late or Too Late?

Springer

John Ure
Hong Kong, Special Administrative
Region, China

ISBN 978-3-032-07618-2 ISBN 978-3-032-07619-9 (eBook)
https://doi.org/10.1007/978-3-032-07619-9

This Springer imprint is published by the registered company Springer Nature Switzerland AG
The registered company address is: Gewerbestrasse 11, 6330 Cham, Switzerland

Foreword

In part this book grows naturally out of *Achieving the UN Sustainable Development Goals: Late or Too Late?* (2024),[1] chapter 3 *Climate Change and the Environment*. However, its motivation is the ever-growing need to convey, in non-technical language as far as possible, the nature and causes of climate change, together with the levels of risk and uncertainty associated with rising average global temperatures. Then to look at Ways Forward. The urgency is clear: global average temperatures since 2023, on a month-on-month basis, have risen consistently to more than 1.5C over pre-industrial levels.

Underpinning the book is a restatement and defence of science and of rational thinking in an age of scepticism, fossil-fuelled all too often by misinformation spread across social media. Yet social media also provides access to some really excellent podcasts and postings which are referenced frequently throughout the book, along with science-based research papers.

The issue here is that science is not about absolute 'proofs' but about probabilities, about degrees of uncertainty, and the consequential levels of risks associated with different outcomes. A high existential risk but with low probability demands at least as much attention as a low level of risk but with high probability. The book does its best to evaluate the work of climatologists, the prescriptions of various bodies for climate change mitigation, the yawning gaps in policies and financial resources required to fund mitigation and adaptation strategies, especially for poorer nations, and finally, to offer ideas for Ways Forward. Over recent years many others, from different angles such as science, policy-making and civil action, have published in books, research papers and the media, their contributions to understanding and addressing the challenges of climate change. This book tries to tie together many of these perspectives in order to provide a more comprehensive foundation for anyone who takes climate change and the associated threats to biodiversity in the natural environment seriously, in acknowledgement that mitigation is late, or is it too late?

---

[1] John Ure (October 2024) Achieving the UN Sustainable Development Goals: Late or Too Late? *Emerald Publishing* https://www.emerald.com/insight/publication/doi/10.1108/9781835494042.

Greta Thunberg

No account of climate change and how to reverse it by net negative GHG emissions would be complete without acknowledging the bold and thoughtful interventions of Greta Thunberg. If only slightly, she has helped move the dial of public debate, which is no mean achievement. Her book *The Climate Book* (2023)[2] is a series of short contributions which offer a good account of what she and her collaborators stand for. But notwithstanding inputs by esteemed economists Nicolas Stern, Lucas Chancel and Thomas Pikerty, there is little said about much needed changes in economic policies towards fighting climate change other than a role for carbon taxes.

This book argues that a shift (a transformation) from neo-liberalism to more government interventions supported by global collaboration that specifically focuses upon GHG-reduction projects in EMDEs (Emerging Markets and Developing Economies) is a necessary and indispensable approach.

As an economist who has spent most of his professional life in academia, and in consulting for UN development agencies and governments as well as in the private sector, upon semi-retirement I decided to devote myself to doing something more important than preaching to the converted. Rather, I wanted to reach some conclusions, provisionally as they may be, as to Ways Forward. I choose to start with the UN Sustainable Development Goals (SDGs) because by definition they cover a broad landscape of social needs. I always include within those needs the balance humans seek to find in their role with the natural world, as part of it rather than as a one-sided obsession with being masters of it. That naturally led me onto the existential issue of our time, climate change. Reaching even provisional conclusions as to Ways Forward required a better understanding of both the problems and the proposed solutions. Extensive reading therefore became a journey of self-education, and as anyone who has travelled a similar journey will be aware, it is endless. Endless sources of enlightenment, endless excitement in learning, endless 'wokeness' in understanding the self-defeating shallow nonsense of so much of human activity. The shallowness is the depressing part, the wokeness is the rewarding exhilarating part because it goes someway to answering the question: why?

Climate Change and Human History

The history of humanity has been a combination of migrations and settlements influenced heavily, and in some cases absolutely determined, by climate change. From ice ages to periods of lush fertility, humans have encountered and reacted to a multitude of environments. One of the most dramatic in 'recent' history is how around 5000

[2] Greta Thunberg (2023) *The Climate Book* Penguin Press, New York.

years ago climate change created civilization in the form of the very first city on Earth, Uruk in southern Mesopotamia in what is modern day Iraq. What is known as the Holocene Climatic Optimum,[3] was an epoch of warmer and wetter climate during which what is today known as the Persian Gulf extended northwards, creating wetlands for fishing and seabird life, alongside a fertile plain between the mighty Euphrates and Tigris rivers. Harnessing these resources resulted in concentrating the growing population into the world's first ever built-up urban area, Uruk. This is lucidly outlined in the podcast, *The Rest is History*.[4] These events took place before the earliest form of writing, cuneiform, so it is pre-history, yet it left a record of itself through ideograms. Without climate change at that time, the rise of a human civilization would have been delayed; yet it was also climate change that subsequently ended the days of Uruk.[5] Sustainability is never guaranteed, and natural climate change is unavoidable, but climate change driven by human activity (Anthropocene) is unique and can be mitigated, and adaptation is possible.

Post-January 2025

The first draft of this book was almost completed by January 2025, around the time the USA installed its 47th President. That event, policy-wise, proved to be cataclysmic, and instantly required a second drafting, a process that would be standard under any normal circumstances, but which took on an entirely new dimension with a President who apparently denied climate change. But did he? In September 2016, he was widely reported as stating "The concept of global warming was created by and for the Chinese in order to make U.S. manufacturing non-competitive."[6] By 2020, to the absurdity of this statement he made a slight concession to reality, but not to the science behind it.[7] By 2025, in an address reported in the *New York Times* to delighted oil and gas executives, the President's Secretary of Energy, Chris Wright promised a "180-degree pivot" on climate policy, saying "The Trump administration

---

[3] Wikipedia (March 2025) *Holocene* https://en.wikipedia.org/wiki/Holocene.

[4] The Rest is History (December 2024) 'The World's First City' *YouTube video* https://www.youtube.com/watch?v=s7u6p9VDl-Y.

[5] ScienceDirect (2016) 'Climatic changes and social transformations in the Near East and North Africa during the 'long' 4th millennium BC: A comparative study of environmental and archaeological evidence' Quaternary Science Reviews V.136, 15 March 2016, Pages 96–121 https://www.sciencedirect.com/science/article/pii/S027737911530127X#:~:text=In%20Mesopotamia%2C%20the%20expansion%20and%20subsequent%20contraction,been%20due%20entirely%20to%20social%20and%20economic.

[6] PolitiFact (27 September 2016) *Stand up for the Facts* https://www.politifact.com/factchecks/2016/sep/27/donald-trump/donald-trump-denies-saying-global-warming-chinese-/.

[7] APnews.com (30 September 2020) *Getting warmer: Trump concedes human role in climate change* https://apnews.com/article/election-2020-joe-biden-donald-trump-technology-politics-b5ae9468286d92f44c6a10d1e6d2cad0.

will treat climate change for what it is—a global physical phenomenon that is a side effect of building the modern world."[8]

A "side-effect" is indeed one way to recognise the Anthropocene. More important than the quixotic reinterpretations of, not so much what President Trump actually said, but what he meant, or might have meant, is the unquestionable 180-degree policy pivot. That has changed the entire landscape of global climate mitigation and how to finance it. What was previously an uphill struggle to fund mitigation and adaptation, especially for poorer nations, has reversed progress. In the following chapters, the gulf between the resources needed and those available will become clear, and crucially the inadequacy of the funding mechanisms. These mechanisms and their direction of travel start at the international level with the United Nations Framework Convention on Climate Change (UNFCCC).[9] In September 2024, the UN General Assembly convened to agree a *Pact for the Future*,[10] including pledges by Heads of State and Governments to reinvigorate their action against climate change. The only dissenting voices were those of Russia and its allies. By March 2025 the USA had joined them, changing everything when the President withdrew from both the UN's funding of climate mitigation and from the UN's Agenda 2023, the 17 *Sustainable Development Goals.*[11]

Perspectives

Climate change is a long-term process. Climatologists usually talk of periods of not less than 30 years of changing weather patterns, patterns of change that can go on for thousands of years. Not to let a good story go to waste, in 1971 it was reported (lost in translation as it turns out) that when Henry Kissinger, then US Secretary of State met Chinese Premier Chou En Lai, he asked the question what did Chou En Lai think was the impact of the French Revolution (1789), Chou En Lai replied "it is too early to say." In reality it was a reference to the student demonstrations of Paris 1969, but true or false,[12] the story as it stands serves to illustrate an important point, that historical perception, and future perspectives, are deeply rooted in culture. China is very conscious and proud of its thousands of years of history. By contrast,

[8] CarbonBrief (14 March 2025) US '180-degree pivot' *Debriefed Weekly Climate Newsletter* https://www.carbonbrief.org/.

[9] UNFCCC (2025) *UN Climate Change* https://unfccc.int/.

[10] UN (22 September 2024) *World Leaders Pledge Bold Action to Protect Present, Future Generations amid Climate Crisis, Conflicts Gripping Globe, as General Assembly Adopts Pact for Future* https://press.un.org/en/2024/ga12627.doc.htm.

[11] Reuters (7 March 2025) *United States quits board of UN climate damage fund, letter shows* https://www.reuters.com/world/us/united-states-quits-board-un-climate-damage-fund-letter-shows-2025-03-07/.

[12] Oxfordreference.com (June 2011) *Zhou Enlai (Chou En Lai) 1898–1976 Chinese Communist statesman, Prime* https://www.oxfordreference.com/display/10.1093/acref/9780191826719.001.0001/q-oro-ed4-00018657.

modern Europe emerged from the events of the Middle Ages 500 years ago. Modern America, post-European migration, took shape during and after the revolutionary wars of the eighteenth-century. Perspectives for the future are highly influenced by perceptions of the past, and that applies just as much to the issues of climate change mitigation and the accompanying loss of biodiversity and species. In other words, is it late or too late to save life on earth.

This question has taken on an immediate urgency following the US withdrawal, for the second time, from the 2015 Paris Agreement on Climate Change, not least because the US is the world's second largest polluter and the most important source of climate finance. The significance of the withdrawal *for America* itself will be determined by what measures in practice the USA does or does not take to reduce Green House Gas emissions. At state and city levels in the US, many local administrations and communities are committed to cleaner air and lowering GHGs, but as the President is a firm advocate of fossil fuels ("burn, baby, burn") to the delight of the Big Oil majors, the outcome will be the net effect of private sector efforts to shift to renewable sources of energy *if* the economics make sense *vs.* official government policies to strengthen the US as an oil and gas super-power.

More impactful, certainly in the shorter term, is the decision of the USA to cease involvement in (despite apparently remaining for the while a member of) the UN IPCC,[13] the body that commissions and assesses scientific climate change modelling, forecasting and data analysis. This puts in jeopardy the next AR-7 (Annual Report-7) scheduled for 2029. Simultaneously, nearly 400 scientists working on the *National Climate Assessment* were dismissed at the end of April 2025.[14] There is none so blind as cannot see, as the saying goes, and one way to create blindness is to purge the collection of data, analysis and online accessibility. The data sets from the NOAA (National Oceanic and Atmospheric Authority)[15] and from the FESAC (Federal Economic Statistics Advisory Committee)[16] are among many that are being erased. Others are racing to preserve them before they and the historical record are forever lost.[17]

---

[13] Blog.ucs.org (27 February 2025) *Trump Blocked Federal Scientists from Attending Latest IPCC Meeting: What Now?* https://blog.ucs.org/delta-merner/trump-blocked-federal-scientists-from-attending-latest-ipcc-meeting-what-now/.

[14] CBSNews.Com (29 April 2025) *Trump administration dismisses nearly 400 scientists working on congressionally mandated national climate report* https://www.cbsnews.com/news/national-climate-assessment-report-scientists-fired/.

[15] TheDailyClimate (13 March 2025) *Government climate data quietly removed as Trump administration reshapes policy* https://www.dailyclimate.org/government-climate-data-quietly-removed-as-trump-administration-reshapes-policy-2671322796.html.

[16] Financial Times (20 March 2025) *The White House war on federal statistics: Economists and scientists worry over disappearing data and advisory groups* https://ft.pressreader.com/article/281840059450972.

[17] E360.Yale.Edu (11 March 2025) *Saving U.S. Climate and Environmental Data Before It Goes Away* https://e360.yale.edu/features/eric-nost-interview#:~:text=Since%20Donald%20Trump%20returned%20to%20the%20White,risk%20from%20climate%20change%20and%20toxic%20hazards.

Consistent with these withdrawals has been the US rescinding its agreements with Indonesia, South Africa, Vietnam to help finance their shift from fossil fuels to renewables under the *Just Energy Transition Partnership* (JETP) programme, an initiative launched in 2021.[18] In addition, the US rescinded its pledge of USD4 billion to the UNFCCC-supported Green Climate Fund, and gave up its seat on the board of the Loss and Damage Fund that helps finance climate change adjustment, such as shoreline defences, for less developed countries.[19] At the same time twelve agencies and their actions under the Inflation Reduction Act (IRA) which had been the cornerstone of the previous administration's climate and environmental policies, providing subsidies and investment incentives to enterprises and communities adopting renewable energy, creating solar and windfarms, and households installing heat pumps, were revoked.[20] With immediate effect billions of dollars of investment on clean energy and manufacturing came to a halt.[21]

The global retreat of the US from areas such as international aid (USAID) and environmental protection has inevitably given rise to speculation that China, the world's largest polluter but also the world's leading developer of renewable energy sources, such as wind and solar farms, and technologies such as 'new energy' (electric) vehicles and infrastructure, will move to fill the void. The most obvious channel would be to enhance the green parts of China's Belt and Road initiative (BRI), especially as the World Bank reckons BRI countries by 2018 were consuming 57% of global energy and emitting 67% of global CO_2, but producing only 38% of global GDP (Gross Domestic Product).[22] Aid of any sort is never unassociated with political influence, but that is not a solid foundation for climate change mitigation. Local communities need to be heavily involved, not a known BRI approach, otherwise it too easily dissipates into a sordid exercise of turning aid into profits for local elites. Rather, there is an alternative to use the current withdrawal of the USA from global efforts to combat climate change as an opportunity for the rest of the world to seriously explore ways to avoid fragmentation into competing trading blocs by using the climate change mitigation and adaptation as a unifying factor. If the opportunity is

---

[18] Climatechangenews.com (7 March 2025) *US withdraws from coal-to-clean JETP deals for developing nations* https://www.climatechangenews.com/2025/03/07/us-withdraws-from-coal-to-clean-jetp-deals-for-developing-nations/.

[19] UNEP (November 2025) *Adaptation Gap Report 2024* https://wedocs.unep.org/bitstream/handle/20.500.11822/46497/AGR2024.pdf?sequence=1&isAllowed=y.

[20] The White House (20 January 2025) *Unleashing American Energy* https://www.whitehouse.gov/presidential-actions/2025/01/unleashing-american-energy/.

[21] YaleEnvironment360 (2 March 2025) *Bill McKibben on Climate Activism in the Age of Trump 2.0* https://e360.yale.edu/features/bill-mckibben-interview.

[22] spj.science.org (March 2023) 'An Evaluation of the Priority Mitigation Technology Needs of Belt and Road Initiative Developing Countries' Ecosystem Health and Sustainability (14 Mar 2023) Vol. 9. Article ID: 0026 https://spj.science.org/doi/10.34133/ehs.0026#:~:text=Climate%20technology%20transfer%20is%20vital%20to%20greening,(BRI)%20and%20strengthening%20the%20BRI%20cooperative%20partnership.&text=Such%20information%20will%20guide%20the%20transfer%20of,BRI%20strategy%20with%20enhanced%20climate%20technology%20cooperation.

lost, it well might be that the force of nature rather than political will determines the outcome.

With the USA out, at least temporarily, and China not yet in, and philanthropic organisations doing their best, the financing and technical aid gap for the world's poorest countries does offer at least one new perspective. Some developing countries could make serious efforts to become more self-reliant in funding by transforming their approach to social issues. This could involve everything from reforming their tax and revenue collection systems, to reducing financial 'leakage' through fraud and money-laundering, to creating national and local state-sponsored employment schemes that pay unemployed people to work on green infrastructure and service projects in partnerships with local enterprises. These could be apprenticeships with skills training. Filling the missing link of importing new energy technologies, from hardware to software monitoring and control systems, would call for collaboration with international companies wherever the shareholders (owners) and stakeholders (staff) retain a commitment to the UN's Sustainable Development Goals. A resetting of alliances based on climate change across the OECD countries would be necessary, and possibly even sufficient, to trigger such reforms in EMDEs. But a more decisive move would be the creation of an International Green Bank, supported by green bonds composed of a basket of currencies from richer countries, including leading roles for China and India, and with an *exclusive* focus upon EMDEs. In addition to climate-mitigation and adaptation project financing a Green Bank should also look towards underwriting climate and environment-related technology transfers from the participating member states.

There is an urgent need for the most vulnerable poorer countries to find ways to combat climate change and local adaptation, and to receive recognition for their efforts by richer nations. There is equally an urgent global need to preserve and restore biodiversity. Currently, environmental issues and climate change are addressed by the UN and world governments in parallel tracks, for example in two sets of COPs (Conference of the Parties), yet while nature provides carbon sinks such as forests and wetlands, oceans and ice-caps, Anthropocene global warming destroys all. Diversity is also the mechanism by which nature ensures resilience and sustainability even during periods of extinction as some plants and animals are more adaptable than others, and more resistance to climate-induced diseases. The central role of diversity is illustrated in the diagram below.

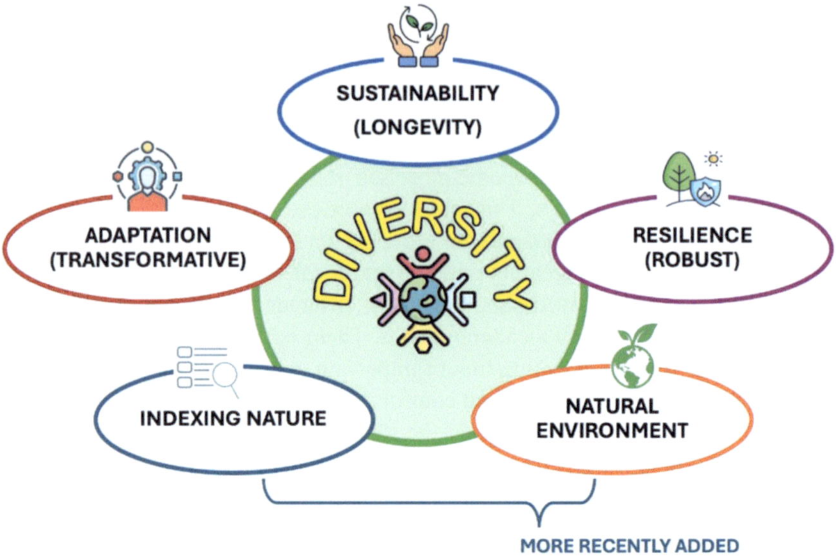

Components of Sustainability. *Design by* Layla Ali

For a long time sustainability as a concept was primarily focused upon issues of resilience and adaptation, and the role and preservation of nature and biodiversity were given short shrift due to the fact that it is difficult to conceptualise nature and biodiversity within a market framework. The value of nature was regarded as being outside normal pricing mechanisms; it was an externality. More recently, as a range of climate-related 'natural' disasters have added to the costs of industry, commerce and society, nature has come out of the cold and been reintegrated into environmentally-conscious thinking. What links each and all of these sustainability components is the concept of diversity. Nothing is forever, change is unavoidable, but the greater the diversity of either nature or of society, the greater the chances of sustainability and survival of both. But of the two, nature may have fewer short-term chances but better long-term chances. Although this book does not focus upon biodiversity, but see Chaps. 4 and 6 in particular, the links between climate change and conservation are too fundamental to ignore. That is deserving of another book.

The Final Draft

This book has chapters on the UNFCCC's meteorological models, the IPCC's different pathways to the Paris goals, what the economic models tell us, or don't, the policies of various national governments worldwide, the nature and problematic operation of carbon markets, where green finance comes from, and why the UN's perspective of using public finance to catalyse private capital will, on its own, remain totally insufficient, and finally the conclusion on where things currently stand, what could be done, and why Net Zero just won't do it. That leaves the biggest challenge of all, how to change policies towards climate change mitigation and adaptation. By past experience, governments only partially wake up to real threats when either their economies are about to tank, without the prospect of a recovery, or/and when society erupts into dissent. If social activists are to be influential, in addition to having a well-founded understanding of climate change science and pathways towards the Paris Agreement, climate change needs to be understood in terms of health issues, in terms of livelihood issues, and in terms of social equity, otherwise civil societies will be sidetracked into alternative populist concepts of ultra-nationalism and xenophobia which divide societies rather than unite them for the struggles ahead. Diversity, not the lack thereof, is the bedrock of sustainability within societies as it is within nature, and a diversity of resources is essential to combat climate change.

Preface

My curiosity about how carbon markets worked started me on the journey to write this book. As an economist I thought I should understand them, but somehow I could not fathom out how the original prices were set. At the time I had moved to Singapore to start up a consulting company after having spent the 1990s and early 2000s in Hong Kong as an Associate Professor at the University of Hong Kong. My specialism was telecommunications and digital development and had nothing to do with climate change or even with the environment as such, despite in my spare time being an avid hiker. The time I had available to do research on carbon markets was limited, and only slowly did I come to understand their variety and complexity, but what I managed to learn fed into Chap. 5 of this book. See my special acknowledgement below.

In 2021 my consulting company (and me with it) was absorbed by a much larger one that, unlike my company, operated timesheets for all staff. This was serious capitalism at work. Independent thinking came second to fulfilling the requests of clients within timelines and budgets. My frustrations grew as I had a deep desire to do more in-depth research on issues that were of greater social value. I graduated as an economist in an era when it was still an exciting discipline with a strong focus upon social welfare and an awareness of its own history and of the competing schools of thought. Commercial consultancy rarely gave scope for any of that, even when I tried to smuggle it in. With the inevitable cost cutting post-COVID-19 I became too expensive as well as increasingly frustrated. We parted company in January 2024, but not before I had started on a new journey, researching the UN Sustainable Development Goals (SDGs) which seemed to me bound to fail on the one hand, but as increasingly important issues on the other. Climate change, environmental degradation (as a hiker I saw at first hand across Asia) and the loss of biodiversity were compelling issues. But as an economist what did I really understand about any of them?

By the time I departed to join my wife in California (she from the Philippines, me an Anglo-Scott from the UK, but I consider myself European) I was well into writing a book about the UN SDGs and what could be done to achieve them. The book grew out of a power point presentation I made for the University of the Philippines in 2023, and was published in October 2024 as *Achieving the UN Sustainable Development Goals: Late or Too Late*?[23] Chapter 3 of that book is devoted to each of the SDGs that cover climate change and the environment on land, in the air and below the sea. My learning about carbon markets first appeared in that chapter, but from that book it was a natural step to start serious research into the entire scope of climate change scenarios as generated by the models used by the UN Intergovernmental Panel on Climate Changes (IPCC), see Chap. 1, and on the pathways towards climate change mitigation and adaptation and the achievement of the Paris Agreement (2015) goals for 2050, 2070 and 2100, see Chap. 2.

Logically it followed I should understand what governments were doing about it, see Chap. 4, and then what role the private sector was playing, see Chap. 5. Besides policy issues such as licensing and carbon taxes, it became clear that at the centre of the efforts of the UNFCCC (Framework for Convention on Climate Change), which acts as the apex intergovernmental body, was the idea that funding carbon mitigation projects, especially in EMDEs (Emerging Markets and Developing Economies as the World Bank calls them) should focus upon public capital catalysing private capital. In other words, public finances would never on their own be sufficient. Examples of this would be 'blended finance' whereby the burden of investment risk is shared or even completely assumed by public capital leaving private capital to make commercial returns. The role of philanthropy was recognised but it remains marginal. I soon discovered that the data shows that this catalytic approach is just not successful. It simply does not fill the yawning gap between what is needed and what is made available. Yet this remains the model pathway. It is completely insufficient.

The basis of that conclusion grows out of Chap. 6 on green finance. That left one hole to fill, what did other economists say on the subject. Besides the concluding Chap. 7 , the last chapter I wrote was Chap. 3 on the economics of climate change. It was the last for two reasons: first, because I had by that time had a chance to reach some evidence-based conclusions of my own; and second, as an economist I was on much more familiar territory, which was a help, but also a danger of drifting too far into the subject. As it turned out, in Chap. 3, I was able to tap into a rich vein of intellectual thought from a variety of economists who disagree with each other. Here I'll just highlight two issues. Economic models often try to quantify a 'damage function' to determine (there is nothing exact about this) how much damage to economic growth (or to nature) climate change will bring about in the future. The debate gets a little esoteric in a discussion about what discount rate to use, a high rate means privileging the present over the future, and a low rate the opposite, giving more weight to the environmental inheritance of future generations. No rate at all is preferred by socio-economic modellers given the many decades that stretch into the distant future. The other issue is how to build models that take account of the fact that greenhouse gases (GHGs) since the industrial revolution have been mostly the result of human activity. In other words, models should treat GHGs as an endogenous

[23] John Ure (2024) *Achieving the UN Sustainable Development Goals: Late or Too Late?* Emerald Publishing.

variable rather than as exogenous as in the tradition neoclassical fashion. For readers less interested in the economics the chapter can be skipped or left to the end.

By the time I came to Chap. 7, I had to declare my position. I decided the best way to do that was to 'follow the money'. If the UN's catalytic model is not working, and if the risk to society and to all species on Earth is so serious, then rather abandon as unachievable the target of average global temperatures rising no more than 1.5C (at max 2C) above pre-industrial levels, which would imply adaptation rather than mitigation as the only way to go, it is the pathway itself that calls for revision. The current pathway is literally not fit for purpose. It is fit for marginal improvements, meaning things could get even worse without it, but it won't save the planet. That is especially true if climate change reaches a runaway tipping point with melting polar ice and consequential changes in ocean currents. In Chap. 7, I outline an alternative pathway which relies upon two developments: (1) a substantial increase in public expenditure (and debt) devoted largely to carbon abatement, including R&D into abatement technologies, and (2) a sustained effort to build international collaboration around such policies to stem capital flight arising from growing public debt and promoting a role for an International Green Bank which would have an exclusive focus upon carbon abatement and adapation across the Global South. This does not eliminate the role of the private sector, especially in carbon avoidance, reduction and removal, but would inevitably squeeze the private sector in areas where its investments are less productive. It is a radical approach that sees economic and social transformation as a dialectal process of change, but also of continuity with the best of both surviving into the next generation. A radical approach sometimes takes a revolution to achieve it, but while in history revolutions may be necessary for change, they can end up with the worst of both worlds. There are no guarantees either way, other than that climate change, like a new movie, is coming to you soon. Whichever pathway is chosen is ultimately a question for an active civil society, not a passive one.

Hong Kong, China John Ure

Acknowledgments

I would like to express gratitude to the following for their generous help in many ways: to climate scientist Prof. Michael Mann of the Penn Center for Science, Sustainability, and the Media at the University of Pennsylvania for giving Chap. 1 a sanity check; and through interviews: Michael Mainelli, then Lord Mayor of London, promoting the City as a carbon trading hub; Simon Puleston Jones, founder of Emral Carbon and Jobs in Carbon, consultant to Philip Lee LLP and author of an excellent *An Introduction to Carbon Markets*—see chapter 5; Sebastien Cross, co-founder and Chief Innovation Officer of BeZero, a carbon ratings agency based in London who shared his views; James Alexander, CEO of UKSIF (UK Sustainable Investment and Finance Association) who shared valuable insights; Vicky Janssens, GM for S.E.Asia, the Klik Foundation (Switzerland) who arranged the world's first ITMO transaction between Thailand and Switzerland and to Bob Fox, Chair of the Digital Economy Committee of the Joint Foreign Chambers of Commerce, Thailand for the introduction; Juan David Duran Hernandez, Executive Director of EcoRegistry, Colombia, for how registries work; Felix Chung, founder and CEO of Hong Kong Carbon Trading Centre who gave me insights into the supply and demand for credits; Sophie Le Clue, CEO ADM Capital Foundation, Hong Kong, for insights into green finance; Grace Cheah, founder and owner of Singapore sustainability consulting company, Sustinere Pte Ltd for good pointers; and to Shivin Kohli, an ex-colleague of mine in Singapore, now at the World Economic Forum, who gave me early guidelines to launch into this project.

Collaborators I wish to thank are Terance Graham and Dr. Jenny Wan who between them read through the entire text with insightful comments (and corrections) and Layla Ali for her help with graphic design. All three have been my colleagues and lasting friends either in Hong Kong or in Singapore. Finally, thanks to my wife, Charleth Luna (Shai) for her constant support and keeping me in great humour all the way through it. I wish to thank them all, noting that I am solely responsible for any remaining errors.

Contents

1 The Science of Climate Change 1
The Science 1
Green House Gases and How They Heat the Planet 2
The IPCC and Its Modelling of Climate Change 7
True or False? 8
The State of Play 10
Denial, Scepticism, Uncertainty, and Risk 11
Is the Record Straight? 14
Climate Change Scientists and the IPCC's Scenarios and Pathways 15
Conclusion 18
References 20

2 Pathways to Net Zero 25
Pathways Towards Net Zero and Lower 25
From Scenarios to Pathways 26
Power Plays 27
Primary and Final Energy Consumption 28
The Future of Renewables in Total Energy Supply (TES)—Models Versus Policies 33
Investing in Renewable Energy 35
The Role of Renewables Going Forward 40
Energy Efficiency and Energy Intensity 43
Technologies 45
Grids 47
Population Trends and the Demand for Energy 48
Conclusions 49
References 51

3 Economics of Climate Change ... 55
Tipping Points ... 55
Incomplete Models ... 56
Discount Rates ... 57
The Concept of Equilibrium *Versus* Tipping Points ... 58
Classical Economics ... 59
Neoclassical Economics ... 60
Exogenous Versus Endogenous ... 63
Exogenous Models ... 64
Endogenous Models ... 66
Innovative Modelling ... 67
Climate Models ... 68
Stern and Stiglitz Versus Neoclassical IAM Models ... 69
Weitzman Versus Nordhaus ... 71
Ecology Economists ... 73
Modern Monetary Theory (MMT) ... 74
System-Based Analysis (SBA) ... 75
Steve Keen versus Neoclassical Economics and Climate Change Modelling ... 76
Keen on Cobb–Douglas Versus Leontief Functional Forms ... 76
Non-linear Damage Functions and Statistical Modelling ... 77
Climate Versus Weather Versus GDP Growth ... 78
Energy and Growth ... 79
Science Versus Economics ... 81
Conclusions ... 83
References ... 85

4 State Policies Towards Greenhouse Gases ... 89
Local Versus Global Reach ... 89
Nationally Determined Contributions (NDCs) ... 90
Government Policies ... 90
Government, Governance, and Civil Society ... 91
Environmental, Social, and Corporate Governance (ESG) ... 96
Article 6 of the Paris Agreement ... 98
United States ... 104
Carbon Trading ... 104
California ... 105
United Kingdom ... 106
Contracts of Difference (CfDs) ... 107
Licences ... 109
Carbon Taxes ... 110
UK Regulation Challenges ... 111
'Natural' Capital Assessment and Investment Standards ... 111

European Union ... 112
EU Carbon Trading ... 112
Licences ... 113
Carbon Taxes ... 114
China ... 114
Coal ... 116
Gas and Oil ... 117
Renewables ... 118
Grid Infrastructure ... 118
Carbon Taxes ... 118
The Voluntary Market ... 119
India ... 120
Carbon Trading in India ... 121
Singapore ... 121
Carbon Tax ... 121
Conclusions ... 122
References ... 124

5 Voluntary Carbon Markets ... 135
Allowances Versus Credits ... 135
Buyers and Sellers and Projects ... 136
Voluntary Market Woes ... 139
REDD+ and Nature-Based Projects ... 144
Nature-Based Removals ... 145
Nature-Based Solutions (NbS) and Natural Capital ... 147
Natural Capital Accounting and the Agro-Business! ... 148
Registries and Credits ... 149
Standards Upgrades ... 150
GHG Protocol ... 150
SBTi ... 151
ICVCM ... 153
Verra ... 153
Gold Standard and the WWF ... 154
ACR Standard ... 155
Carbon Action Reserve (CAR) ... 156
Plan Vivo ... 156
Puro.Earth ... 157
The Global Carbon Council ... 158
ICAO/CORSIA ... 158
ICROA ... 159
IETA ... 159
Voluntary Carbon Markets Integrity Initiative (VCMI) ... 160
Climate, Community and Biodiversity Alliance (CCBA) ... 161

The Changing Carbon Credit Ecosystem ... 161
Conclusions ... 162
Fragmentation ... 163
The Future for the VCM ... 165
References ... 165

6 Green Finance ... 177
Mind the Gap ... 177
A Just Transition ... 179
Green Finance and the Conference of Parties (COPs) ... 179
China ... 180
Climate Finance Landscape ... 180
Mitigation and Adaptation ... 182
Loss and Damage Fund ... 183
New Collective Quantified Goal (NCQG) and COP-29 Outcomes ... 184
COP-29 ... 184
Article 6 as 'Shred of Progress' ... 185
The Anomalies of Public Sector Green Finance ... 186
Political Economy of Financing Zero Emissions ... 186
Public Sector as a Catalyst ... 188
Public–Private Finance into Nature-Based Projects, Carbon Removal, Nuclear Power ... 190
Nuclear Power ... 191
Multilateral Green Finance and EMDEs ... 192
Commercial Sectors ... 194
Financial Taskforces ... 195
COPs 26 and 27 ... 196
GZANZ: Banks and Climate-Related Finance ... 198
Not Fit for Purpose ... 199
Private Capital Markets ... 201
Natural Capital and the Pricing of Public Goods ... 202
Innovation and New Business Models ... 203
Financing Biodiversity ... 206
Climate-Biodiversity-Environment ... 207
Summary of Funding Estimates ... 210
The Role of the Multinational Companies (MNCs) ... 214
The Role of Banks and Scope 3 Emissions ... 216
Financing Nature-Based Projects by Debt Swaps ... 218
Concluding Remarks ... 219
Another Intelligent Approach ... 220
References ... 224

7 Ways Forward 239
Global Warming and Cooling 239
Chapters 1–3 239
Chapters 4–6 241
Out of the Bag: Evidence of the Anthropocene 242
The Yawning Green Gap in Finances 245
The Financial Ecosystem of Tackling Climate Change 246
Public and Private Financing of Carbon Abatement 249
Areas of Carbon Urgency 253
Après Trump, Le Déluge? 255
Après Le Déluge? 256
The Polluters 257
Ghost-In-The-Machine 258
What is Fit for Purpose? 259
The Basis of Action Beyond Slogans 259
An Alternative and Transformative Perspective 260
Technologies, Climate Change, Health, and the Environment 260
Facing Realities 261
References 262

Index 269

Chapter 1
The Science of Climate Change

The Science

Learning from the past is a good place to start, not least because it helps in the design of climate models. Paleoclimate science is the study of climates and climate change in eons past. One of the surest ways is to examine tree rings as each ring within a tree trunk represents a year of growth, and trees annually store carbon. Measuring a tree's carbon content provides evidence of carbon levels in the surrounding atmosphere at the time of growth, but this method can only take scientists back at most a few thousand years. Examining cores driven into the ice in places like Greenland can take science back a further 200,000 years, and cores near the Earth's poles a further 900,000 years or more, with 'time resolutions' of maybe between 10- and 50-year intervals. Isotopes found within these samples, such as deuterium, a hydrogen isotope, and oxygen18, are a useful way to record the temperatures of the air, land, or sea as lighter isotopes evaporate faster than heavier ones as temperatures rise. A brief explanation is provided in the informative series of Climate Emergency Forum videos [1].

Isotopes

An isotope is a kind of chemical fingerprint that identifies different atoms with different physical properties within substances [2]. They comprise the *atomic number of an element*, which is given by the number of protons which attach to neutrons in the nucleus of any particular atom, a number that never varies, while the number of neutrons, which does vary, gives *the mass or weight of the element* such that the *atomic number x mass* produces different isotopes of the same element. For example, ^{1}H, ^{2}H, ^{3}H are three hydrogen isotopes of heavier or lighter weight [3]. Lighter isotopes decay faster than heavier isotopes giving clues as to temperatures millions of years ago.

J. Ure, *Climate Change and Carbon Markets: Late or Too Late?*,
https://doi.org/10.1007/978-3-032-07619-9_1

Similarly with carbon isotopes. C^{12} and C^{13} are stable (and therefore last longer in the atmosphere) owing to the number of neutrons being equal to the number of protons in the atom, whereas with other carbon isotopes neutrons outnumber protons and need decay before stability is established, which renders them temporarily radioactive, such as C^{14}. C^{12} isotopes emitted by plants during photosynthesis are lighter than C^{13} emitted by the burning of fossil fuels, but as the latter are diluted in the atmosphere the ratio of $C^{13}:C^{12}$ decreases which is an indicator of by how much the burning of fossil fuels has increased atmospheric CO_2. Currently estimates suggest that human (anthropogenic) activities using fossil fuels have added 50% to CO_2 in the atmosphere [4].

Examining marine and oceanic sedimentary cores can take scientists back many millions of years, as can the forensic analysis of stalactites and stalagmites preserved in caves, although with varying degrees of precision. Together evidence from all of these sources can give a picture of temperatures and greenhouse gas levels, such as carbon dioxide, as they were thousands and millions of years ago.

Green House Gases and How They Heat the Planet

Green House Gases (GHGs) comprise compounds such as carbon dioxide, methane, nitrous oxide, water vapour, and various hydrofluorocarbon aerosols such as ozone. They are so called because they imitate the warming effects of a greenhouse. But the popular view that GHGs create a blanket around the Earth which reflects back the radiation is only an approximation and not quite right,, which is a pity as it would give rise to less contentious arguments.

The essential feature of climate physics and of climate modelling is one of *system equilibrium*; heat from the Sun's radiation is balanced by heat reflected back into Outer Space from the Earth to maintain an average *equilibrium* temperature on the surface of the planet of 15 °C, 59 °F, or approximately 290 K [5]. (1 degree Kelvin = 1 °C, but inconveniently zero degree Kelvin starts at −273.15 °C) [6]. In climate science, global warming is therefore based upon models in which the release of GHGs disturbs the equilibrium which triggers physical mechanisms by which a warming of the Earth's surface re-establishes an equilibrium but at a higher average temperature. This is referred to as *radiative forcing* [7]. The existential part is that it does not guarantee the survival of any particular species. It is entirely indiscriminate.

Although the role of GHGs in climate change was first uncovered by Joseph Fourier in the late eighteenth century [8], and more GHGs were revealed in the nineteenth century [9], discovering exactly how all that worked was pioneered in a ground-breaking research paper in 1967 by two scholars, Syukuro Manabe and Richard T. Wetherald. The paper focused on the role of temperatures in *different layers* of the atmosphere [10]. As a bio-article points out in the *New Yorker* magazine on Manabe, who was awarded the Nobel Prize in 2021, despite carbon dioxide being only three parts per thousand of the Earth's atmosphere, its effects upon life on Earth through photosynthesis is defining, and so too it turns out are its effects on the atmospheric conditions that give rise to the warming of the Earth's surface [11]. The central methodological breakthrough was to plot points representing the sensitivity of the Earth's temperature to GHGs at different altitudes and the atmospheric effects of GHGs that absorb the radiation emitted by the surface of the Earth.

This needs some explanation, and a short detour into particle physics will help. Radiation consists of particles, including the light particle, the proton, which has no mass but has kinetic energy which gives it speed (the speed of light). All other known particles (except gluons) have mass which can transform into energy according to Einstein's famous equation $e = mc^2$, where e is energy, m is mass, and c^2 is the speed of light (about 300000 kms per second) [12]. Particle physics is the study of sub-atomic matter and has revealed that when atoms in molecules collide with each other, they either gain negatively charged electrons or lose them, a process known as ionisation [13]. This means energy in the form of mass and/or heat is exchanged, resulting in atoms either warming or cooling. For example, different aerosols, which exist in the atmosphere as minute solid particulates suspended within liquid droplets inside gases [14], either absorb the heat from the Sun's ultra-violet (UV) radiation and heat the atmosphere, or reflect it back into Outer Space with a cooling effect [15]. This is exactly what occurs when GHGs in the atmosphere encounter radiation rising from the Earth's surface. Figure 1.1 shows radiation along the entire electromagnetic spectrum, ranging from 30-mm radio wavelengths carrying low-energy particles to wavelengths of nanometres (a nanometre is one billionth of a metre) carrying extremely high-energy particles that would easily kill a living organism if exposed for any length of time.

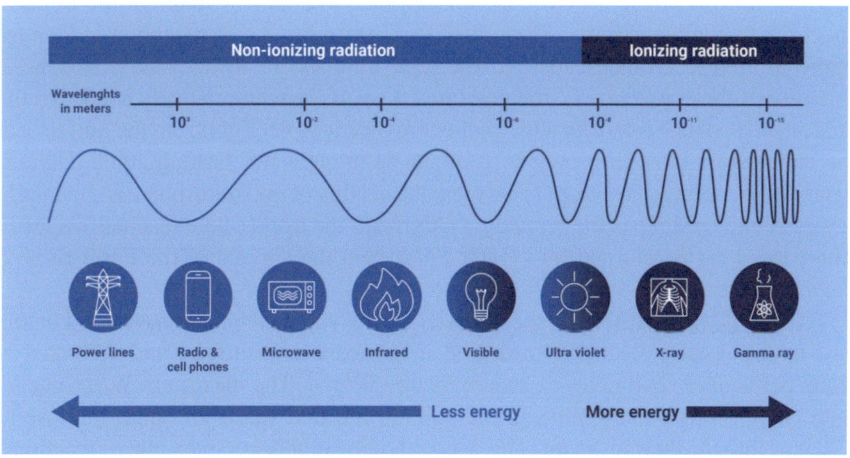

Fig. 1.1 Ionizing and non-ionizing radio spectrum frequencies. *Source* IAEA [13]

But the explanation is not yet complete. Infrared is the heat-producing part of the spectrum [16], most evident as summer sunshine, but if that heat is ultimately captured by GHGs above the Earth, instead of the heat escaping into Outer Space, the surrounding atmosphere will be cooled as energy is transferred to the GHG atoms. From this comes a definition of what is a GHG. Other gases which contain only one atom, such as oxygen, hydrogen, and nitrogen, do not interact with the infrared radiation, and are not GHGs [17]. Their atoms, although never at rest, do not get ironized. By contrast, gas compounds such as water vapour, carbon dioxide, methane, and ozone that contain two or more atoms do interact with radiation and thereby become GHGs.

But it gets even more complex because different GHGs have different molecular structures and react with different parts of the electromagnetic spectrum. Carbon dioxide, water vapour, and methane gas each react to infrared radiation, while ozone reacts with both infrared and ultra-violate radiation. As infrared spectrum wavelengths are longer and carry less energy-intensive particles than shorter waves of radiation, when GHG molecules come into contact with molecules of other gases in different layers of the atmosphere, such as nitrogen, oxygen, and argon, their exchange of energies either cools the surrounding atmosphere in the case of infrared rays/light, or heats it up in the case of ultra-violet rays/light. Where temperatures in the atmosphere fall as a result, the temperature of the Earth's surface has to rise to maintain the natural balance. Should atmospheric temperatures rise, then the opposite at the surface level would need to occur to maintain system equilibrium.

The first and lowest layer of the atmosphere is the troposphere, see Fig. 1.2 [18]. Here carbon dioxide will naturally lower the temperature and thereby push up global warming. But there is a natural limit to how far the temperature of the troposphere can go down, which climate deniers point to as evidence that global warming has its limits.

However, the effects of CO_2 are *not* confined to a narrow range of the infrared spectrum that influences the troposphere because some CO_2 in neighbouring ranges of the infrared spectrum escapes into the upper atmosphere, towards the stratosphere. In-between these two layers, temperatures it seems remain relatively stable. In the stratosphere, ozone interacts with high-energy ultra-violet light from the Sun to heat the atmosphere [19]. In principle, it might be thought this heating could counterbalance the cooling of the troposphere, but the heating of the stratosphere is eventually reversed into cooling as the remaining CO_2 from the troposphere reaches this level. Thus, less heat is returned into Outer Space and the surface of the Earth rises to restore the heat-exchange equilibrium.

Each GHG has a different 'specific irradiation' which is radiation (emitted in this case by the agitated electrons) measured on (atmospheric) surfaces and calibrated as watts per square centimetre or square metre (W/m^2). The maximum W/m^2 occurs with carbon dioxide emissions in the 700 nm infrared spectrum range, at close to 290 kms into the thermosphere. This account of the highly complex mix of reactions, atmospheric cooling, heating and cooling again, is highly simplified and a helpful but also somewhat simplified introductory video on YouTube is recommended [20].

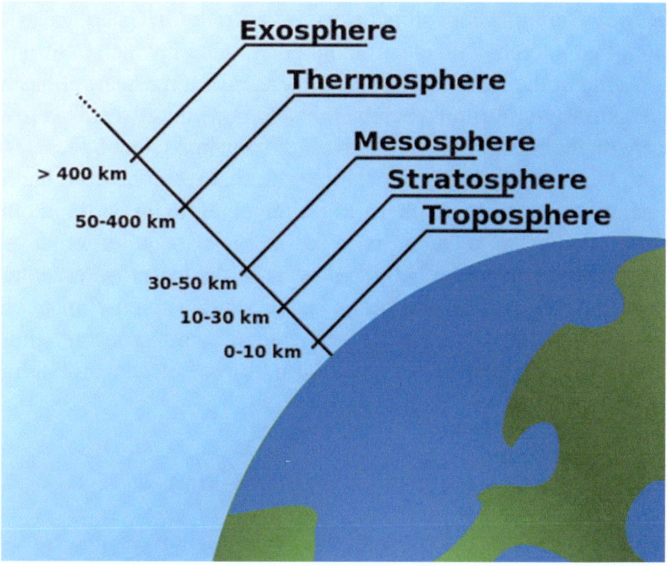

Fig. 1.2 The Earth's atmospheric layers. *Source* NIWA [18]

Why Carbon Dioxide and not Water Vapour?

Why is CO_2 the focus of most attention yet it only constitutes around 0.04% of the atmosphere, and the next gas/element/particle that receives the most focus is methane, which is only around 0.00017%, while water vapour accounts for anything up to 3% of the total atmosphere and 95% of the troposphere? [21]. The explanation is that water vapour molecules only hang around in the sky for up to 2 weeks before turning into rain or snow. According to NASA, CO_2 can stay in the atmosphere for between 300 and 1,000 years [22]. Methane, from volcanoes, fissures in the Earth's surface, from wetlands, farm animals and rotting organic matter in landfills [23], while it traps 28 times more heat than carbon dioxide in the atmosphere, lasts only between 7 and 12 years. While water vapour is the major GHG, it is constantly recycled and is not the driver of climate change [24].

Nitrous Oxide (NO_2), which globally comes mainly from fertilisers in agricultural land use, is another potent GHG. Its lifespan in the atmosphere is over 120 years, and its Global Warming Potential (GWP) which is a measure of how much energy the emissions of 1 tonne of a gas will absorb over a given period of time, relative to the emissions of 1 tonne of carbon, is 273 times that of carbon [25]. However, while CO_2 makes up over 80% of GHGs, NO_2 accounts for only 6% [26].

The takeaway from this convoluted story is twofold: GHGs are gases with more than one atom that interact with electromagnetic radiation which in turn produces a heating or cooling of the surrounding atmosphere, and at the heart of the story is the concept of a natural equilibrium whereby the radiation energy received from the Sun is in balance with the energy that the Earth sends back into Outer Space. If the energy sent back into Outer Space is not in balance with that received from the Sun then a new equilibrium is created at a higher surface temperature on Earth. In other words, if the radiation sent back into Outer Space falls below the equilibrium level, the heat of the Earth rises to compensate. All the models of the Intergovernmental Panel on Climate Change (IPCC) [27], the world's leading authority on climate change, are based upon the working of this mechanism as their central concept, although they differ widely in their estimations of what the multiple between a doubling of CO_2 and global warming may be.

Radiative Equilibrium and the Earth's Energy Budget

Keeping the global temperature of the Earth in a relative stable equilibrium, meaning its fluctuations of heating up and cooling down are not too variable, requires the amount of energy (radiation or heat) incoming from the Sun to be matched by a similar outgoing of energy (radiation or heat) from the Earth. The distance between the Sun and the Earth is approximately 150 million kilometres, and the solar energy reaching the Earth is about 1,360 watts per square metre. Because the Earth is spherical, the sunlight reaching the top of the Earth's atmosphere is around one-quarter of that solar radiance, or 340 watts per square metre. Because the Earth is tilted in its rotating orbit around the Sun, the solar heat that reaches the Earth's surface is distributed unevenly, giving rise to seasons alongside day and night, and with oscillations in the rotation of the Sun (plasma) and the Earth (plasma + rock), there are natural fluctuations in average global temperatures. 29% of solar radiation is reflected directly back by the Earth's upper atmosphere. Of the remaining 71%, 23% is absorbed by water vapour, dust, and the ozone layer, and 48% reaches the surface of the Earth. The reflectiveness of the atmosphere and the Earth's surface of radiated heat back into Outer Space (its albedo) is influenced by various factors, such as the scale and composition of GHGs, and the size of snowfields, ice caps, sea ice, and deserts. Keeping the equilibrium between incoming and outgoing radiation is called the Earth's energy budget [28].

If the budget is in deficit, owing more back to the Sun than receiving energy from the Sun, global temperatures will rise as a result of radiation forcing. Different GHGs have different impacts at different levels of the Earth's atmosphere, but each is transparent to the sunlight, but opaque to infrared arising from the heat captured or generated by the Earth itself. Excessive heat from the Earth will then force radiation levels back through the atmosphere to restore the energy budget, but at the expense of global warming. The alternative, if the energy budget failed to rebalance would be runaway temperature rises, exactly the current fear.

The IPCC and Its Modelling of Climate Change

Established in 1988 jointly by the United Nations Environment Programme (UNEP) [29] and the World Meteorological Organization (WMO) [30], the IPCC is hosted by the WMO out of Geneva, Switzerland. In 2023, it published its 6th Annual Report (AR-6) [31]. Due to its ambitious research programme AR-7 is not due out until 2029 [32], however the withdrawal of US participation in 2025 raises doubts.

IPCC

'The IPCC is the United Nations body for assessing the science related to climate change'. This is an important statement because it prioritises the science of climatology, and related sciences such as particle physics, statistical mechanics, and mathematical modelling over other issues. However, if the *objective* is to achieve a scientific understanding of climate change and the phenomena of global warming, the *aim* is to have governments and societies-at-large adopt evidence-based policies to stop global warming before it becomes runaway, irreversible, and too late. To achieve that, the science must be seen to be credible, understandable, and widely adopted. That is the basic task of the three working groups. Working Group I [33] deals with the physical science basis of climate change; Working Group II [34] deals with impacts, adaptation, and vulnerability; and Working Group III [35] deals with the mitigation of climate change. It also has a Task Force on National Greenhouse Gas Inventories [36] that develops methodologies for measuring emissions and removals [37].

To avoid the worst-case scenarios for humanity, it becomes imperative that climate science fully uncovers the processes at work, and currently that is controversial. Research either centres upon *Equilibrium Climate Sensitivity* (ECS) models, which try to estimate the long-term impact of GHGs once a computed *new* equilibrium has been reached, which is where paleoclimate data is necessary, or *Transient Climate Response* (TCR) models—otherwise known as Global Climate Models (GCM) [38]—which analyse pre-equilibrium impacts of more recent events measured over just a few years or decades [39]. *These models measure by how much average temperatures will rise following an annual doubling of carbon dioxide in the atmosphere.* Where methane is being measured a conversion rate of 1 kg of methane to 29.8 kg of

CO_2e is used [40]. The resulting estimations of climate sensitivity of these models are poles apart. TCR models produce estimated multiples of around 1.5° temperature increases for an annual doubling of CO_2, whereas ECS models tend to estimate increase multiples of between 2.5 and 4.5 degrees, and some go way beyond that. It is important to keep in mind that a small GHG volume at a multiple of 5 could be less damaging than a large GHG volume at a multiple of 2. An accessible popular science discussion graphically explaining these research positions, and other climate change topics, can be found in the excellent YouTube video by Sabine Hossenfelder in 2024, *I wasn't worried about climate change: now I am*, but should be viewed with a rejoinder by 'Climate Adam' [41]. The implications are serious and to better understand why, the World Climate Research Program (WCRP) [42] is running the *Coupled Model Intercomparison Project* (CMIP) to find out.

True or False?

No-one who keeps abreast of world news can seriously doubt that extreme weather events appear to be on the rise. As NASA's website [43] puts it (assuming it survives) 'Record-breaking heat waves on land and in the ocean, drenching rains, severe floods, years-long droughts, extreme wildfires, and widespread flooding during hurricanes are all becoming more frequent and more intense'. NASA provides the following summary, Fig. 1.3.

Not all damaging environmental events are a result of extreme weather. Wildfires may be started in all sorts of ways, although the dryness of the land will be a factor in their spread. In China, as in several other countries, towns and cities are literally sinking [44], not due to flooding but to the collapse of the water table as intensive urban development sucks more water out of the ground for drinking, washing, and other purposes. Golf courses and data centres add to the demand for water outpacing the supply, despite hotter and more humid weather resulting in a wetter climate in many parts of the world. In many Latin American countries and in Africa, deforestation has made land vulnerable to weather changes involving heavy rains resulting in the appearance of deep gullies, with dwellings, people and animals collapsing into their abyss [45]. Another example is the voluminous discard of single-use plastics which finds its way into the oceans. As the oceans warm, sea creatures already at risk face this additional challenge to their existence. In such circumstances, climate change reduces the resilience of the environment to withstand other catastrophes.

But is extreme weather really an indicator of climate change, rather than of a period of weather fluctuations, and how would we know the difference? Most climatologists and scientists in related fields have little doubt about it, but it is never sufficient in science to deny doubt. As the great philosopher scientist of the European Enlightenment, René Descartes [46] famously put it, 'doubt everything'. But science only doubts what is uncertain in the sense that the *mathematical probability* of something being true or untrue is so small that the hypothesis that something is true or false can, on the evidence tested, be rejected. Nothing is ever literally 100% certain for

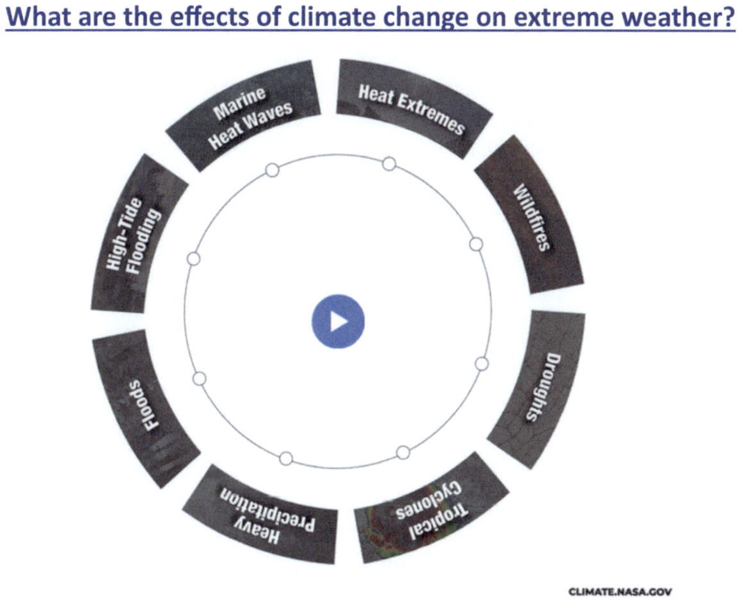

Fig. 1.3 NASA's listing of extreme weather events. *Source* NASA

all kinds of statistical and philosophical reasons. Statistically, the evidence sample must be large enough for a probability to be assigned to it, and only if the sample was infinite could certainty be infinite. Philosophically, and *computationally* [47], humans are restricted in our ability to conceive of, for example, dimensions beyond the four: height, width, length, and time. Maths can offer 'n' dimensions, but no-one really understands what 'n' dimensions in time and space can be or can mean.

What scientific investigation has taught us is that we are, like everything else in the material world, composed of atoms and they are composed of sub-atomic particles. What scientific investigation has further taught us is that the laws of sub-atomic particles, a field known as quantum physics, are unlike the laws of the visible natural world, the world of Newtonian physics and of Albert Einstein's *General Theory of Relativity*, as we experience them, such as the law of gravity. The weirdness of the sub-atomic particle world includes phenomenon known as Heisenberg's 'uncertainty principle' [48]. This states that the position and the movement of a particle cannot both be measured at the same time. In the natural world that we observe every day, the location and movement of a hurricane can both be observed and measured, but understanding exactly what determines how and why that hurricane is where it is, and moving as it does, calls for an understanding of how the particles that make up the chemistry of that hurricane are reacting with each other across the atmosphere, in other words, the laws of particle physics. It follows that uncertainty is an inevitable outcome of scientific investigation, and probability is the best that can be expected of it. If real events consistently turn out according to their probability then all is well and good. If they don't, then either the theory that predicts their probability has to be revised to take into account new evidence, or a different theory is required. A guiding principle of science is to explain more with less, and a theory in trouble is one that explains less with more. In the world of climate change, there is just so much more to be explained, so there is always room for some degree of doubt in the theory or the climatology models used. But doubt cannot be arbitrary. Doubt itself must prove itself, must live up to the same scientific standards of what is being doubted. Only super-natural explanations can claim a pass.

The State of Play

Climate change is nothing new in paleoclimatology, but its driving forces may be. Without GHGs to trap some of the Sun's radiation, the Earth's surface would be 255 K (−18 °C or 0 °F) [49]. But the Earth also has a store of heat of its own as 84% of the planet's volume is a mantel of molten minerals about 2,900 kms (1,802 miles) thick at temperatures ranging from 3700 °C (6692 °F) at the core to around 1,000 °C (1832 °F) close to the crust [50]. Whenever the mantle bursts through the crust in volcanic eruptions the Earth's surface temperature rises, as does the volume of GHGs. Over eons the equilibrium temperature, meaning a temperature *maintained* over very long periods of time, of the Earth's surface has changed.

These changes in climate have naturally affected the types of flora and fauna that could sustain themselves. For example, the dinosaurs survived during, but not beyond, the Mesozoic Era, which lasted 186 million years and includes three periods: the Cretaceous, Jurassic, and Triassic [51]. In palaeontology, eons sub-divide into eras, then into periods, and then into epochs, and climate change can be detected within each epoch, from 30 years to a million years or more. Differentiating climate changes from changes in the weather is a basic necessity as weather change can be scattered and erratic over decades without implying a change in climate. This is the realm of statistical analysis of the available data, when sustained rates of change can be detected; at the quantum level, statistical mechanics studies the properties of sub-atomic particles in the compounds that make up GHGs.

Mathematically modelling these changes is as much an art as it is a science, with so many parameters to be taken into account, such as the temperature sensitivities at different altitudes, the irradiation properties of different particles in different GHG ionised molecular compounds, and fluctuations in the strength of radiation from the Sun and from the surface of the Earth. The room for uncertainty, scientifically expressed as levels of probability, is wide and manifold, and all too often grist to mill of doubters and climate change deniers. The riposte that 97% of climatologists seem to agree with the findings of the IPCC models sounds convincing but may not be insofar as science has a long history of minority opinions later becoming widely accepted [52]. Galileo could be an historical example if it were not for the fact that his critics were not scientists. A better example is Luis Walter Alvarez, a Nobel Prize winner who first suggested the dinosaurs were wiped out by an asteroid [53].

Asteroids would be another parameter/variable to be taken into account. But put into perspective, it is highly unlikely that blind peer-reviewed published scientific research papers could all be wrong on the basics, although care needs to accompany that statement. Confirmation bias by reviewers cannot be excluded. Nevertheless, the real controversy has to be around the issue of probabilities, especially but not only in forecasting.

Denial, Scepticism, Uncertainty, and Risk

It is helpful to make a distinction between climate denial and climate scepticism. Denial is often based upon a predetermined ideological stance, or complete stupidity, or the denier is a hired-gun speaking and writing on behalf of a far-Right organisation or fossil-fuel company. Readers can check for themselves as there is no end of propaganda of this sort online. The term 'sceptic' is different, because a scientist can only be a credible denier if the probabilities are so small as to justify the term. In this sense, it is similar to the difference between being an atheist and an agnostic; a rational choice should be based upon an assessment of probabilities. If the probability of a super-natural power seems within reason, then doubt is best suited to agnosticism, but if the probability seems too low, then doubt will tend towards atheism.

Of course, ideology can equally come from the Left whenever an acceptance of climate change and its consequences is more of an assertion than a debate about the science. Appeals to examples of more violent storms, rising tides or exceptional heatwaves may sound common sense—see NASA's list above—but how can anyone be absolutely sure these are instances of long-lasting climate change rather than short-lived and often dramatic changes in weather patterns. Scientists may argue over the models, with sceptics among them emphasising the uncertainties involved, and they may argue over the validity of the data put into the models and the analytical methods and interpretations of the data, but ultimately the issue comes down to the degree of *estimated risk*. The worst-case scenario is low uncertainty and high risk, the case supporting scepticism would be high uncertainty and low risk. That argument goes to the heart of the scientific element of the dispute. Of course, for the fossil-fuel industry it is all about risk to them. Also see Table 3.1, Chap. 3.

But risk does not only apply to the effects of climate change, but to those associated with remedial action, i.e. the policies to be pursued. On the one hand is the argument that climate change risk is too great not to take remedial action now to contain and ultimately reduce global warming as agreed by the Paris Accords. On the other hand, is the sceptics' argument that the risk of the cure outweighs the risk of the disease, such as energy insecurity should fossil fuels be abandoned; the destruction of communities, jobs, and livelihoods that rely upon the production and use of fossil fuels; and the threat to economic growth in developing countries that rely upon fossil fuels. If the science of climatology is not absolutely clear on the risks of climate change, then how does a society make its choices? Assuming, that is, that society and not the fossil-fuel lobby determines the choice.

For these reasons, scientists on both sides of the argument take very strong and often uncompromising stands from their respective positions. To call the debate 'robust' would be an understatement, vehemence is characteristic, and toxic is often an appropriate way to describe it. This is confirmed by any sampling of the many public videos on social media. Egregious cases arise. The most polluting country after China is the USA because of the size of its fossil-fuel industry, its consumption of fossil fuels, and the size of its agricultural sector. These represent Scopes 1, 2, and 3 emissions under the Greenhouse Gas Protocol [54]. This has created many powerful and well-funded vested interests opposed to efforts to reduce GHG emissions if it means reducing fossil-fuel usage, which it inevitably does, notwithstanding arguments for carbon capture and ways to make sectors using fossil fuels 'greener'.

Until the end of the twentieth century, public policy debates in the US were mostly within the bounds of bi-partisan politics. Democrat and Republican Presidents alike signed onto international climate change conventions [55]. That was true until the 45th Presidency (2017–2021) when Donald Trump tweeted that 'The concept of global warming was created by and for the Chinese in order to make U.S. manufacturing non-competitive' [56]. By 2020, to the absurdity of this statement he made a slight concession to reality, but not to the science behind it [57]. By 2025, in an address reported in the *New York Times* to delighted oil and gas executives, the President's energy secretary Chris Wright promised a '180-degree pivot' on climate policy, saying 'The Trump administration will treat climate change for what it is—a global physical phenomenon that is a side effect of building the modern world' [58].

In Europe much the same bi-partisanship occurred. That now seems like a past/bygone era as the rise of popularism and the spread of conspiracy theories, especially the vortex of misinformation on social media, has marked a massive and dangerous retreat from rational thinking and from a respect for science. Climate change has become weaponised [59]. For example, two conservative writers in 2012 made accusations against leading climatologist and professor at the University of Pennsylvania, Michael Mann comparing him to a child molester [60]. In subsequent, but ongoing litigation at the time of writing., he was awarded USD1million damages in February 2024 [61].

On the other side, climatologist sceptic and professor at the University of Alabama, John Christy reported a rifle being fired at him [62]. The weaponisation of climate change is not just a metaphor, assuming the shooting was science gone mad. John Christy specialises in satellite technology to monitor the Earth's atmospheric pressure and surface heating and argues that satellite and air balloon data show much less warming of the Earth's surface than the models suggest is the case [63]. Responses include a 50+ point-by-point rebuttal on a website Skeptical Science [64], and a response from Michael Mann that John Christy's satellite data sets had been debunked [65]. The overtly political attacks against climate change scientists continues unabated. In 2023, to a question as to why so many climate change scientists managed to get their peer-reviewed papers published in leading journals, Prof. John Clauser, a Nobel-Prize-winning physicist, but not a climatologist, who as a self-proclaimed climate change denier [66] joined the right-wing CO_2 coalition [67] gave as his response 'dishonesty'. His claims have been very publicly rebuked in another *YouTube* video by Sabine Hossenfelder [68].

Skirting around the toxicity is not easy. However, claims that climate change is a hoax and that scientific findings are deliberately falsified are not worth spending time on. As mentioned, there are plenty of videos online along those lines for anyone interested in viewing them. To be taken more seriously are critiques by scientists themselves, even when their thinking may be propelled by ideological bias. Ideally, science should be an open book, but there are three realities to be taken into account: (1) researchers frequently fall in love with their own models and theories, and often see the prospects of their promotion within universities as dependent upon the success of their models, which has at least two consequences, first they may choose to play safe and avoid more controversial fields of study or methodologies, and second they fall victim to *confirmation bias* or the tendency to give preference to things that seem to support their models; (2) scientists also have their ideological biases and again confirmation bias becomes a challenge; and (3) scientists may be sponsored by commercial or ideological interests. Consultants in any number of fields will have experienced the pressures that are often exerted by those putting up the money.

There are three questions that need answering about models, theories and predictions of climate change.

- What evidence is there for climate change and what is the evidence that human activity is now the major driver of it?
- How good are the models at forecasting the future of climate change and its consequences?
- What can be done to slow down or reverse climate change and/or mitigate the consequences?

Scientists who doubt the first are sceptical of the second and see the third as a cure that could be worse than the disease. Although this book is by-passing the deniers in favour of the sceptics, the most convincing bullshit stories are those that have some grounding in truth. One such is the fact that the rotating blades on windfarms are a threat to birds such as high-flying eagles, and that solar farms are bereft of flora and fauna, claims that unfortunately have value. But it does not follow that fossil-fuel extraction is less environmentally unfriendly. What it might be is a logical argument for a greater use of nuclear fuel, but that would not help the fossil-fuel lobby, nor the ultra-conservative producers of many such propaganda videos [69]. Nevertheless, it is estimated that over one million birds are killed in the USA alone by wind turbines and while redesigning and colouring the blades, adding sonar warning signals and avoiding the migratory path of birds are actions that can mitigate the dangers, and putting the numbers into perspective by noting that cats are the major cause of bird death [70], it remains a problem, as a more balanced YouTube video shows [71]. Needless to say, climate change poses an even greater and sustained threat to wild life.

Is the Record Straight?

Climate change sceptics claim the models typically used by climatologists have too much uncertainty attached to them to lead to firm conclusions [72]. One prominent climate sceptic is Steven Koonin, a theoretical physicist rather than a climatologist. His book *Unsettled: What Climate Science Tells Us, What It Doesn't, and Why It Matters* was published in 2021, and in 2023 he outlined his arguments at the conservative Hoover Institute in a YouTube-recorded public lecture [73]. He summarises his arguments in three high-level statements: (i) the 'climate crisis' has scant scientific support; (ii) arguments for an 'overly rapid' *global* decarbonisation is immoral as it penalises poor countries the most; and (iii) 'overly rapid' national decarbonisation will be disruptive, expensive, and will degrade national security.

The first point is clearly scientifically contentious, as a critical book review of *Unsettled* appearing in *Scientific American* fact-checks [74]. Often the dispute arises from cherry-picking. For example, while Steven Koonin cites heat waves as not becoming more frequent in the USA, the book review points out they are becoming hotter and longer. Given that even hotter and longer is insufficient to declare for certain climate change, it remains the case that the many models used by the IPCC predict those outcomes (AR6 WG 3) [75], and on that basis they need to be heeded as they point to a growing high risk to society. The questioning of the accuracy of the models will never disappear because the physics of climate is so complex, as is the physics of weather and its forecasting. The issue ultimately comes down to standards of variance or deviation (designated as 'sigma' σ) [76], which is a measure of how dispersed the data is in relation to the mean, and degrees of error in the models. These measures are used to obtain levels of confidence in the data analysis of the models. In many sciences, especially in the social sciences, a 95% confidence level in the model's analytical findings is often used [77], whereas in theoretical physics, such as in astrophysics or cosmology, a 5-sigma (i.e. 0.00005) is demanded before a result can be accepted beyond reasonable doubt [78]. The reason is that in the natural world quanta, such as energy particles, are measurable, although only if and when they have been correctly identified. Given the complexity of factors influencing climate, and variance of the data, climatology models can never reach such absolutely high levels of confidence. So, criticisms of climatology models can be made regarding their data and the standard deviations arising, and to compensate for that climatologists compare the predictions of many models with recordable weather events over time. That is where the fact checking comes in.

The second and third points Koonin makes are policy issues, not science issues as such, other than the science predicting catastrophes if policies are not adjusted in time. Notwithstanding the debates over the models from which the predictions are made, or the data they use, the policy issues are important for two reasons. First, however necessary it is to reduce GHGs, there will inevitably be costs and consequences involved. These need to be understood and acknowledged, and where possible mitigated. There is no magic wand. Second, those costs and consequences will impact upon people and there will be political and social resistance if the policies fail to compensate those impacted/affected. Developing countries who are offered no assistance in making a 'just transition' (see Chap. 6) to green energy sources will simply not make it, and nations that are threatened with energy shortages will look to fossil fuels if there are no accessible and affordable alternatives. These are not grounds for scepticism of climate change, but of ways to combat it. As noted in the Introduction, the USA in 2025 rescinded its 'just transition' commitments to Indonesia, South Africa, and Vietnam. Whether Prof Koonin considers this policy reversal as immoral is unknown.

Climate Change Scientists and the IPCC's Scenarios and Pathways

The mainstream of climate change scientists working on models that variously predict temperature rises by the end of the twenty-first century of 3–4 °C, well beyond the Paris Accords' target of 1.5–2 °C above pre-industrial levels, are best represented by the IPCC and its Annual Reports, especially AR6 (2023) [31]. AR6 consolidates the findings of a multitude of models: ECS models drawing upon long-run palaeo-climatology data, and TCR/GCM models drawing upon more recent data, typically from before the industrial revolution to modern times [79]. Combinations of these models help to narrow-down the uncertainties of data and measurement. But in 2025 a report from the University of Exeter (UK) and the Institute and Faculty of Actuaries (IFoA) entitled *Planetary Solvency* [80] was far more pessimistic than the IPCC, warning of *impending* tipping points for runaway climate change with catastrophic socio-economic impacts leading to 50% loss in GDP between 2070 and 2090—see YouTube video discussion [81].

However, the IPCC works within a consensus approach, which means that models that generate predictions that are outliers to the normal distribution of findings are not included. While climate sceptics often seize upon this to suggest the IPCC ignores findings that minimise the effects of climate change, it is also the case that models of researchers that generate more alarming findings are similarly treated as outliers. One of the best-known researchers is James E. Hansen, Adjunct Professor at Columbia University, who in 2023 published with collaborators a seminal paper titled *Global warming in the pipeline* [82]. The term 'pipeline' is meaningful because the message in the paper is that the forecast of climate feedback mechanisms as being studied in the Equilibrium Climate Sensitivity (ECS) models (see above) are poised to accelerate the rate of global warming towards a horrifying 10 °C, that is, 7 °C beyond the 'average' forecast temperature rise in the IPCC models. In a video, James Hansen explains that 'We live on a planet with a climate characterized by *delayed response* and *amplifying feedbacks*' [83]. Naturally, predicting 'pipeline' events, that is to say events currently yet to be observed such as ice cap melting tipping points, is an invitation to sceptics, even those who are not deniers.

While the paper is highly technical, including its review of IPCC models, it helps lay readers by summarising its findings in plain English. Without delving too far into the technical details, the following may be noted. The paper develops a Global Climate Model (GCM) with a heavy emphasis upon melting ice caps and rising sea levels, drawing upon longer-term paleoclimatic evidence, such as isotopes dating back 800,000 years from boreholes in ice. The authors note that while the IPCC attribute 'high impact, low probability' to dramatic changes in ocean currents resulting in rising sea levels during the twenty-first century [84], their own assessment is one of 'high impact, high probability'. They argue that 'most climate models are unrealistically insensitive to freshwater injected by melting ice and that ice sheet models are unrealistically lethargic in the face of rapid, large climate change'.

They continue that the 'slow feedback' mechanisms such as melting ice caps, unlike the 'fast feedback' mechanisms such as above-ground GHGs, are accorded insufficient significance, hence the dangers of, as yet foreseen, climate changes 'in the pipeline'. At the technical heart of these models is the concept of 'forcing', the impact that GHGs, such as the doubling of CO_2 have upon the atmosphere and the Earth's surface as measured by watts per square centimetre or square metre (W/m^2)—see above. Once that is estimated, the heating or cooling effects of changes in GHGs are assessed, but because the processes involved are non-linear, while possible scenarios can be speculated, all projections are subject to doubt. Everything hinges upon the probabilities of the estimates being correct, and the chemistry of the atmosphere and the volatility of the inner core of the Earth not changing too radically.

Of the IPCC's three working groups, WG1 is focused upon the physical science models of climate change from which it derives possible *scenarios* for global warming going forward, and WG2 on the likely impacts on life on Earth, whereas the focus of WG3 is upon risk mitigation and future *pathways* towards that. The debates only schematically outlined above hinge upon the findings of WG1 and the wisdom and practicality of the proposals of WG3. By contrast, the findings of the *Planetary Solvency* report seriously question the less alarmist outputs of WG2—see also Chap. 3. It is therefore time to review the key findings of AR6 (2023). Section 3 '*Long-Term Climate and Development Futures*' of AR6 in the summary of its findings, draws an important line of distinction with AR5.

IPCC AR6 Section 3

'The uncertainty range on assessed future changes in global surface temperature is narrower than in the AR5. For the first time in an IPCC assessment cycle, multi-model projections of global surface temperature, ocean warming and sea level are constrained using observations and the assessed climate sensitivity. The *likely* range of equilibrium climate sensitivity has been narrowed to 2.5–4.0 °C (with a best estimate of 3.0 °C) based on multiple lines of evidence, including improved understanding of cloud feedbacks. For related emissions scenarios, this leads to narrower uncertainty ranges for long-term projected global temperature change than AR5'.

Source IPCC AR6 Synthesis Report [85]

The findings of AR6 are 'narrower' than those of AR5, which arises from displaying on a graph all the data points from all relevant models, and running a 'best-fit' line through them. Some data points will lie above the line and some below, and maybe with a few actually on the line. The idea is to minimise the aggregated (squared) distances between the data above and below the line. This is known as least-squares regression analysis, in other words, a form of averaging. The procedure minimises the chance/possibility that residual data is not captured by the average. Outliers are data that lie well beyond the average, either above or below. In this sense, the model of James Hansen and his colleagues (see above) is seen as an outlier. That does not invalidate it, and the fact that the recorded average global temperature for every month since July 2023 has exceeded 1.5C suggests that outliers may be closer to the forthcoming mid-century reality. This is consistent with the *Planetary Solvency* conclusion that the pace of temperature increases is rising.

It is entirely natural to attribute the increasing frequency of turbulent weather across the world, such as heat waves, torrential and unseasonal rain, violent winds, floods, and droughts, to climate change. That does not prove that they are… until it does. That is the risk. Defining climate change as lasting at minimum 30 years and possibly accelerating, if it becomes runaway, over an entire period of a million years or more, is an exercise in empirical data collection at the scientific level, and in human misery and ecological disaster at the practical level. If it is too early to tell definitively, that opinion is not shared by researchers such as the World Weather Attribution (WWA) among many others. In the first quarter of 2024, extreme weather events hit Africa, the Middle East, and Southeast Asia. Across the Sahel and West Africa temperatures rose to records of 45 °C in Burkina Faso and to 48.5 °C in Mali [86]. The desert countries of United Arab Emirates (UAE) and neighbouring Oman were devastated and inundated with the heaviest and longest rains since records began [87]. Deaths rose as a result of a heatwave in the Philippines where temperatures hit an insufferable 47 °C, a combination of actual temperature and humidity [88]. The heatwaves affected most of Southeast Asia [89]. In each of these cases, analysis by WWA found that, in addition to other factors such as the El Nino effect and changes in the tilt of the Earth's axis, climate change to be responsible for the excessive nature of these events.

Their research is careful. For example, in analysing the unusually severe flooding of Mindanao, the Southern Province of the Philippines, during the seasonal Northeast Monsoon, they cautiously concluded that given 'the trend in the observations is large, there is probably an aspect of the atmospheric circulation that is systematically misrepresented by the models. This prevents us, without further detailed assessment of underlying processes and their representation in the climate models, to draw an overarching attribution conclusion that quantifies the influence of climate change on

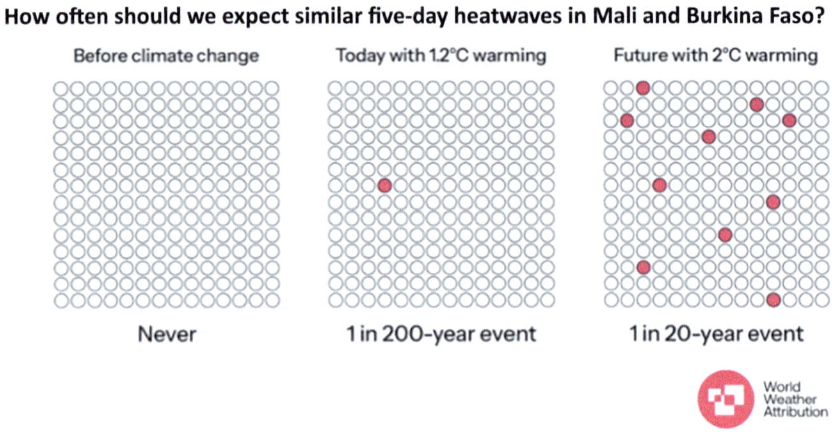

Fig. 1.4 The frequency of extreme heatwaves projected for two African countries

this event'. It adds to credibility when research stops short on the basis of evidence [90]. In their analysis of the Sahel however (see above) they are able to reach a conclusion on the impact of climate change should climate warming be 1.2 °C above pre-industrial levels or 2 °C above. As can be seem from Fig. 1.4, at 2 °C, they estimate a 1-in-20-year chance of similar heatwaves for the future.

The approach of WWA is what has been referred to as 'conservativism' with a small 'c', basically meaning not going beyond what the data says, as in the case of Mindanao (see above). In an interesting YouTube video, Dave Borlace, host of the UN Climate Change channel *Just Have a Think* [91] cites research that contrasts 'conservativism' by data scientists with how actuaries in the field of finance and risk management have been conservative when it comes to an assessment of climate risk, yet when it comes to financial risk in businesses such as insurance they lean in the opposite direction of risk aversion. While the IPCC uses a regression analysis (see above) to work out the average forecasts of the climate change models, such averages for insurance purposes are entirely misleading. If WWA, for example, forecasts one really bad heatwave 'on average' every 20 years, that does not exclude the risk of two or three successive years of heatwaves with disastrous ecological and financial implications. Awareness of climate change risk is not just a 2050 or 2100 issue; it is a very real growing annual issue.

Conclusion

Climate change is usually identified with changes in weather patterns that persist for 30 years or more, but the scientific study of understanding the factors involved, especially the causes and effects of changes at different strata in the Earth's atmosphere, is very much an ongoing project. It involves climatologists building and testing Equilibrium Climate Sensitivity (ECS) models to estimate the impacts of GHGs. At its analytical core is testing what would happen to global warming if, *hypothetically*, CO_2 were to be doubled over pre-industrial levels. The leading 50–60 models cited by the IPCC predict a warming of between 2 and 4.5 °C. Taken together these models are somewhat clumsily called the Coupled Model Intercomparison Project (CMIP) which provides the IPCC with the basis for projections of how much global warming there will be at different levels of GHG emissions. But in recent years so-called 'hot models' have begun predicting temperature rises of 5 °C or substantially higher. As these are incompatible with the historical records, they have been treated as outliers by the IPCC, but they are based upon a revised understanding of the chemistry of cloud cooling, and this remains an unresolved debate among climatologists. Actual average global temperatures have already *consistently* exceeded the 1.5C level, and the *Planet Solvency* report using actuarial forecasts sees the pace of change quickening with substantial damage to GDP by the mid-century.

A separate, but related debate is whether turbulent weather conditions are becoming more frequent and/or becoming more severe and if so, is this an indicator of climate change or simply of variable weather. The issue here seems at first to be less about modelling and more about statistical analysis of the historical record. The problem with statistical analysis is that it is influenced by the interpretation and weighing analysts place upon different variables and measurements, which is where climate deniers and sceptics find room to cast their doubts, and whose arguments might be more persuasive were it not for the support they receive from the well-funded fossil-fuel lobby and politicians who have a stake in that worldview. Global experience of violent weather events during 2024 has certainly reinforced *the perception* of these amounting to climate change. The National Oceanic and Atmospheric Administration (NOAA) measures annual temperature anomalies [92], meaning temperatures above or below the long-term average temperature or reference value—see Fig. 1.5.

NOAA's National Center for Environmental Information (NCEI) recorded the 10 warmest years on record since 1940 as being the years between 2010 and 2023 [94]. One would need to be a denier or sceptic not to be seriously concerned, and ongoing evidence of global average temperatures exceeding 1.5C above pre-industrial levels adds to that concern. Although the Paris Agreement does not specify when the dating of pre-industrial levels should be, it is widely understood by climatologists that the baseline is the 1850–1900 average, even though this is at least one century later than the dating of the industrial revolution by economic historians. The IPCC uses 20-year averages to make its measurements, 10 years before and 10 years after the measurement data. On the basis of single months, 1.5 °C was systematically breached for all months from July 2023 onwards [95].

For public understanding and debate, the statistical analysis seems confusing: is the Paris Agreement target already 'dead as a dodo' or can it be asserted that these extreme weather events are themselves anomalies? It will take another 10 years from 2025 to know using the IPCC ruler, but clearly if the risk is understood to be close to calamitous, then it is better to be wrong after 10 years and discover things are not so bad after all, than to be right and yet wait until it is proved to be the case. The

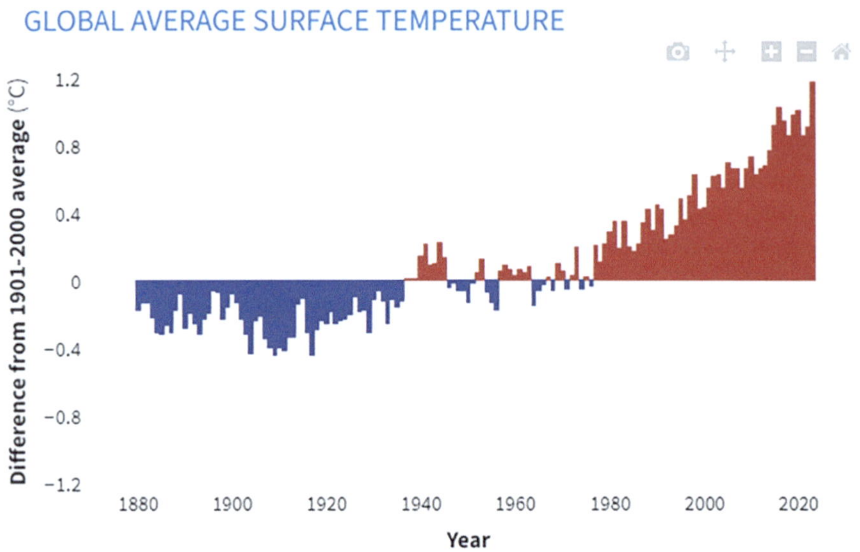

Fig. 1.5 The 2010–2023 were the warmest years on record. *Source* NOAA: Climate.Gov [93]

challenge now is therefore twofold: how to mitigate GHG production and dispersal into the atmosphere to achieve a net-negative CO_2, because net zero will not suffice to bring down average global temperatures to levels that will protect and preserve the environment, and how to maximise the engagement of political and civil society in tackling climate change. What level of crises will it take to accomplish either? The IPCC reports are not read by the general public, but teachers, civil society, and political leaders have the capability, and responsibility, for disseminating the knowledge to a wider audience. But if the value of IPCC conservatism in the past was to gain credibility, that value is rapidly diminishing, and the focus now needs to shift to a Planetary Solvency approach which starts directly from *the value of nature* in sustaining life on the planet Earth. This is discussed in more detail in Chap. 3.

References

1. Climate Emergency Forum (2023) Global Warming in the Pipeline YouTube https://www.youtube.com/watch?v=v-ArA_xYxfs
2. IAEA (2024) What are Isotopes? https://www.iaea.org/newscenter/news/what-are-isotopes
3. Yale University Isotope Evidence for Climate Change https://oyc.yale.edu/geology-and-geophysics/gg-140/lecture-26
4. NASA (June 2025) Carbon Dioxide https://climate.nasa.gov/vital-signs/carbon-dioxide/?intent=121#:~:text=Since%20the%20onset%20of%20industrial,ice%20age%2020%2C000%20years%20ago

5. NASA Solar System Temperatures https://science.nasa.gov/resource/solar-system-temperatures/#:~:text=Mean%20Temperatures%20on%20Each%20Planet&text=The%20mean%20temperatures%20of%20planets,%C2%B0F%20(15%C2%B0C)
6. NIST Kelvin: Introduction https://www.nist.gov/si-redefinition/kelvin-introduction
7. Science Direct (2022) Radiative Forcing https://www.sciencedirect.com/topics/earth-and-planetary-sciences/radiative-forcing
8. Rutgers (2020) The Discovery of Global Warming http://mpe.dimacs.rutgers.edu/2013/01/19/the-discovery-of-global-warming/
9. Iowa State University (2021) The Greenhouse Effect is proven science https://www.extension.iastate.edu/agdm/articles/hof/HofJul21.html
10. Syukuro Manabe and Richard T. Wetherald (1967) Thermal Equilibrium of the Atmosphere with a Given Distribution of Relative Humidity Journal of the Atmospheric Sciences https://journals.ametsoc.org/view/journals/atsc/24/3/1520-0469_1967_024_0241_teotaw_2_0_co_2.xml
11. The New Yorker (2021) The Man Who Predicted Climate Change https://www.newyorker.com/news/persons-of-interest/the-man-who-predicted-climate-change
12. Nova Einstein's Big Idea https://www.pbs.org/wgbh/nova/einstein/lrk-hand-emc2expl.html
13. IAEA (2023) What is Radiation? https://www.iaea.org/newscenter/news/what-is-radiation
14. CAICE What is an aerosol? https://caice.ucsd.edu/introduction-to-aerosols/#what-is-an-aerosol
15. National Geographic (2015) Aerosols, explained https://www.nationalgeographic.com/environment/article/aerosols
16. Herschel Types of Infrared Heat https://www.herschel-infrared.co.uk/how-do-infrared-heaters-work/types-of-infrared-heat/
17. Ahachemistry.com Why are nitrogen and oxygen not greenhouse gases? https://www.ahachemistry.com/2709-why-are-n2-and-o2-not-greenhouse-gases.html
18. NIWS Layers of the atmosphere https://niwa.co.nz/education-and-training/schools/students/layers
19. UK Air Ozone Depletion and Climate Change https://uk-air.defra.gov.uk/research/ozone-uv/moreinfo?view=deleption-climate-change
20. Sabine Hossenfelder (2023) I Misunderstood the Greenhouse Effect. Here's How It Works YouTube https://www.youtube.com/watch?v=oqu5DjzOBF8
21. NASA Steamy Relationships: How Atmospheric Water Vapor Amplifies Earth's Greenhouse Effect https://science.nasa.gov/earth/climate-change/steamy-relationships-how-atmospheric-water-vapor-amplifies-earths-greenhouse-effect/
22. NASA The Atmosphere: Getting a Handle on Carbon Dioxide https://science.nasa.gov/earth/climate-change/greenhouse-gases/the-atmosphere-getting-a-handle-on-carbon-dioxide/
23. SoCalGas (2019) Sources of Methane Emissions https://www.socalgas.com/stay-safe/methane-emissions/sources-of-methane-emissions
24. MIT Climate Portal (2023) Why do we blame climate change on carbon dioxide, when water vapor is a much more common greenhouse gas? https://climate.mit.edu/ask-mit/why-do-we-blame-climate-change-carbon-dioxide-when-water-vapor-much-more-common-greenhouse
25. UCAR (2024) Some Greenhouse Gases Are Stronger than Others https://scied.ucar.edu/learning-zone/how-climate-works/some-greenhouse-gases-are-stronger-others
26. EPA (2024) Understanding Global Warming Potentials https://www.epa.gov/ghgemissions/understanding-global-warming-potentials
27. IPCC Intergovernmental Panel on Climate Change https://www.ipcc.ch/
28. NOAA (June 2023) The Earth-Atmosphere Energy Balance https://www.noaa.gov/jetstream/atmosphere/energy
29. UNEP United Nations Environment Program https://www.unep.org/
30. WMO World Meteorological Organization https://wmo.int/
31. IPCC (2023) AR6 Synthesis Report: Climate Change 2023 https://www.ipcc.ch/report/sixth-assessment-report-cycle/

32. IPCC (2024) IPCC agrees on the set of scientific reports for the seventh assessment cycle https://www.ipcc.ch/2024/01/19/ipcc-60-ar7-work-programme/
33. IPCC (2024) Working Group I The Physical Science Basis https://www.ipcc.ch/working-group/wg1/
34. IPCC (2024) Working Group II Impacts, Adaptation and Vulnerability https://www.ipcc.ch/working-group/wg2/
35. IPCC (2024) Working Group III Mitigation of Climate Change https://www.ipcc.ch/working-group/wg3/
36. IPCC (2024) The Task Force on National Greenhouse Gas Inventories https://www.ipcc.ch/working-group/tfi/
37. IPCC (2024) The Intergovernmental Panel on Climate Change (IPCC) is the United Nations body for assessing the science related to climate change https://www.ipcc.ch/about/
38. Wikipedia Transient Climate Simulation https://en.wikipedia.org/wiki/Transient_climate_simulation
39. Carbon Brief (2018) Explainer: How scientists estimate 'climate sensitivity' https://www.carbonbrief.org/explainer-how-scientists-estimate-climate-sensitivity/
40. Ecometric (2023) Greenhouse Gases, CO2, CO2e, and Carbon: What Do All These Terms Mean? https://ecometrica.com/assets/GHGs-CO2-CO2e-and-Carbon-What-Do-These-Mean-v2.1.pdf
41. Sabine Hossenfelder (2024) I wasn't worried about climate change. Now I am YouTube https://www.youtube.com/watch?v=4S9sDyooxf4 and reply by ClimateAdam (Adam Levy; 2024) Climate Scientist responds to Sabine Hossenfelder on Climate Sensitivity https://www.youtube.com/watch?v=q4EuvpDzlUY
42. WCRP (2024) W CRP Coupled Model Intercomparison Project (CMIP) https://www.wcrp-climate.org/wgcm-cmip
43. NASA (2024) Extreme Weather and Climate Change https://science.nasa.gov/climate-change/extreme-weather/
44. Nature (2024) Nearly half of China's major cities are sinking — some 'rapidly' https://www.nature.com/articles/d41586-024-01149-7
45. Tony Guerra and Harry Hofman (2006) Urban gully erosion in Brazil Geography Review v.19.3, January 2006 https://www.researchgate.net/profile/Antonio-Guerra-7/publication/264167273_Urban_gully_erosion_in_Brazil/links/5536bbf00cf218056e9531b9/Urban-gully-erosion-in-Brazil.pdf
46. Stanford Encyclopaedia of Philosophy René Descartes https://plato.stanford.edu/entries/descartes/
47. Stephen Wolfram (2023) The Mystery of Entrophy FINALLY solved after 50 Years? YouTube https://www.youtube.com/watch?v=dkpDjd2nHgo&t=2805s
48. CalTech What Is the Uncertainty Principle and Why Is It Important? https://scienceexchange.caltech.edu/topics/quantum-science-explained/uncertainty-principle
49. Energy Education Earth Temperature without GHGs https://energyeducation.ca/encyclopedia/Earth_Temperature_without_GHGs
50. National Geographic Mantle https://education.nationalgeographic.org/resource/mantle/
51. University of Kansas Mesozoic Era https://geokansas.ku.edu/mesozoic-era-252-million-years-ago-to-66-million-years-ago
52. Arvin Ash (2023) Is Climate Change Man Made? The Surprising Data! https://www.youtube.com/watch?v=-skE4jCuf-w
53. Natural History Museum How an asteroid ended the age of the dinosaurs https://www.nhm.ac.uk/discover/how-an-asteroid-caused-extinction-of-dinosaurs.html
54. GHG Protocol (2024) What is GHG Protocol? https://ghgprotocol.org/about-us
55. Michael Mann (2021) Climate Change YouTube https://www.youtube.com/watch?v=ydtmJ8RwxHs
56. PolitiFact (27 September 2016) Stand up for the Facts https://www.politifact.com/factchecks/2016/sep/27/donald-trump/donald-trump-denies-saying-global-warming-chinese-/

57. APnews.com (30 September 2020) Getting warmer: Trump concedes human role in climate change https://apnews.com/article/election-2020-joe-biden-donald-trump-technology-politics-b5ae9468286d92f44c6a10d1e6d2cad0
58. CarbonBrief (14 March 2025) US '180-degree pivot' Debriefed Weekly Climate Newsletter https://www.carbonbrief.org/
59. Michael Mann Shorts YouTube https://www.youtube.com/shorts/3Fg4md1pEgA
60. @democracynow Climate scientist Michael Mann wins $1 Million defamation case against right-wing #climate deniers https://www.youtube.com/shorts/3Fg4md1pEgA
61. CBS News (2024) Renowned climate scientist Michael Mann awarded $1 million in suit over his work being likened to child molester's acts https://www.cbsnews.com/news/michael-mann-1-million-suit-vs-writers-likened-his-work-to-child-molester-jerry-sandusky/
62. John Christie (2023) Lecture YouTube https://www.youtube.com/watch?v=kDAXzcMD2Ng
63. Academic Influence (2021) John Christy on making sense of data in the climate change debate YouTube https://www.youtube.com/watch?reload=9&app=desktop&v=ULpGDnuz308
64. Sceptical Science (2024) Climate Misinformation by Source: John Christy https://skepticalscience.com/John_Christy_quote.htm
65. ThinkProgress (2014) Quoting John Christy On Climate Change Is Like Quoting Dick Cheney On Iraq https://archive.thinkprogress.org/quoting-john-christy-on-climate-change-is-like-quoting-dick-cheney-on-iraq-414c7bc89cac/
66. Wikipedia August 2024) John Clauser https://en.wikipedia.org/wiki/John_Clauser
67. Wikipedia (June 2024) CO2 Coalition https://en.wikipedia.org/wiki/CO2_Coalition
68. Sabine Hossenfelder (8 August 2024) "Climate Change is a Myth"—A Nobel Prize Winner's Embarrassing Ideas https://www.youtube.com/watch?v=_kGiCUiOMyQ
69. Wikipedia Prager U https://en.wikipedia.org/wiki/PragerU
70. Hannah Ritchie (2024) How many birds do wind farms kill? https://www.sustainabilitybynumbers.com/p/wind-power-bird-deaths
71. Rosie (2023) Blades vs Wings: Wind Turbine Technologies to Save Bats and Birds YouTube https://www.youtube.com/watch?v=XQa0PPHfFz0
72. Hoover Institution (2017) Flawed Climate Models https://www.hoover.org/research/flawed-climate-models
73. Steven Koonin (2023) Unsettled: What Climate Science Tells Us, What It Doesn't, and Why It Matters YouTube https://www.youtube.com/watch?v=acyErLNL7kQ
74. SciAm (2021) A New Book Manages to Get Climate Science Badly Wrong https://www.scientificamerican.com/article/a-new-book-manages-to-get-climate-science-badly-wrong/
75. IPCC 6th Assessment Report (2022) Climate Change 2022: Mitigation of Climate Change https://www.ipcc.ch/report/ar6/wg3/downloads/report/IPCC_AR6_WGIII_FullReport.pdf
76. National Library of Medicine Standard Deviation https://www.nlm.nih.gov/oet/ed/stats/02-900.html
77. StatSig (2022) Understanding the role of the 95% confidence interval https://www.statsig.com/blog/95-percent-confidence-interval
78. CERN (2024) Why do physicists mention "five sigma" in their results? https://home.cern/resources/faqs/five-sigma
79. USDA Basics of Global Climate Models https://www.climatehubs.usda.gov/hubs/northwest/topic/basics-global-climate-models
80. IFoA (16 January 2025) Planetary Solvency—finding our balance with nature https://actuaries.org.uk/planetary-solvency
81. Climate Emergency Forum (January 2025) 'Planetary Solvency' YouTube video https://www.youtube.com/watch?v=DHXx93nfRU0
82. James E. Hansen, et al. (2023) 'Global warming in the pipeline' Oxford Open Climate Change, Volume 3, Issue 1 https://academic.oup.com/oocc/article/3/1/kgad008/7335889
83. UN Climate Change (2023) An Intimate Conversation with Leading Climate Scientists To Discuss New Research on Global Warming YouTube https://www.youtube.com/watch?app=desktop&v=NXDWpBlPCY8

84. IPCC (2021) Climate Change 2021: The Physical Science Basis, https://www.ipcc.ch/report/ar6/wg1/
85. IPCC AR6 (2023) AR6 Synthesis Report: Climate Change 2023 https://www.ipcc.ch/report/sixth-assessment-report-cycle/
86. WWA (2024) Extreme Sahel heatwave that hit highly vulnerable population at the end of Ramadan would not have occurred without climate change https://www.worldweatherattribution.org/extreme-sahel-heatwave-that-hit-highly-vulnerable-population-at-the-end-of-ramadan-would-not-have-occurred-without-climate-change/
87. WWW (2024) Heavy precipitation hitting vulnerable communities in the UAE and Oman becoming an increasing threat as the climate warms https://www.worldweatherattribution.org/heavy-precipitation-hitting-vulnerable-communities-in-the-uae-and-oman-becoming-an-increasing-threat-as-the-climate-warms/
88. Reuters (April 2024) Heat wave in Southeast Asia closes schools, triggers health alerts https://www.reuters.com/world/asia-pacific/philippines-dangerous-heat-prompts-shift-online-classes-power-crunch-2024-04-29/?ref=hans.news
89. CNA (April 2024) CNA Explains: What can we do about the extreme heat across Asia—including in Singapore? https://www.channelnewsasia.com/asia/hot-weather-asia-heatwave-high-temperatures-climate-change-cna-explains-4300206
90. WWW (2024) Reducing vulnerability and improved land management needed with increasing heavy rainfall in Mindanao Island, southern Philippines https://www.worldweatherattribution.org/reducing-vulnerability-and-improved-land-management-needed-with-increasing-heavy-rainfall-in-mindanao-island-southern-philippines-2/
91. UN Climate Change (2023) The money men know the truth about planetary boundaries! https://www.youtube.com/watch?app=desktop&v=X-FJvzgrM00
92. NOAA.GOV Global Surface Temperature Anomalies https://www.ncei.noaa.gov/access/monitoring/global-temperature-anomalies/#:~:text=The%20term%20temperature%20anomaly%20means,cooler%20than%20the%20reference%20value.
93. Climate.Gov (2024) Climate Change: Global Temperature https://www.climate.gov/news-features/understanding-climate/climate-change-global-temperature#:~:text=Earth%27s%20temperature%20has%20risen%20by,0.20%C2%B0%20C)%20per%20decade
94. NCEI.NOAA.GOV (2024) Annual 2023 Global Climate Report https://www.ncei.noaa.gov/access/monitoring/monthly-report/global/202313/supplemental/page-1
95. NCEI.NOAA.GOV (September 2024) Earth had its hottest August in 175-year record https://www.noaa.gov/news/earth-had-its-hottest-august-in-175-year-record#:~:text=by%20the%20numbers-,August%202024,in%20the%20global%20climate%20record.

Chapter 2
Pathways to Net Zero

Pathways Towards Net Zero and Lower

Just as Chap. 1 was designed to understand the various models used by climatologists and the UN's Intergovernmental Panel on Climate Change (IPCC) to construct different climate change scenarios, Chap. 2 is designed to understand the pathways that arise from those models using different policy and market-led assumptions that could ultimately result in permanently limiting global warming to no more than 1.5 °C. The reason this remains important is because *sustained* average global temperatures exceeding 1.5 °C will have disastrous consequences for all life in Earth, and threaten to turn into a cascade of tipping points and runaway global warming. For example, according to the IPCC's 2019 *Special Report on Global Warming* [1], if the new normal became 2 °C rather than 1.5 °C, the IPCC estimates 14% of the world's population would be exposed more often to severe heatwaves, that ocean levels by 2100 would rise an additional 10 cm above those forecast under 1.5 °C, that twice as many plants, insects, and vertebrates would lose their habitat, and whereas under 1.5 °C, 70–90% of corals would disappear, under 2 °C, close to 99% would disappear. Further, hundreds of millions of people would be exposed to severe drought, and agricultural yields would fall even more drastically exacerbating food insecurity. Even with major regional variations, the consequences will be dire.

Further, and even more serious if it happens, and there are alarming signs that it is beginning to happen, is the irreversibility of melting ice caps at the poles and the effects this has upon the Atlantic Meridional Overturning Circulation (AMOC) which determines the Atlantic Ocean currents [2]. This does not just raise the levels of oceans, it exposes land that has been acting as a carbon sink for eons, releasing more carbon dioxide and methane into the atmosphere that has to be extracted by human effort. The loss of the ice caps also reduces the reflection of solar energy, adding to the average equilibrium global temperature. The result will be an inflexion point, accelerated runaway global warming. These are the reasons why humanity should not give up on the target of 1.5 °C even if this means years of >1.5 °C before

J. Ure, *Climate Change and Carbon Markets: Late or Too Late?*,
https://doi.org/10.1007/978-3-032-07619-9_2

a reversal can take place. This is also the reason net zero is not good enough, it has to be net negative.

This chapter is inevitably quite dense with data as it dives into the outcomes of the modelled pathways. Readers can treat the data as indicative of the urgency of measures that are needed to increase the capacity of electricity and 'green' hydrogen from renewable sources, to reduce the use of fossil fuels, to build a green grid infrastructure, and to invest at the required levels. The most important takeaway is why, as argued above, the net-zero target of a global average temperature of 1.5 °C by 2050 is the only one under consideration, despite its inadequacy.

The *World Energy Outlook 2023* from the International Energy Authority (IEA) makes for bleak reading.

> Today, the global average surface temperature is already around 1.2 °C above pre-industrial levels, prompting heatwaves and other extreme weather events, and greenhouse gas emissions have not yet peaked. The energy sector is also the primary cause of the polluted air that more than 90% of the world's population is forced to breathe, linked to more than 6 million premature deaths a year. Positive trends on improving access to electricity and clean cooking have slowed or even reversed in some countries [3].

From Scenarios to Pathways

There are numerous bodies or 'modelling shops' as Nolan Lindquist of CAS (**C**enter for **A**ctive **S**tewardship) calls them [4] that map out possible pathways towards carbon and GHG reductions. He distinguishes between 'incumbent modelers' using 'prescriptive scenarios' such as the IEA (above) [5], ExxonMobil [6], OPEC [7], Shell [8], to derive future pathways, and others who focus upon an interpretation and extrapolations of contemporary trends, new emerging technologies, and the most recent data such as icecap melt rates, or policy and investment initiatives either by the public or private sectors. Examples include the 'Net-Zero Scenario' (NZS) and the 'Economic Transition Scenario' (ETS) from Bloomberg's Net Energy Finance (BNEF) [9], and Rystad Energy's 1.6 °C and 1.8 °C pathways [10]. The CAS takes the view that 'incumbent modelers' are in danger of 'status-quo modelling', but equally is the case that using very contemporary data and trends can over-estimate their importance and duration. The spectrum runs between cautious optimism that things can be done, things can improve, and alarmism. Caution places an emphasis upon pragmatism and the art of the possible, alarmism upon the jolt of wake-up calls and urgency.

The various pathways are primarily based upon a comparison of existing policies towards global warming and alternative more robust approaches towards net-zero carbon by 2050 as produced by different think-tanks, including the IEA. Species extinction is the risk involved, unless climate mitigation and species adjustment can be managed.

Power Plays

All known life forms on Earth are carbon based. This does not imply, as some climate change deniers suggest—a search of the Internet will reveal more than enough—that there cannot be an excess of carbon beyond what is necessary to sustain life. Carbon, along with other GHGs released into the atmosphere, is well in excess of what is needed for life. It also lingers in the atmosphere for hundreds or even thousands of years, depending upon the different paths taken by CO_2 molecules once they are airborne [11]—see Chap. 1.

Non-animal power generation has, since the industrial revolution, shifted from wind power (windmills), water power, and steam power to electricity. The generation, and in particular the *ease of transmission and storage* of electricity, propelled it into becoming what economists call a General-Purpose Technology (GPT). Electricity is to power what the Internet is to communications. Every aspect of economic and social life uses it, and one measure of a country's level of development is how universally available it is. This poses two dilemmas: how to assist less developed countries catch-up in their development, which implies more power consumption, and how to reduce the carbon emissions of *all* countries including those of the most developed and most polluting.

Ideally the answer is, and both will be discussed in this chapter, through (i) a more efficient use of energy and (ii) renewable energy sources, notably solar, wind, hydro, biomass, and, arguably, nuclear, replacing the extraction and combustion of fossil fuels. If this is insufficient in the time available, say by 2050, as a means to limit global warming to 1.5 °C or 2 °C above pre-industrial levels, then, as the argument goes, abatement through technologies, such as carbon capture, utilisation and storage (CCUS) from the ground or carbon dioxide removal (CDR) from the air, is proposed as a way to take carbon out of the atmosphere or prevent more of it from being added to the atmosphere. This line of reasoning sets the year 2050 as net zero, which means no more carbon is added than is taken out. But net zero is *not* net negative which implies carbon reduction. The Paris scenario is that carbon dioxide stabilises at net zero by 2050 and then starts to decline by 2070, with other GHGs in decline by 2100. If by 2050 nuclear fusion has become commercially available, the world can move into an energy-abundant era without global warming. Maybe 'utopian' is another term for such a scenario.

The stark reality of the 2020s is that the bulk of energy supply comes, directly or indirectly, from fossil fuels: directly as burned fuel, and indirectly as an indispensable input into renewable energy outputs. Examples would be fossil-fuel-generated electricity that is then used in hydrolysis to produce 'green' hydrogen, or fossil-fuel consumption in the manufacturing of solar panels, heat pumps, wind turbines, and blades, accompanied by carbon emissions. In other words, there are supply chains which may result in renewable energy but start with fossil fuels. The terms scope 1, scope 2, and scope 3 are used to identify sources of GHG generated along all such supply chains [12].

- Scope 1 refers to the GHG emissions of organisations (enterprises, governments, households) which use the electric power generated under scope 2.
- Scope 2 refers to the generation of energy and how much GHG is generated in electricity production and transmission.
- Scope 3 refers to the chain of economic and industrial activities which feed into scope 1 activities such as agriculture, manufactured products, logistics including warehousing, packaging and transport, and various services, for example, Internet and web-services from energy-hungry data centres.

For these reasons, any analysis of climate change caused by global warming and how to combat it, how to slow it down, and how to reverse it requires a wholistic approach covering all three scopes, even if the actions of different participants until recently focused mostly upon scope 1 and maybe scope 2.

Primary and Final Energy Consumption

In 2023, the UN General Secretary, António Guterres [13], referred to global warming as 'global boiling' and excessive heat has indeed been a feature of global weather conditions since that statement [14]. To fully grasp the nature of the problem, and given that generating and consuming electricity lies at its heart, it is necessary to start with global energy production and consumption—see also by this author, Chap. 3 in *Achieving the UN's Sustainable Development Goals: Late or Too Late?* [15] The concept of primary energy consumption [16] measures *final energy consumption* [17] which is limited to end users, agriculture and industry, *plus* the energy consumption of the energy sector itself and losses in transformation (energy used to produce energy) and distribution from leakages and evaporation: in other words, scopes 1, 2, and 3 together. Figure 2.1 from Our World in Data [18] shows that in 2022, fossil fuels, coal, oil, and gas were over 80% of primary energy consumption. Also see the Energy Institute which hosts interactive maps.

The measure of global energy consumption reaches into quadrillions or 1,000 TWh (Trillion-Watt hours) which is a measure of total demand over the period of 1 h. By 2022 that was 167.8 Quad TWh [20], just over 604 exajoules of primary consumption, up from 587 exajoules or 162.2 Quad TWh in 2019 [21]. *Final* energy consumption in 2019 was 161.1 Quad TWh or 418 exajoules, or 99% of the total, with the remainder dissipated (leakages). To explain some of the esoteric metrics of electric energy flows and power, see Box 2.1.

Box 2.1 The Metric of Electric

Measures of Electricity: Watts, Btu (British Thermal Unts), and exajoules (xj)

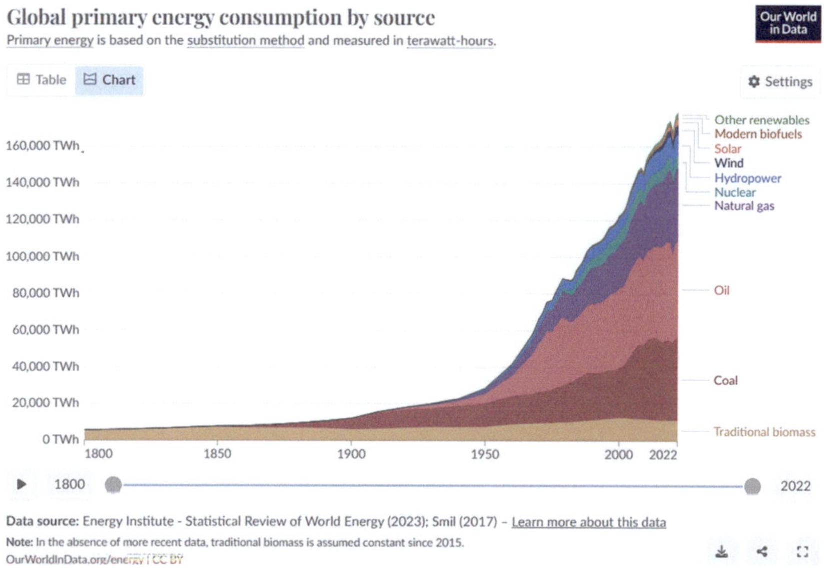

Fig. 2.1 Global primary energy consumption by source up to 2022. *Note* Primary energy is calculated by the 'substitution' method, that is, by taking into account energy losses from the conversion of non-fossil fuels using the same conversion loss rate as for fossil fuels. *Source* Our World in Data [19]

Measures of energy become confusing to those outside the industry.

Joules [22] (or exajoules, EJ) are units of *the force of the energy* produced, where according to the international system of units (SI) [23], a joule is *the equivalent of* one watt-second.

A **Watt** is the *unit of power*, being *the equivalent of* one joule per second, W = J/s, corresponding to the power of one **volt**.

A **Volt** is the *difference in power* between two points in an electric circuit of one **ampere** which is the *power of the circuit*, or its flow, while its *specific or actual power* is measured in **coulomb** [24].

A **Btu** is a British thermal unit, often used to categorise household appliances, as a *measure of heat* generated by a unit of energy at work, where rather unhelpfully 1 Btu = 1055.06 J.

Watts are measured in units such as TW (terawatt), GW (gigawatt), MW (megawatt), and kW (kilowatt) of power in electricity generation, and measures of TWh, GWh, MWh, and kWh are *units of energy* produced or consumed over X number of hours (h). For example, in a city of 1 million households, during the course of 1 h the *per capita* kWh consumption (by household) might be 30 kWh, or a total city consumption of 3 MWh, then over 5 h would be 150 kWh, or a total city consumption of 15 MWh, and so on. Hence TWh/GWh/MWh/kWh implies *energy produced over X many hours* by a multitude of generators or consumed simultaneously by multiple households.

Source John [15] *Achieving the UN Sustainable Development Goals: Late or Too Late?* Emerald Publishing

It is notable that renewable energy sources in 2022 accounted for less than 20% of total energy consumption. This is partly because major areas of consumption, such as the transport sector and general heating and cooling units, are currently low users of renewable energy sources. For example, renewables only accounted for 17% of heat consumption in 2023 [25], and although electric vehicles (EVs) are being heavily promoted as zero emitters of carbon monoxide—notwithstanding that tyre wear adds particulates to the roadside atmosphere and the production of automobiles itself emits GHGs—global take-up has been constrained by factors such as time to recharge, uncertain battery life and distance anxiety, the cost of battery replacements, and a lack of expensive charging infrastructure to support longer journeys. In the absence of easily available alternatives, coal, oil, and gas have all risen in volume despite climate change alarms. Figure 2.2 from the IEA shows in greater detail the composition of total energy supply (TES) by source, 2019 [26]. Solar and wind power were both less than hydro which was less than nuclear, although by 2023 *the capacity* of solar for the first time exceeded that of hydro [27]. In the use of biofuels and waste, the standout countries are the USA at 452 TWh and Brazil at 254 TWh, followed by China a long way back at 41 TWh [28]. But the IEA forecasts solar and wind will constitute over 90% of renewable energy sources by 2030. It should be noted that estimates of TES and demand vary somewhat according to sources. From the IEA the TES of *final consumption* in 2019 was 606 EJ, slightly higher than the Energy Institute estimate of 587 EJ *primary consumption*—see above. Figure 2.2 shows the major sources of energy by type.

'Energy security' has been one factor driving a continuing reliance upon these fossil fuels, together with a political society often severely compromised by the donations and lobbying of the fossil-fuel industry, and a civil society driven by the general public's reaction against rising energy costs, especially for home heating and fuel prices for family cars. See below for fossil fuel and renewable energy trends and cost comparisons. Estimates and certainly *projections* of the supply and demand for energy tend to vary widely according to the data sources used. In practical terms this means working with ballpark figures. Box 2.2 provides an illustrative example of this taken from the work of CAS's Lindquist in June 2024 (see above).

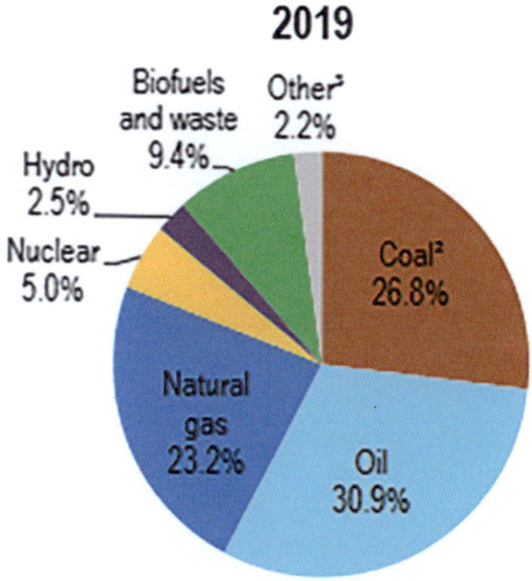

Fig. 2.2 World total energy supply (TES) by source, total final consumption, 2019. *Source* IEA (2021)

Box 2.2 Variance Depends upon Data Sources and Model Assumptions

An illustration of this variance was provided by Nolan Lindquist of the Center for Active Stewardship (CAS) in June 2024. He compared the range of 'pathways' towards net zero arising from five sources providing the 'base case' scenarios (meaning things continue as they are) and seven sources providing 'policy-driven' scenarios (meaning things change towards more ambitious targets). Fuel levels are measured at the point of production as primary energy rather than at the point of delivery and consumption. The table shows the ranges of energy demand reductions in the three base cases for Coal, Gas, and Oil in the first column, and the same for policy-driven reductions in the second column, and in column three the difference in the *implied* emission reductions between the two scenarios.

Base-case demand—range of fossil-fuel decline 2023–2030	Policy-driven demand—range of fossil-fuel decline 2023–2030	Policy-driven emissions *minus* base-case emissions by 2030
Coal—2023 emissions ~15Gt		
−5% to −12% (Av = −6%) or +2%	−15% to −40% (Av = −28%)	~3.4Gt
Gas—2023 emissions ~7Gt		
+3% to +10% (Av = +6%)	−16% to +2% (Av = −6%)	~850Gt
Oil—2023 emissions ~12Gt		
+3% to +11% (Av = +4%)	−18% to −1% (Av = −7%)	~1.4Gt

Source Author's summary of Nolan Lindquist (June 2024) *Climbing down from the fossil peak?* Center for Active Stewardship (CAS)

All but one of the sources (Shell at +2%) sees the demand for coal collapsing by 2030, so although the average declines range from 6% in the base case to 28% in the policy-driven case, the difference in *implied* emissions from Coal by 2030 is small, ~3.4Gt. For gas, the variance is very wide, ranging from a 10% increase in demand to a fall of 16%, with the averages being a 6% increase and a 6% decrease, and a significant difference in the *implied* emissions for Gas by 2030 of ~850Gt. The differences of pathways are greatest in the continuing demand for oil and its products, ranging from an 11% increase to an 18% decline, the averages being an increase of 4% and a decline of 7%. The *implied* difference in emissions from burning oil is ~1.4Gt. The variations highlight the sensitivity of these models to the assumptions made, but they also suggest that the higher emission reductions are within reach if, but only if, policies are climate change driven.

The rest of this chapter will explore the role of renewable energy in energy consumption, the global forecast models of the IEA, the geographical or international spread of energy consumption by sources, and the costs and price of different forms of energy consumption.

The Future of Renewables in Total Energy Supply (TES)—Models Versus Policies

The IEA uses Global Energy and Climate (GEC) modelling to illustrate the estimated capacity of different renewable sources of energy to grow over time [29]. Note: First, the GEC models are *not forecasts but potential pathways* based upon varying assumptions; second, capacity implies the generation of electricity but does not guarantee it. Shifts in prices, and the cost of inputs and/or distribution, or the emergence of new technologies, will influence producers in their decisions to generate electricity. The GEC models are bottom-up using data in the following three areas:

Final energy demand	Industry, transport, buildings, agriculture, and other non-energy use driven by detailed modelling of energy servicing and material demand
Energy transformation	Includes energy generation, heat production, refineries, production of biofuels, hydrogen, hydrogen-based fuels, other energy-related processes, transmission, distribution systems, storage, trade
Energy supply	Fossil-fuel exploration, extraction, trade, availability of renewable energy sources

Box 2.3 summarises the IEA's model scenarios.

Box 2.3 IEA Pathway Models

In its *World Energy Outlook* series, the IEA uses modelling for three different pathways (not forecasts): the **Stated Policies Scenario (STEPS)**, the **Announced Pledges Scenario (APS),** and the **Net-Zero Emissions by 2050 (NZE) Scenario**.

- **STEPS**: Based on prevailing policy settings—global energy demand growth of around 1% per year to 2030 is met in aggregate almost entirely by renewables; coal demand peaks in the mid-2020s, natural gas demand reaches a plateau by the end of the decade, and oil demand reaches a high point in the mid-2030s before falling slightly; the share of fossil fuels in the global energy mix falls to less than 75% by 2030 and to just above 60% by mid-century; carbon emissions reach a plateau around 37Gt before falling slowly to 32Gt in 2050, a trajectory that would lead to a 2.5 °C rise in global average temperatures by 2100.
- **APS:** Based upon stated commitments—demand for all fossil fuels declines up to 2030; announced global manufacturing capacity for electrolysers (for green hydrogen) and solar PV modules in 2030 is sufficient to reach and go beyond APS commitments; carbon emissions peak in the mid-2020s falling to 23Gt in 2030 and to 12Gt by 2050, a trajectory that would see the global median temperature rise by 2100 to 1.7 °C.

- **NZE**: Based upon the Paris targets and an accelerating scenario—global renewable energy capacity triples from 2022 levels to 11,000 GW by 2030; carbon emissions fall to 23Gt in 2030 and to zero in 2050, a trajectory consistent with limiting the temperature increase to less than 1.5 °C in 2100.

Differences in the assumed energy efficiency gains by 2030 account for nearly 50% of the total energy savings in the APS scenario over the STEPS scenario, 'reflecting additional retrofit and updated energy codes, more stringent standards in transport, and upgrades to industrial processes efficiency. In the NZE scenario, year-on-year energy intensity improvements double by 2030' [30].

Source IEA *World Energy Outlook 2022* [31]

But note, according to the IEA (2024) [27].

'Under existing policies and market conditions, global renewable capacity is forecast to reach 7,300 GW by 2028. This growth trajectory would see global capacity increase to 2.5 times its current level by 2030, falling short of the tripling goal. Governments can close the gap to reach over 11,000 GW by 2030 by overcoming current challenges and implementing existing policies more quickly. These challenges fall into four main categories and differ by country: (1) policy uncertainties and delayed policy responses to the new macroeconomic environment; (2) insufficient investment in grid infrastructure preventing faster expansion of renewables; (3) cumbersome administrative barriers and permitting procedures and social acceptance issues; and (4) insufficient financing in emerging and developing economies. This report's accelerated case shows that addressing those challenges can lead to almost 21% higher growth of renewables, pushing the world towards being on track to meet the global tripling pledge'.

Figure 2.3 shows the share of each of the fuels under each of the scenarios. The major shift from STEPS to APS to NZE is the declining share of oil and the rising share of electricity. However, electricity only overtakes oil in the APS and NZE scenarios. The shares of coal and natural gas decline in all three scenarios, but the shares of hydrogen and hydrogen-based fuels rise along with 'modern renewables' which the IEA defines as 'all uses of renewable energy with the exception of traditional use of solid biomass'.

In the IEA's modelled scenario for 2028, over 40% of electricity will be generated by renewable energy sources, with wind and solar panels (photovoltaic panels or PV) accounting for 25% of that. In fact, that point was reached by the end of 2024 [32]. Wind and solar PV surpass hydropower in the generation of electricity in 2024, coal in 2025, and nuclear power in 2026. Wind and solar PV will account for 95% of global renewable expansion 2022–2028 [33].

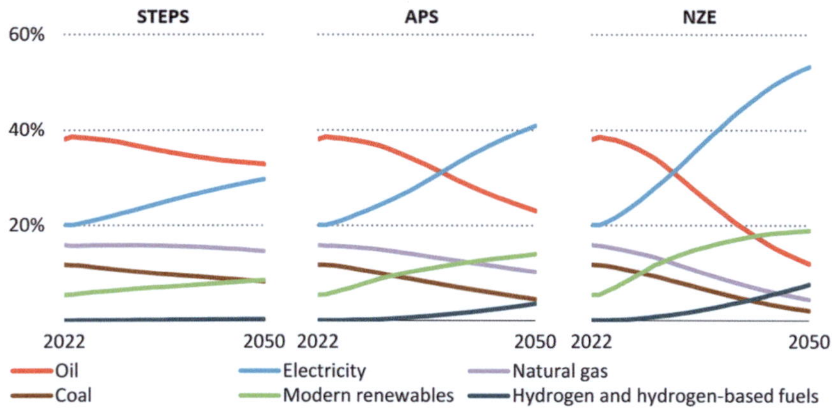

Fig. 2.3 Share of fuels in each of the scenarios, 2022–2050 [29]. *Source* IEA (2024)

Investing in Renewable Energy

Realising these targets involves investments, and most important for the future in fighting climate change are the factors influencing the decisions to invest in *additional capacity* followed by investments in *increased efficiency* in the generation, delivery, and use of energy. Despite the growth in renewables, overall levels of investment are insufficient, as the intergovernmental *International Renewable Energy Agency* (IRENA) has pointed out [34]:

> Global investment in energy transition technologies, including energy efficiency, reached a record high of USD 1.3 trillion in 2022. However, annual investments need to at least quadruple to remain on track to achieve the 1.5°C Scenario in IRENA's World Energy Transitions Outlook 2023 [35]. Investment in renewable energy was also unprecedented—at USD 0.5 trillion—but represented less than one third of the average investment needed each year. Investments are also not flowing at the pace or scale needed to accelerate progress towards universal energy access; investments in off-grid renewable energy solutions in 2021–at USD 0.5 billion–fell far short of the USD 2.3 billion needed annually in off-grid solar products alone (not including mini-grids).

IRENA has developed two scenarios, the *Planned Energy Scenario* (PES) and the *1.5 °C Scenario* [36]. The PES assumes strong global economic growth from 2023 to 2050, based upon a macro-model developed by *Cambridge Econometrics*. Somewhat inevitably, strong economic growth (2.8% per annum) results in high-energy requirements which are offset by more efficient energy production and consumption but not enough to achieve 1.5 °C above pre-industrial levels by 2050. The second scenario adds into the model 'policy baskets' (PB-A and PB-B) where PB-A starts with a low level of international financial cooperation but introduces a high tax on carbon, USD300 per tonne for high-income countries and USD150 per tonne for low-income countries. This is projected to produce an annual average global financial flow of USD918 billion which is three times higher than the 2023 international commitments, and could open the door to greater international financial cooperation. PB-B has a lower carbon tax, USD60 and USD30, respectively, but much higher levels of international cooperation. In gross terms, the PES pathway foresees cumulative investments 2023–2050 at USD103 trillion, although 40% remains focused on fossil fuels, whereas the cumulative total in the net-zero 1.5 °C scenario pathway reaches USD150 trillion, of which 17% is directed into fossil fuels.

International cooperation is basically gilt-edged, that is to say, governments commit whatever the risks involved, but for private sector investors the nature of uncertainty and risk varies with each type of investment, including the cost of investment, or what business economists call the Weighted Average Cost of Capital (WACC) [37]. Natural renewables will be influenced by weather cycles, but also by the costs of competing energy sources, such as a rise or fall in the price of oil. Fossil-fuel exploration for oil and gas is subject to the vagaries of geography and geology, and to changing terms and conditions of licensing agreements which are heavily influenced by geopolitics. Forecasting investment scenarios is therefore fraught with uncertainty, nevertheless, making sensible judgements about the levels and effectiveness of investment in energy production, distribution, and end-use (industrial, commercial, public service, and residential) is dependent upon the methods and accuracy of the measurement. Without these, the judgement of IRENA above would be purely subjective and certainly open to challenge by the fossil-fuel industry and likely by the banks that provide the finance.

A key source that provides detailed measurements is the IEA's *World Energy Investment* (WEI) publications. Table 2.1 provides a methodological overview used by the WEI breaking the energy sector into *power generation* and *fuel supply* sub-sectors [38].

Not included in the WEI calculations but separately treated by the IEA are standalone costs such as operating expenses, maintenance and repairs, research and development, financing costs, and market operations such as merger and acquisition costs. Investments in *power generation* fall very broadly into two categories:

- Investment in projects, such as solar or wind farms, in hydrolysis to produce 'green hydrogen' or in nuclear plants that have a defined lifecycle.
- Investments in exploration of coal, oil, and gas reserves which are on-going.

Investment in *energy efficiency* is more difficult to determine. It falls into three broad categories:

- Household spending on more power-efficient equipment and consumables, such as heat pumps, more energy-efficient washing machines, and energy-efficient lightbulbs. As these transactions are distributed across hundreds of thousands of households, often with no external financing, WEI makes estimates based upon the total sale of such goods and services.
- Building retrofits and general upgrades are accounted by WEI by comparing the difference in the costs of compliance under new regulations governing commercial and residential buildings with those under previous regulations.
- Industrial and freight transport investments are modelled by the *World Energy Model* (WEM) including the IEA's *World Energy Outlook* [40], using an average

Table 2.1 Sub-sectors and assets included in power sector investment

Sub-sector power generation	Assets
Fossil-fuel-based power generation	Coal-fired power with or without CCUS, gas-fired power with or without CCUS, oil-fired power
Nuclear power generation	Nuclear power plants (greenfield), refurbishments, and upgrades of existing plants for long-term operations
Renewable power generation	Bioenergy, hydropower, wind (onshore and offshore), geothermal, solar PV (utility-scale and buildings), solar thermal, marine
Electricity grids	Transmission, distribution, public EV chargers
Battery storage	Utility-scale and buildings
Sub-sector fuel supply	Assets
Oil and gas	Upstream oil, upstream gas, midstream oil (pipelines), midstream gas (pipelines and LNG), refining (greenfield), refining (upgrade and maintenance)
Coal supply	Coal mining, coal transportation
Low-carbon fuels	Biogases, liquid biofuels, hydrogen production

Source World Energy Investment, 2022 [38]; Note: CCUS = Carbon Capture, Utlisation and Storage [39]; nuclear plant investment calculated as from the initial decision to connection to the grid, thereafter investment captured as refurbishment and upgrades; electricity grids include smart meters, automation, etc.; as the largest component of storage, investment in pumped-hydro storage is included under renewable power generation

technology-efficiency base year and projected energy savings cost curves, *plus* published data on energy-efficient management systems.

- Light vehicles such as electric vehicles (EVs) and plug-in hybrid electric vehicles (PHEVs) are a sector on their own. The energy efficiencies of vehicles for personal use are estimated from market price data given that EVs remain more expensive than petrol-driven vehicles, but fleet vehicle costs are estimated using commercial data from various industry sources.

It should be noted that throughout their reports the IEA stresses the need for *energy flexibility* in addition to efficiency, meaning an industry able to adapt quickly to changes in energy supplies and market demand. IEA/WEI scenario modelling revolves around contemporary financial practices (STEPS), assumed future practices (APS), and climate-driven assumptions (NZE), noting that different projects will attract different sources of funding according to levels of risk and issues of project scale. For example, public funds are more likely in hydropower and geothermal projects, while windfarms have expensive upfront costs before making revenues and often financial concessions are made to the private sector in the initial stages. For the NZE model, greater support from international funds is assumed. The real issue is not so much the scenario as such but more the level of commitment by all financial stakeholders, from governments and banks to corporate interests to public sentiment. If data and reason alone were sufficient, then there would be no problem. Unfortunately, data and reasoning are not sufficient to turn commitments into realities; nevertheless, the IEA/WEI have to be convincing about their methodology. The WEI goes into considerable detail and there is no room here to summarise other than to take note of the following (Table 2.2).

The overall picture of annual global investments in renewables versus fossil fuel over the decade to 2023 shows 'clean energy' new investment overtaking that for fossil fuels from 2018 onwards, and while investments in fossil fuels continue to rise after a decline to 2020, the gap between the two grows. See Fig. 2.4.

From this data, it is possible for the IEA to forecast the overall investment needs if the NZE objective is to be met in advanced economies (Advanced) and in emerging markets and developing economies (EMDE) by 2050. See Fig. 2.5.

The left side of Fig. 2.5 measures the estimated investment in the power sector (generation and transmission) in 2022 (2022E), the compound average growth rate (CAGR) of that investment from 2019 to 2022 projected into the future (2015 to 2022 for investments in energy storage), and the annual average investment necessary (pathways) to achieve the APS and the NZE scenarios by 2050. What the data suggests is that the CAGR of investment in renewables needs to be doubled, in fossil fuel needs to be reduced substantially, needs to be maintained in nuclear, but in electricity grids it needs to grow by 75%. In other words, between 2022 and 2050 investment in renewables needs to rise from around USD600 billion annually to around USD1.2 trillion annually, and in electricity grids from around USD400 billion to around USD700 billion annually. Neither CCUS nor battery storage technologies are given much significance in the two scenarios.

The right side of Fig. 2.5 compares for the Advanced economies and the EMDEs the estimated actual investment in 2022 in each, and the CAGR derived for each,

Table 2.2 WEI methodology breaks down the following areas

Funding	Financial actors
Funding sources	Household, retained earnings, banks, public finance, capital markets
Funding instruments	Equity, debt, corporate finance, project finance, grants, guarantees
Capital markets	1. **Institutional investors**: Asset managers, infrastructure funds, insurance companies, pension funds, private equity and venture capital, sovereign wealth funds *Note: WEI noted 90% of energy investments come from the balance sheets of companies and consumers, and a smaller amount from project finance, mostly from banks* 2. **Fixed-income investments**: Sustainable finance (debt securities) including green bonds, green loans, etc., certified under schemes such as the Green Bond Principles [41] or the Climate Bonds Standard and Certification Scheme [42] 3. **Equity markets**: A combination of institutional investors (see above) and ESG (Environmental, Social, and Governance) investment such as ESG exchange-traded funds designed to focus upon corporate responsibility and sustainability [43]

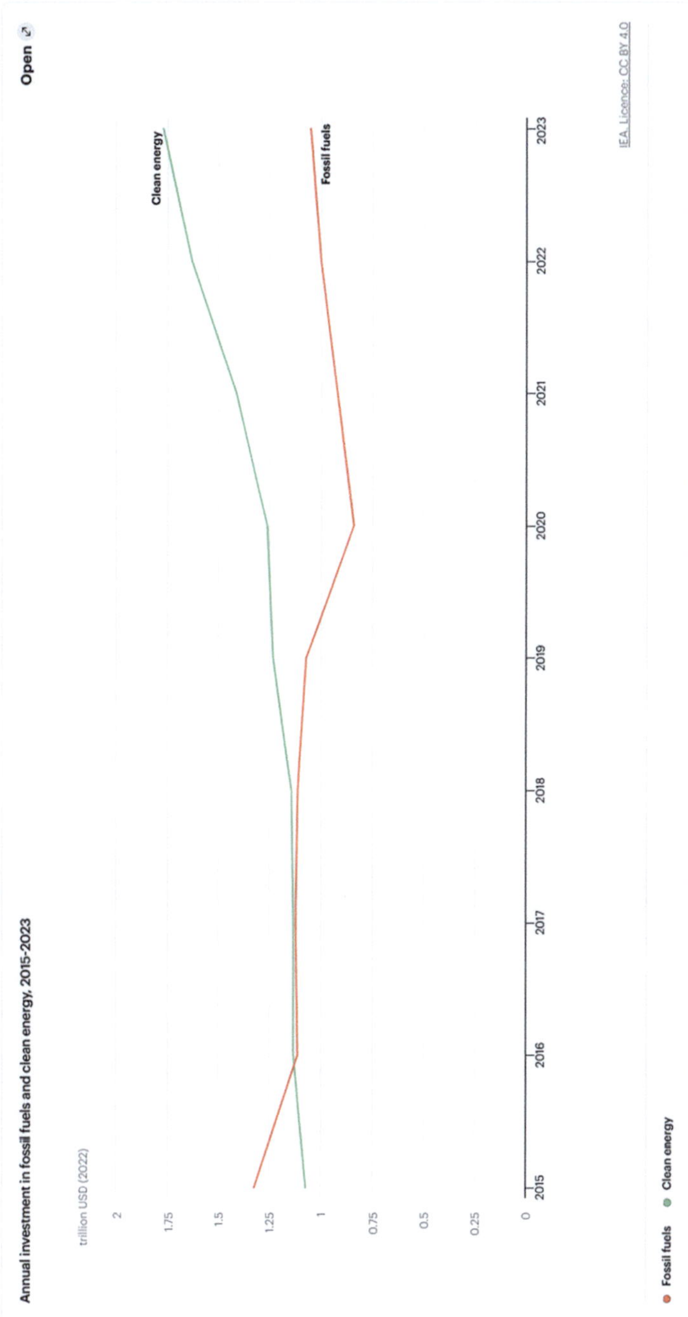

Fig. 2.4 Investment in renewables outstrips investment in fossil fuels [44]. *Source* IEA World Energy Outlook 2023

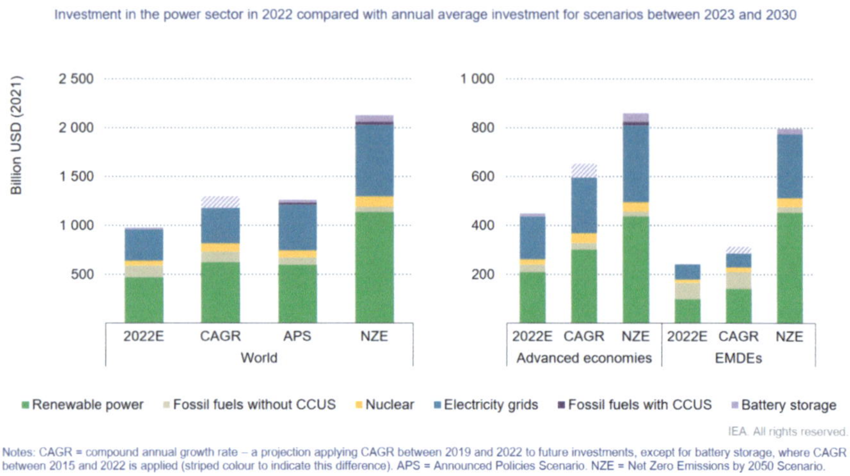

Fig. 2.5 Investment in energy sources and storage estimates for 2022 compared with APS and NZE and between advanced economies and EMDEs [45]. *Source* IEA World Energy Investment 2022

with the NZE investment requirements for each. As expected, the difference between the CAGR and the NZE requirements is less for the Advanced economies than for the EMDEs. Investment in renewables in Advanced economies needs to ramp up from USD300 billion to over USD400 billion annually and in electricity grids from little over USD200 billion to over USD250 billion annually. By contrast, in EMDEs the figures are, respectively, from under USD 200 billion to over USD400 billion for renewables and from less than USD50 billion to upwards of USD250 billion for grids. These are massive numbers for the EDMEs.

The Role of Renewables Going Forward

Ploughing through the highly technical reports of IEA, WEI, and IRENA among others is a painstaking process, and one which illustrates the challenge ahead for slowing, if not reversing, climate change and its impact upon the environment. Just as the climate change models discussed in Chap. 1 are only approximations, with varying degrees of uncertainty but which demand to be taken seriously if only because of the very high existential risks associated with them, so too the modelling of pathways towards net-zero emissions by 2050 (as discussed above) probably have higher degrees of associated uncertainty. This is because they are based upon assumptions regarding investments, especially by the corporate sector, and the policies to be adopted, such as whether or not carbon taxes, are introduced, and if so are they sufficiently high to reflect the costs of negative externalities, such as health costs and loss of species. If they are not they will be ineffective other than as ways to raise funds.

The rest of this chapter is devoted to how effective renewables can be in reducing overall emissions in the atmosphere.

The key metrics are (i) the carbon intensity of electricity generation given that renewables will not account for 100% of energy production and consumption, (ii) the contribution renewables will make to total energy generation and consumption, and (iii) increased energy efficiency in both generation and usage. An additional key factor is how flexible will renewable energy supplies be, for example, will battery storage compensate for periods when there is a lack of solar energy or wind power due to weather conditions, and is the grid system capable of managing bursts of renewable high energy and of redistributing energy supplies to areas temporarily depleted of energy.

Carbon Intensity is measured as grams of carbon dioxide per kilowatt-hour (g CO_2/kWh), heavily influenced by the amount of unabated coal in the mix [46]. In 2022, the global average in electricity generation was 460 g CO_2/kWh [45]. An interactive map is available showing carbon intensities in 2023 by countries [47]. The IEA's NZE scenario sees carbon intensity fall below zero 'as negative emissions in the power sector offset residual emissions in industry and transport'. It might be noted here that the International Maritime Organization (IMO) aims to reduce the carbon intensity of international shipping, a major user of unabated bunker coal, by 40% by 2030 and by 70% by 2050 compared with 2008 levels [48]. By 2050 in STEPS by contrast, carbon intensity only falls to 160 g CO_2/kWh. The IEA provides a figure showing reductions in carbon intensity necessary by different countries to 2040 under a Sustainable Development Scenario (SDS) [49]. The IEA also notes that currently the majority of power plants are located in industrial areas where the carbon intensity of the grid is very high, 'above 600 tonnes of carbon dioxide per-gigawatt-hour (t CO_2/GWh)' and only around 30 percent are located in areas with high potential supplies of solar PV. Further, even when CCUS is used to abate emissions from the use of coal, the average distance globally 'over which CO_2 is currently transported by pipeline between existing CCUS facilities is around 180 km'. Therefore, the technical and commercial challenges of abatement and of shifting to renewable sources of energy are significant, but it does become clear what needs to be done and where it needs to be done.

Renewables and the Total Energy Supply—COP28 (Dubai, 2023) targeted tripling renewable energy supply from 2023 to 2030 to 11,000 GW, a scenario in line with achieving NZE by 2050. The IEA's estimate of STEPS is only 7,300 GW by 2028, a way off from the 2030 target [50]. IRENA [51] provides Fig. 2.6 to illustrate the progress of renewables and what is required to meet the NZE target of 1.5Gt CO_2 by 2050. It can be seen that only the share of renewables in final electricity generation is on track, while total electricity generation and the contributions of both solar and wind power remain behind schedule.

In each of the three IEA scenarios and two IRENA scenarios, the roles of solar and wind power dominate the contribution of renewable energy up to 2050. In APS, assuming governments lived up to their commitments, the global *annual additions* of solar power should be 370 GW and of wind power 210 GW, with China, India, and the USA leading on solar and the European Union on onshore and offshore wind power

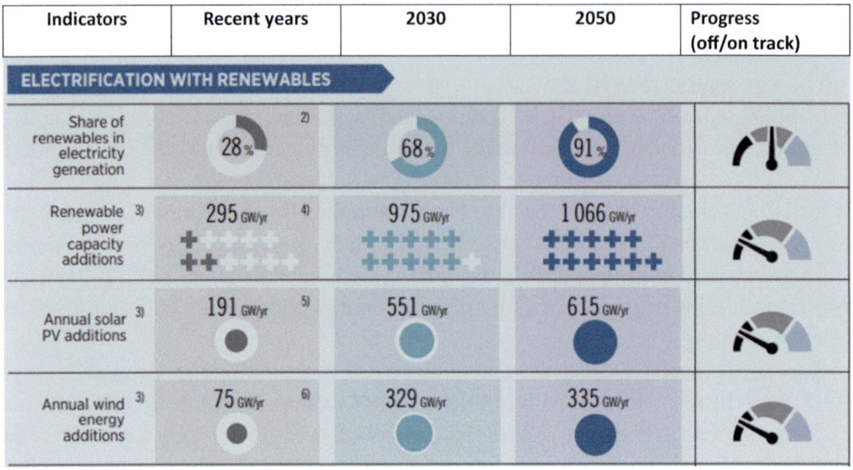

Fig. 2.6 Tracking progress of key energy system components to achieve the 1.5 °C scenario. *Source* IRENA World Energy Transitions Outlook 2023

[52]. It may be noted that 1 MW of electricity is sufficient to simultaneously power around 750 homes in the USA [53], but it should also be noted that the average consumption per person in the USA is around 100 times greater than in some of the poorest countries [54]. Global forecasts suggest that by 2050, if countries meet their Nationally Determined Contributions (NDCs), which are due for to be revised every 5 years [55], and pledges, electricity generation will reach 70,000 TW [56]. Forecasts for hydropower are much more varied. Hydrogen is less efficient than electricity by about one-third. Due to its inflammable nature, production, storage and pipeline or cylinder distribution, the handling of hydrogen requires careful regulation. Currently, most hydrogen is 'blue' hydrogen using natural gas in its production, which is less costly than electricity, but under the NZE scenario 'green hydrogen' using only renewable energy will reach over 6 MW, mostly for use in the transportation sectors. IRENA estimates 25% of electricity by 2050 could be used to produce green hydrogen. IRENA also forecasts a capacity of 6,000 GW of green hydrogen by 2050 compared with less than 500 GW in 2023 [51].

The most controversial energy source that is sometimes classified as clean if not exactly renewable is nuclear. The by-product of nuclear fission [57] is radioactive waste which needs underground sealed storage for hundreds of thousands of years. The Nuclear Energy Institute (NEI) reports that total waste accumulated since the 1950s would cover a football field 10 m high [58]. In addition to a nuclear capacity of 50 GW in 2022, the IEA's STEPS scenario sees this growing by 2050 to 120 GW, and in the APS scenario to 160 GW. However, there are emerging nuclear plant technologies that are smaller than traditional plants, meaning cheaper, which use liquid metal (silicon) instead of water for cooling at much lower temperatures. Microsoft has received regulatory approval for one such plant, the Natrium reactor in Wyoming, USA [59]. In the words of the manufacturer, TerraPower, the use of silicon makes it 'simpler and safer than any of today's plants' [60].

To summarise, under IRENA's 1.5 °C scenario, by 2050 renewables would contribute 91% of power *generation* (82,148 TWh), with solar and wind power accounting for 70% of this, and with smart digital technologies providing energy *flexibility* between renewable sources. Other renewables in the mix would include nuclear, hydrogen (mostly 'green') and bioenergy (used mostly by industry, transport and in buildings and construction), concentrated solar-thermal power (CSP) which uses mirrors to reflect and concentrate sunlight onto a receiver [61], geothermal, and ocean, tidal and wave power. Renewable power capacity would reach 33,216 GW [62]. But these pathways are periodically revised upwards as a consequence of new developments. For example, in the USA, in large part due to the US Inflation Reduction Act (IRA), and before the Trump administration halted it, the IEA foresaw annual solar and wind capacity *additions* as growing 250% by 2030 over 2022 levels [52].

There are several trade-offs involved in securing renewable sources of energy. Green hydrogen is perhaps the prime example, as it requires electricity from renewable sources such as solar and wind, to power the electrolysis in making green hydrogen. Electricity generated using fossil fuel creates blue hydrogen and its hydrogen-fuel spinoffs. Synthetic kerosene produced from a mix of hydrogen and CO_2 provides around 25% of energy used in the aviation sector, and ammonia and hydrogen provide more than 40% of energy use in shipping [52]. The installed capacity for electrolysis by 2050 according to IRENA's 1.5 °C pathway will be 3.7 GW up from 500 MW in 2022, generating over 14,800 TW/h of electricity and accounting for nearly 75% of low-emission hydrogen—the rest using fossil fuel with CCUS—mostly used in land, sea, and air transport, and in the power and industry sectors. The remainder of this chapter looks briefly at three areas: energy efficiency, emerging technologies, and the longer-run impact of population decline in countries with falling birthrates.

Energy Efficiency and Energy Intensity

Energy efficiency 'is the use of less energy to perform the same task or produce the same result' [63]. Energy intensity is the measure of how much energy is required to do the job, so an increase in efficiency will reduce the intensity [64]. IRENA published a study in 2017 of the effects of energy efficiency upon the demand for renewables, setting its base year in 2010 for projections to 2030 [65]. It first identified the use of renewables under three different assumptions, namely, a no-change 'freeze' on the use of renewables, a policy commitment to increase renewables in the total energy mix towards 2030 plus the adoption of additional policy measures; second, it then factored in assumptions about increased energy efficiency which would lower costs and therefore increase the demand for renewables. The report notes a synergy that the growing use of renewables will help drive up their efficiency, and even more so if advanced technologies drive up the use of renewables. It concludes: 'In general, the effect that energy efficiency has on the share of renewables is greater than the effect renewables have on energy intensity'. It is also the case that increased energy efficiency will dampen aggregate demand for different sources of power. Overall, IRENA apportions achieving a reduction in carbon from 45 $GtCO_2$/yr in 2015 to just over 15 $GtCO_2$/yr in 2050 to the use of more renewables (44%) and increased energy efficiency (32%) with electrification accounting for 14%.

The IRENA study looks at five countries: China and India as fast developing economies, and the USA, Germany, and Japan as advanced industrial economies. In general, it found that 'Greater renewable energy reduces the demand for energy, and greater energy efficiency results in higher shares of renewable energy'. The way this works is that increased energy efficiency reduces the need for energy intensity so that the share of renewables in the energy mix increases even when the use of renewables remains unchanged. In China and India, increased energy efficiency due to a switch to electricity was the major driving factor in reducing carbon, whereas renewable technologies, such as the switch from fossil-fuel-generated electricity to solar, were the principal drivers in the USA, Germany, and Japan. The major reductions were in the power sector followed by buildings and transport. The report found that in 2017 the energy intensity (measured in terms of Megajoules (MJ) of primary energy per unit of gross domestic product (GDP) in US dollars) was highest in the two developing economies, China and India, at 7.5–9.0 MJ per USD of GDP, compared with the advanced economies where energy intensity ranged between 4.2 and 6.8 MJ per USD of GDP. This is akin to saying production levels are higher in China and India, but productivity levels are higher in the USA, Germany, and Japan.

Table 2.3 is extracted from the report to show how, *in the projections using IRENA's Global Renewable Energy Roadmap (Remap)*, rising energy efficiencies can drive up the share of renewables in the total energy mix. The Remap projection for 2030 (from 2010) in column 2 shows the share of renewables *before* energy efficiency is taken into account, column 3 includes the increase in energy efficiency *before* its effect upon the share of renewables is taken into account, and column 4 the share of renewables *after* the effects of increased energy efficiency are taken into account. In all cases, the *percentage* impact of rising energy efficiency is substantial.

Table 2.3 The Share of Renewables Before and After Rising Energy Efficiencies

Economy	Renewables share of total final energy consumption		
2030–1	Remap reference case-2 (%)	With energy efficiency (EE)-3 (%)	With remap + energy efficiency-4 (%)
China	19.1	28.1	32.0
Germany	25.9	35.6	38.4
India	22.2	25.9	30.9
Japan	8.2	15.5	18.2
USA	9.0	26.6	30.0

Source Extracted from IRENA [65]

Efficiency technologies helping to drive down energy intensity levels include on the demand side, the electrification of domestic appliances, such as heat pumps and electric cookers, and the gradual adoption of electric vehicles (EVs) including E-Buses, and on the supply side, renewable energy technologies, such as improvements in electrolysis, battery storage, and solar PVs. An important point the report makes is that planning is at its best when it becomes holistic, what it names as 'sector coupling' joining up plans for power, heating (or air conditioning) and mobility, such as electrified vehicles and a recharging infrastructure to support them. But as an aside, IRENA also notes that in international energy statistics, *all* renewables are regarded as 100% efficient compared with efficiency levels of 25%–85% for fossil-fuel power plants, adding that the 'measured magnitude of this phenomenon can change depending on the primary energy accounting methodology'.

Energy efficiency, *as measured by the rate of decline in energy intensity*, is central to the IEA's NZE scenario, being the 'single largest measure to avoid energy demand' up to 2050. Accompanying it are 'closely related measures of electrification, behavioural change, digitalisation and material efficiency' [66]. In the NZE scenario, global energy efficiency needs to grow by around 4% per year compared 'with 1.7% over the last 10 years'. However, during periods of volatile demand for fuel, such as that following Russia's invasion of Ukraine, the annual rate of energy efficiency can change quickly. The IEA is typically optimistic about new energy technologies, noting annual energy efficiency investment 'is up 45% since 2020, with particularly strong growth in electric vehicles and heat pumps' [67]. Figure 2.7 lists the areas of greatest growth.

But achieving an overall annual rate of energy efficiency growth of 4% is, to put it mildly, challenging. Nevertheless, the optimism of the IEA (see above) shines through: 'In the past 10 years, 90% of countries have achieved the 4% rate at least once, and half have done so at least three times. However, only four G20 countries—China, France, the United Kingdom, and Indonesia—have done so over a continuous 5-year period within the last decade, though several others have come close'.

Technologies

There is plenty of literature on solar panels and wind turbines, but there are also less well-known emerging technologies and the repurposing of existing technologies, including recycling such as using spent batteries for power storage, which can help boost renewable power generation and energy consumption. On the supply side, two novel technologies under development can be mentioned. In Sweden an aircraft manufacturer has pioneered 5-m-wide underwater 'tidal kites'—the next generation will have 12-m wingspans—which move underwater in a figure-of-eight pattern, absorbing energy from the running tide. They are tethered to the fjord seabed by 40-m metal cables. Like aircraft they rise and fall as pressure builds up and lowers beneath the wings. They are computer driven into the areas of the ocean with the highest waves and connected by cable to transit electric power back to the shore. They

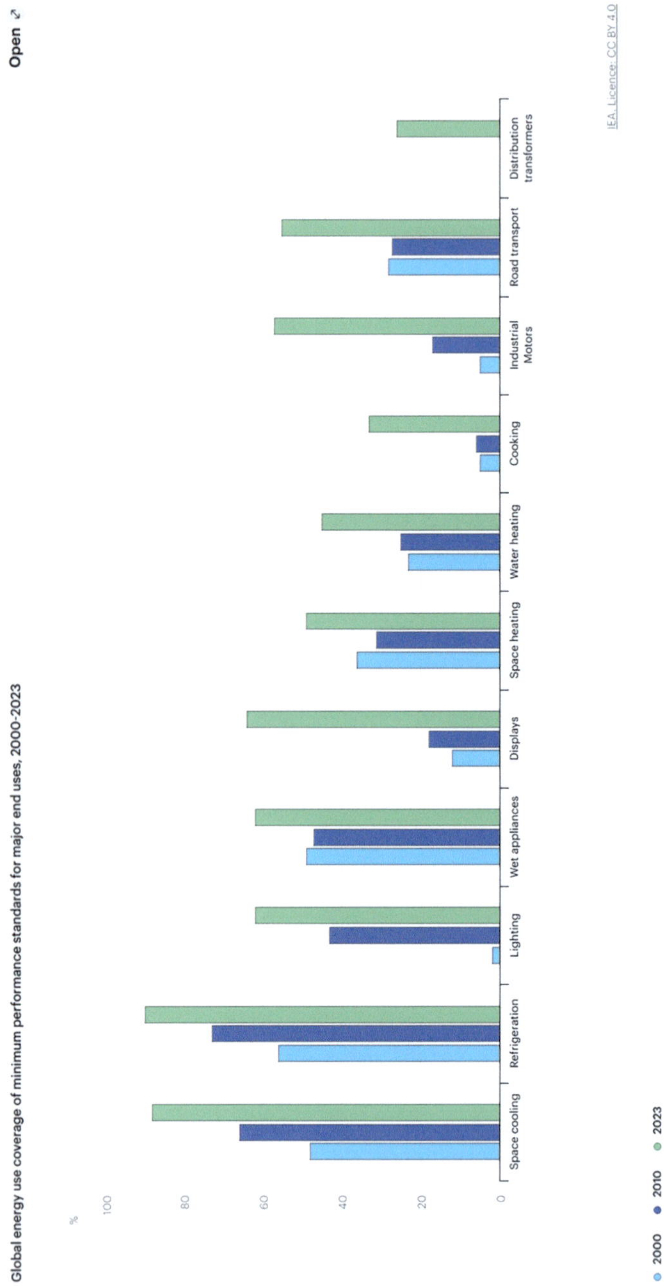

Fig. 2.7 Energy performance standards increase

are designed to achieve net zero for the Faroe Islands off the coast of Iceland by 2030 [68]. However, typical of tidal technologies, they require an upgraded onshore grid to manage the generated capacity.

An example of repurposing is a teleport in Switzerland where an array of specially angled solar panels has been fitted across the dish of an old C-band telecommunications satellite earth station. A satellite dish with a diameter of 32 m contains 307 solar panels with a rated power of 116 kW, producing upwards of 110,000 kWh annually [69]. The power output of a typical grounded solar panel during sunlight is estimated between 250 and 400 watts [70], generating 1.5 kWh, so 307 grounded solar panels would generate between 77 and 123 kW, of comparable power output. A satellite earth station dish plus the cabling and control buildings should take up far less land than a solar farm which has a negative ecological externality.

Recycling is another important and, for the moment, an underutilised option to reduce the demand for the minerals that are critical to batteries and other electrical devices and components. One report notes that research studies 'conducted on the topic of recycling solar panels have resulted in numerous technologies. Some even reach an astonishing 96% recycling efficiency, but the aim is to raise the bar higher in the future' [71]. Currently, most panels end life in refill land sites. The problem remains how profitable can it be for commercial companies to undertake the recycling process. That is unresolved.

Grids

Feeding renewable electricity back into a national or regional grids is a problem akin to providing the infrastructure to recharge EVs. Due to its expense, it is rarely undertaken by the private sector without state subsidy [72]. One model for any networked industry, such as telecoms or gas or water, is a structural separation, one company operating the network on a wholesale basis, while competing retailers buy capacity. This normally requires a regulatory framework to prevent collusion or anti-competitive practices. But the grid is crucial to the success of renewable sources of energy such as wind, wave power, or solar because they are often location bounded, while other areas have limited access. According to one source, USD3.1 trillion of investment is necessary by 2030, adding 18 million kilometers of grid network worldwide [73]. It is difficult to see low-income countries financing nationwide grids. A World Bank report in 2024 covering over 180 utilities in more than 90 countries shows that 'fewer than 40 percent of utilities are able to collect sufficient revenue to meet their annual operating and debt servicing costs, the bare minimum for financial sustainability' [74].

Population Trends and the Demand for Energy

By 2050, two-thirds of countries will fall below the total fertility rate (TFR) of 2.1 children per person giving birth, which is the minimum rate required for population replacement [75]. Half of all countries are already below this rate, but as projected by the United Nations, the global population will rise to reach over 9 billion persons by 2050, and then peak at around 11 billion by 2100 [76], or otherwise by 2080 at below 11 billion, according to the source. On its own that would suggest, other things being equal, the demand for energy will continue to increase until the end of the century. But other things are not equal. Increased energy efficiency and new technologies may reduce the levels of energy intensity. But so may the distribution of population growth. According to the IMF, 'nearly 2 billion net additions to world population projected over the next three decades will occur in less developed regions' [77]. But that in turn could increase the demand for electricity in countries that currently lack it. More than 8 out of 10 people will live in Asia or Africa by 2100 [78]. Figure 2.8 shows the expected world population by 2100 and population growth rates since 1700, peaking in 1963 [79].

For the planet this trend reversal of population growth must be good news, and especially for other species, although it poses major economic and social problems for countries with diminishing workforces. But unless there is a dramatic turn-around in the *per capita* incomes of poor countries and/or significant investment in power

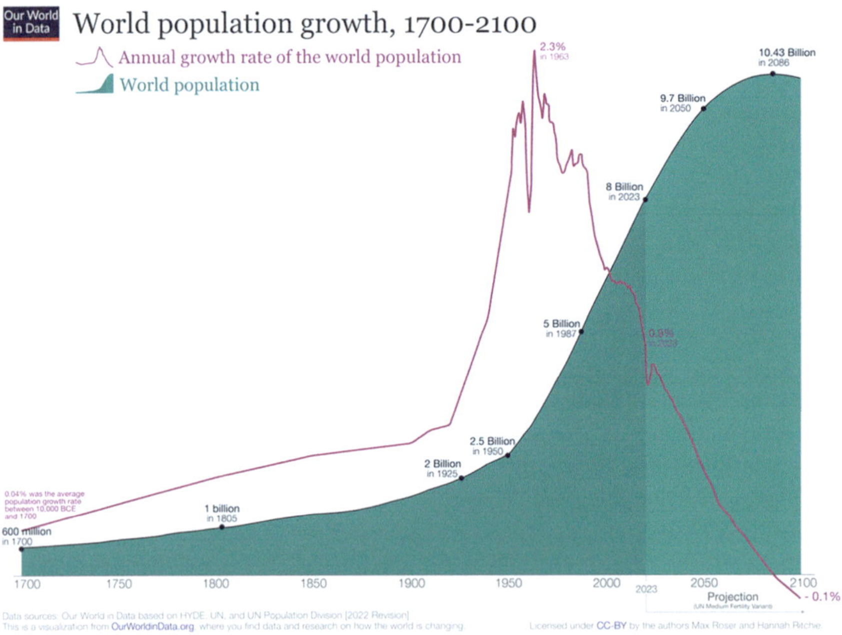

Fig. 2.8 World population by 2100 peaks at 10.43 billion. *Source* Our World in Data

supplies in those countries, growth in demand for energy per unit of GDP is likely to fall, and could even do so in absolute terms by 2100. For example, China's population growth has already turned a corner, with some projections to 2100 seeing a fall from above 1.4 billion people to 800 million or less [80]. This should speed China's journey to net zero. India's population growth is expected to continue into the 2060s and could peak between 1.5 billion and 2 billion [81]. The number of children born in India peaked in the first decade of the 2000s [82]. Indonesia is expected to reach 288 million people by 2050, and then decrease to 243 million by 2100 [83]. In the USA, the population is expected to peak at 367 million by 2080 and then begin to decline [84]. By contrast, population growth across Africa is expected to reach a staggering 4.2 billion, just behind 4.8 billion in Asia [85], countries and populations highly vulnerable to climate change.

These figures herald an enormous shift in world demographics and society, each having profound global economic, cultural, and political effects. Migration within countries and across borders, between continents will be aggravated not only by economic disparities between the Global North and the Global South, but by the challenge of global warming and boiling hot temperatures and droughts (or water deluges) in countries too poor to take mitigating and adaptive action. It is usually the potentially productive younger and more able-bodied who migrate but that will be a drain on the productivity of poorer countries. It will also change the geographical patterns of demand for energy, as it will for consumption, much of which is already online and dependent upon energy-hungry logistics and transport. Ageing populations in poorer countries will require more healthcare and special housing and a supporting energy infrastructure for those facilities. The need for international help with investment and finance will not be confined to achieving 1.5 °C by 2050 or by 2100, it will also be needed to avoid humanitarian emergencies in low- and many middle-income economies.

Conclusions

Derived from the models of the IPCC which establish a baseline from which global warming scenarios of 1.5 °C or above are derived, pathways towards keeping global warming within those limits are the guardrails of the Paris Agreement of COP-15. Pathways are just that, ways ahead modelled upon various assumptions. The global average temperature for the 10-year period, from 2014 to 2023, was around 1.2 °C above the 1850–1900 average [86]. From July 2003 it exceeded above 1.5C and has remained above 1.5C ever since. But the fact that 1.5 °C has been finally breached does not alter the urgency of slowing down and ultimately reversing the process of global warming. But net zero is unlikely to achieve that. Policies need to recognise the need for net negative.

The consequences of failure are seriously colossal. The pathways are illustrative of what could be done were there a collective international will to do so. The importance of the pathways is they identify the various components of what needs to be achieved, such as the levels of energy-source replacement by renewables, the levels of energy efficiencies required, in which sectors of the economy they can be achieved, and the estimated costs and benefits involved. The latter is crucial because it is clear that driving down levels of energy consumption is unlikely to be the complete and achievable answer. That poses unwelcome trade-offs, such as a continued reliance upon fossil fuels for 'energy security' unless alternatives like nuclear power and various other technologies such as CCUS or CDR can bridge the gap. However, in 2025, serious doubts were cast on the effectiveness and cost of immature CCUS technologies, such as CDR (Carbon Dioxide Removal), by a UK Parliamentary Committee [87], and even if they work and prove cost-effective, such a progression would be not much better than a step towards net zero, and not towards net negative which is what is needed to reverse the trend of GHG emissions.

If there is an answer it probably lies in governments mandating and progressively lowering carbon ceilings on polluters, using carbon taxes which genuinely reflect the social costs of emissions, meaning rates well above USD150 per tonne of carbon, and finding ways to mobilise the private sector and capital markets to invest in and adopt clean technologies and outputs across Scopes 1, 2, and 3 combined. But markets respond to profitability, and the reality is, in the language of economists, markets generally fail to account for social costs or externalities. The economics of climate change is examined in Chap. 3.

An additional problem, a massive one and one which is politically unavoidable, is that markets are dominated by vested interests, partisan interests, highly influential and controlling interests, and none more so than the fossil-fuel interests. A story in The Guardian reveals the big oil companies were aware of the global warming effects of fossil fuels from as early as 1954 [88]. But that does not negate the importance of the enterprise sector, especially the major multinational companies, responding to public pressure to reduce their emissions, especially the 57 companies seemingly responsible for 80% of global emissions as suggested by research in 2024 [89]. That would be a good place to start.

The pressure has to come ultimately from civil society and from families suffering ill-health from the effects of GHG pollution and feeling the loss of the local habitat and the degradation of the environment. There are numerous books and journals covering the key issues of biodiversity and environmental stewardship, and this book only touches on the point where climate mitigation and environmental protection meet. There is a series of COPs on biodiversity in parallel to the COPs on climate change, and the two areas are interdependent. COP-16 (2024) on biodiversity recognised the importance of local communities and the need for their voices to be heard. These issues, the role of government, the role of the private sector, the role of the financial sector, and the role of civil society will each be the subject of the following chapters.

References

1. IPCC (2029) Special Report on Global Warming https://www.ipcc.ch/site/assets/uploads/sites/2/2022/06/SR15_Full_Report_HR.pdf
2. Climate Portal (7 November 2024) What would happen if the Atlantic Meridional Overturning Circulation (AMOC) collapses? How likely is it? https://climate.mit.edu/ask-mit/what-would-happen-if-atlantic-meridional-overturning-circulation-amoc-collapses-how-likely#:~:text=Such%20a%20collapse%20would%20be,would%20play%20out%20once%20again.
3. IEA (2023) World Energy Outlook 2023 https://www.iea.org/reports/world-energy-outlook-2023/executive-summary
4. Nolan Lindquist (June 2024) Climbing down from the fossil peak? Center for Active Stewardship (CAS)—by subscription
5. IEA (2024) Latest news, commentaries and reports https://www.iea.org/
6. ExxonMobil (2024) Advancing Climate Solutions Report https://corporate.exxonmobil.com/sustainability-and-reports/advancing-climate-solutions
7. OPEC (2024) Press Releases https://www.opec.org/opec_web/en/press_room/28.htm
8. Shell (2024) World Energy Model https://www.shell.com/news-and-insights/scenarios/what-scenario-planning-models-does-shell-use/world-energy-model.html
9. BNEF (2024) Opportunities generated https://about.bnef.com/?utm_medium=Adwords_SEM&utm_source=pdsrch&utm_content=amer_bnef_2024&utm_campaign=833878&tactic=833878&gad_source=1&gclid=CjwKCAjwmrqzBhAoEiwAXVpgopsZnEMA5N_8nkk0pPeDWaRuwq0_o7IguME0j4Yz5INLY7huEhjS1RoCNw4QAvD_BwE&gclsrc=aw.ds
10. Rystad Energy (2024) Navigating the Future of Energy https://www.rystadenergy.com/?gad_source=1&gclid=CjwKCAjwmrqzBhAoEiwAXVpgokuD3TQ6ceE8yQKMTL4m_r-Npjpa3bea2cvt_wcFuRTLPf9l9YBHSxoC6w8QAvD_BwE
11. MIT Climate Change Portal (2024) How do we know how long carbon dioxide remains in the atmosphere? https://cgcs.mit.edu/how-do-we-know-how-long-carbon-dioxide-remains-atmosphere
12. GHG Protocol (2024) What is GHG Protocol? https://ghgprotocol.org/about-us
13. Sustainability for All (2023) The era of global boiling: the latest twist in the climate crisis https://www.activesustainability.com/climate-change/global-boiling/?_adin=02021864894
14. Science Alert (March 2024) Record Heat Expected Again in 2024 as UN Warns Earth Is 'On The Brink' https://www.sciencealert.com/record-heat-expected-again-in-2024-as-un-warns-earth-is-on-the-brink
15. John Ure (2024) Achieving the UN Sustainable Development Goals: Late or Too Late? Emerald Publishing, Leeds, UK
16. Eurostat (2018) Glossary: Primary energy consumption https://ec.europa.eu/eurostat/statistics-explained/index.php?title=Glossary:Primary_energy_consumption#:~:text=Primary%20energy%20consumption%20measures%20the%20total%20energy%20demand,energy%2C%20and%20the%20final%20consumption%20by%20end%20users
17. Eurostat (2018) Glossary: Final energy consumption https://ec.europa.eu/eurostat/statistics-explained/index.php?title=Glossary:Final_energy_consumption
18. Hannah Ritchie and Pablo Rosado (2020) 'Energy Mix' Our World in Data https://ourworldindata.org/energy-mix
19. Our World in Data (2024) Energy Production and Consumption https://ourworldindata.org/energy-production-consumption
20. Energy Institute (2023) Statistical Review of World Energy https://www.energyinst.org/statistical-review
21. IEA (2019) Key World Energy Statistics 2019 https://www.iea.org/reports/key-world-energy-statistics-2021/final-consumption
22. Dictionary.com Joule https://www.dictionary.com/browse/joule
23. The Metric System The International System of Units (SI) https://metricsystem.net/
24. Techtarget (2023) What is a Watt? https://www.techtarget.com/whatis/definition/watt

25. IEA (2023) Renewables 2023 https://prod.iea.org/reports/renewables-2023/executive-summary
26. IEA (2021) Key World Energy Statistics 2021 https://iea.blob.core.windows.net/assets/52f66a88-0b63-4ad2-94a5-29d36e864b82/KeyWorldEnergyStatistics2021.pdf
27. IEA (2024) Renewables: Executive summary https://prod.iea.org/reports/renewables-2023/executive-summary
28. Our World in Data (2024) Renewable Energy https://ourworldindata.org/renewable-energy#article-citation
29. IEA (2024) Global Energy and Climate Model: Documentation 2023 https://iea.blob.core.windows.net/assets/ff3a195d-762d-4284-8bb5-bd062d260cc5/GlobalEnergyandClimateModelDocumentation2023.pdf
30. IEA World Energy Outlook 2023 https://iea.blob.core.windows.net/assets/86ede39e-4436-42d7-ba2a-edf61467e070/WorldEnergyOutlook2023.pdf IEA
31. IEA World Energy Outlook 2022 https://www.iea.org/reports/world-energy-outlook-2022/key-findings#abstract
32. Weforum.org (17 April 2025) Clean energy's share of electricity reaches 40%, and other nature and climate news https://www.weforum.org/stories/2025/04/clean-energy-electricity-nature-and-climate-stories-this-week/
33. IEA (2023) Renewables 2023: Executive Summary https://prod.iea.org/reports/renewables-2023/executive-summary
34. IRENA (2023) Global Landscape of Renewable Energy Finance, 2023 https://www.irena.org/Energy-Transition/Finance-and-investment/Investment
35. IRENA (2023) Worl d Energy Transitions Outlook 2023: 1.5°C Pathway https://www.irena.org/Publications/2023/Jun/World-Energy-Transitions-Outlook-2023
36. IRENA (2023) World Energy Transitions Outlook 2023: Volume 2 https://mc-cd8320d4-36a1-40ac-83cc-3389-cdn-endpoint.azureedge.net/-/media/Files/IRENA/Agency/Publication/2023/Jun/IRENA_World_energy_transitions_outlook_2023.pdf?rev=db3ca01ecb4a4ef8accb31d017934e97
37. BDC Weighted average cost of capital https://www.bdc.ca/en/articles-tools/entrepreneur-toolkit/templates-business-guides/glossary/weighted-average-cost-of-capital
38. WEI (2022) World Energy Investment 2022: Methodology Annex https://iea.blob.core.windows.net/assets/79c65e95-baf6-4cc3-b16f-9ea7c7e6be42/WorldEnergyInvestment2022MethodologyAnnex.pdf
39. IEA (2024) Carbon Capture, Utlisation and Storage https://www.iea.org/energy-system/carbon-capture-utilisation-and-storage
40. IEA (2023) World Energy Outlook, 2023 https://www.iea.org/reports/world-energy-outlook-2023
41. ICMA (June 2022) Green Bond Principles https://www.ifc.org/content/dam/ifc/doclink/2022/the-green-bond-principles-202206.pdf
42. Climate Bonds (2024) The Climate Bonds Standard https://www.climatebonds.net/standard/the-standard
43. RBC What are ESG ETFs? https://www.rbcgam.com/en/ca/learn-plan/types-of-investments/what-are-esg-etfs/detail
44. IEA (2023) World Energy Outlook 2023: Executive Summary https://www.iea.org/reports/world-energy-outlook-2023/executive-summary
45. IEA (2022) World Energy Investment 2022 https://iea.blob.core.windows.net/assets/b0beda65-8a1d-46ae-87a2-f95947ec2714/WorldEnergyInvestment2022.pdf
46. E3G (2021) Explained: what does 'unabated coal' mean? https://www.e3g.org/news/explained-what-does-unabated-coal-mean/
47. Our World in Data (2023) Carbon intensity of electricity generation, 2023 https://ourworldindata.org/grapher/carbon-intensity-electricity
48. IRENA (2023) World Energy Transitions Outlook, 2023 https://mc-cd8320d4-36a1-40ac-83cc-3389-cdn-endpoint.azureedge.net/-/media/Files/IRENA/Agency/Publication/2023/Jun/IRENA_World_energy_transitions_outlook_2023.pdf?rev=db3ca01ecb4a4ef8accb31d017934e97

49. IEA 50 Carbon intensity of electricity generation in selected regions in the Sustainable Development Scenario, 2000–2040 https://www.iea.org/data-and-statistics/charts/carbon-intensity-of-electricity-generation-in-selected-regions-in-the-sustainable-development-scenario-2000-2040
50. IEA (2023) Renewables 2023: Executive Summary https://www.iea.org/reports/renewables-2023/executive-summary
51. IRENA (2023) World Energy Transitions Outlook 2023 https://www.irena.org/Publications/2023/Jun/World-Energy-Transitions-Outlook-2023
52. IEA (2022) World Energy Outlook 2022 https://iea.blob.core.windows.net/assets/830fe099-5530-48f2-a7c1-11f35d510983/WorldEnergyOutlook2022.pdf
53. California ISO (2023) Understanding electricity http://www.caiso.com/about/Pages/OurBusiness/Understanding-electricity.aspx
54. Hannah Ritchie, et al. (2020) Energy Production and Consumption OurWorldInData.org. Retrieved from: 'https://ourworldindata.org/energy-production-consumption' Our World in Data (2024) Energy Production and Consumption https://ourworldindata.org/energy-production-consumption?ref=ai-infra-link.com
55. United Nations Climate Action https://www.un.org/en/climatechange/all-about-ndcs
56. Enerdata Total electricity generation https://eneroutlook.enerdata.net/total-electricity-generation-projections.html
57. Department of Energy DOE Explains...Nuclear Fission https://www.energy.gov/science/doe-explainsnuclear-fission
58. NEI Nuclear Waste https://www.nei.org/fundamentals/nuclear-waste
59. Wyofile.com (15 January 2025) Natrium 'advanced nuclear' power plant wins Wyoming permit https://wyofile.com/natrium-advanced-nuclear-power-plant-wins-wyoming-permit/
60. TerraPower (2024) Natural Balance https://www.terrapower.com/
61. Department of Energy Concentrating Solar-Thermal Power Basics https://www.energy.gov/eere/solar/concentrating-solar-thermal-power-basics
62. IRENA (2023) WORLD ENERGY TRANSITIONS: OUTLOOK 2023 https://mc-cd8320d4-36a1-40ac-83cc-3389-cdn-endpoint.azureedge.net/-/media/Files/IRENA/Agency/Publication/2023/Jun/IRENA_World_energy_transitions_outlook_2023.pdf?rev=db3ca01ecb4a4ef8accb31d017934e97
63. Energy.Gov Energy Efficiency: Buildings and Industry https://www.energy.gov/eere/energy-efficiency-buildings-and-industry
64. Energy.Gov Energy Efficiency vs. Energy Intensity https://www.energy.gov/eere/analysis/energy-efficiency-vs-energy-intensity
65. IRENA (2017) Synergies between renewable energy and energy efficiency https://www.irena.org/publications/2017/Aug/Synergies-between-renewable-energy-and-energy-efficiency
66. IEA 50 (2024) Energy Efficiency 2023 https://iea.blob.core.windows.net/assets/dfd9134f-12eb-4045-9789-9d6ab8d9fbf4/EnergyEfficiency2023.pdf
67. IEA (2023) Energy Efficiency 2023 https://iea.blob.core.windows.net/assets/dfd9134f-12eb-4045-9789-9d6ab8d9fbf4/EnergyEfficiency2023.pdf
68. BBC (2021) The underwater 'kites' generating electricity as they move https://www.bbc.com/news/business-59401199
69. Sonnenallee Giving old satellite dishes a new lease on life—a great upcycling idea for parabolic antennas https://sonnenallee.sma.de/en/solar-energy/giving-old-satellite-dishes-a-new-lease-on-life/
70. MarketWatchGuides (2024) How Much Power Does a Solar Panel Produce? https://www.marketwatch.com/guides/solar/power-solar-panels-produce/
71. GreenMatch (2024) The Opportunities of Solar Panel Recycling https://www.greenmatch.co.uk/blog/2017/10/the-opportunities-of-solar-panel-recycling
72. Energy.Gov Federal Financing Tools https://www.energy.gov/gdo/federal-financing-tools
73. RystadEnergy (2024) Enable or inhibit: Power grids, key to the energy transition, require $3.1 trillion in investments by 2030 https://www.rystadenergy.com/news/power-grids-investments-energy-transition-permitting-policies

74. World Bank (2024) The Critical Link: Empowering Utilities for the Energy Transition https://documents1.worldbank.org/curated/en/099061724143528903/pdf/P50045015689f70be1b3cf1f6f94e20ac35.pdf
75. IHME (March 2024) The Lancet: Dramatic declines in global fertility rates set to transform global population patterns by 2100 https://www.healthdata.org/news-events/newsroom/news-releases/lancet-dramatic-declines-global-fertility-rates-set-transform
76. UN Shifting Demographics https://www.un.org/en/un75/shifting-demographics
77. IMF (2020) Population 2020 https://www.imf.org/en/Publications/fandd/issues/2020/03/changing-demographics-and-economic-growth-bloom
78. Our World in Data (2019) More than 8 out of 10 people in the world will live in Asia or Africa by 2100 https://ourworldindata.org/region-population-2100
79. Our World in Data (2023) Two centuries of rapid global population growth will come to an end https://ourworldindata.org/world-population-growth-past-future
80. Pew Research Center (2022) Key facts about China's declining population https://www.pewresearch.org/short-reads/2022/12/05/key-facts-about-chinas-declining-population/
81. Pew Research Center (2023) Key facts as India surpasses China as the world's most populous country https://www.pewresearch.org/short-reads/2023/02/09/key-facts-as-india-surpasses-china-as-the-worlds-most-populous-country/
82. The World in Data (2019) India's population growth will come to an end: the number of children has already peaked https://ourworldindata.org/indias-population-growth-will-come-to-an-end
83. The World Counts The population of Indonesia is estimated to grow from 272 million people in 2020 to 298 million people in 2030 https://www.theworldcounts.com/populations/countries/indonesia
84. USA Facts (2024) What will America's population look like by 2100? https://usafacts.org/articles/what-will-americas-population-look-like-by-2100/
85. SAIS Review of International Affairs (2024) How a Population of 4.2 Billion Could Impact Africa by 2100: The Possible Economic, Demographic, and Geopolitical Outcomes https://saisreview.sais.jhu.edu/how-a-population-of-4-2-billion-could-impact-africa-by-2100-the-possible-economic-demographic-and-geopolitical-outcomes/
86. Climate Council (2024) Have we already gone past 1.5 degrees of warming? https://www.climatecouncil.org.au/resources/have-we-already-gone-past-1-5-degrees-warming/
87. UK Parliament (February 2025) Carbon capture: High degree of uncertainty whether risky investment by Govt will pay off https://committees.parliament.uk/committee/127/public-accounts-committee/news/205139/carbon-capture-high-degree-of-uncertainty-whether-risky-investment-by-govt-will-pay-off/
88. The Guardian (12 November 2024) Big oil firms knew of dire effects of fossil fuels as early as 1950s, memos show https://www.theguardian.com/us-news/2024/nov/12/big-oil-fossil-fuel-warning
89. Development Aid (30 April 2024) Just 57 companies accounted for 80% of the world's CO_2 emissions in the last 6 years https://www.developmentaid.org/news-stream/post/178528/57-companies-companies-responsible-80-of-the-worlds-co2-emissions

Chapter 3
Economics of Climate Change

Tipping Points

Of the many fundamental challenges facing economic modelling, the most limiting is their inability to handle tipping points or their consequences, that is, (mathematically) their exponential. Standard economic models are typically not designed to include tipping points. Tipping points, which could lead to runaway climate change—'runaway' could be a hundred years or more—undermine the concept of a system of self-adjusting capitalist markets which always return to an equilibrium, which is the norm for conventional economic theory and its models. Tipping points are events such as the melting of the Artic ice cap which not only raises ocean tides to life-threatening levels, but also reduces the reflection of solar energy back to into Outer Space and exposes buried carbon and methane to release yet more GHGs into the atmosphere.

A concomitant impact of the melting is a change in the direction of the Gulf Stream which brings warm water up from the Gulf of Mexico into the Atlantic Ocean, permanently altering weather patterns. But the more impactful part of the story is what happens to the Atlantic Meridional Overturning Circulation (AMOC) which brings warm water up from the equator to the poles. Changes in the AMOC interrupt the wet-warming effects of high humidity and temperature of the Gulf Stream as it approaches the North Pole. The National Oceanic Service of the NOAA describes the AMOC (see Fig. 3.1) as the slow conveyor belt of the oceans and cites research that suggests it may be slowing down even further than forecast as a result of global warming. If that happens, 'freshwater from melting ice at the poles would shift the rain belt in South Africa, causing droughts for millions of people. It would also cause sea level rise across the U.S. East Coast' [3].

Due to their runaway potential, tipping points would run riot over the aim of economic models which is primarily to estimate the costs of climate change as it occurs. The basic assumption is that, on the one hand, warming could raise agricultural productivity in some regions because photosynthesis responds well to higher

J. Ure, *Climate Change and Carbon Markets: Late or Too Late?*,
https://doi.org/10.1007/978-3-032-07619-9_3

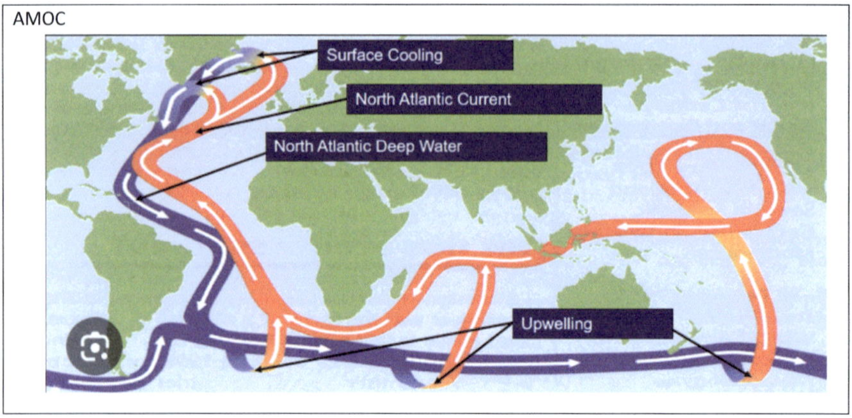

Fig. 3.1 Changes in the Atlantic Meridional Overturning Circulation (AMOC) disrupt the Gulf Stream. *Source* CarbonBrief [4]

levels of CO_2 in the atmosphere, even if humans do not. On the other hand, in other regions, the weather becomes too hot and too dry, or the regions suffer hurricanes/typhoons and inundation from torrential flooding, which will result in lower output and higher costs. The benefits and costs are additions to and deductions from the anticipated growth in GDP, although GDP is widely recognised as a very poor baseline. For example, if more has to be spent on air conditioning, or buildings need to be retrofitted to keep them cool using additional energy, these costs are usually recorded as benefits due to the income they generate to those making and installing them. By contrast, if a forest is cut down to extend an urban area, there may be no recorded lost benefits attributed to the destruction of the forest, despite it being a natural carbon sink; indeed, it may be that GDP goes up by the value paid to the landowner. These economic models are not models of welfare, they are models of market prices, including shadow (attributed) prices, where, as in the wilds of nature, there are no markets.

Incomplete Models

Most economic models treat climate change as an externality, as something that happens outside of an enterprise or the economy but which has an impact upon the economy, rather like a large asteroid hitting the planet, or a very severe volcanic eruption. But this is insufficient. Just as wars are the result of human activity, so is much of climate change, and that means models should attempt to treat them as such, as endogenous rather than as exogenous. The act of creating GDP is also the act of creating GHGs unless entirely green methods are used. In terms of mathematics, endogenous modeling is difficult because it adds considerably to the complexity of the model—see below. It is easier and cheaper to treat possible climate changes as

Table 3.1 Levels of climate change risk versus precautionary action

Risk level	Probability of global warming impact high	Probability of global warming impact low
Risk high	Red alert for immediate action	Orange alert for precautionary action
Risk low	Orange alert for precautionary action	Green alert for precautionary action

Source Author

exogenous, as outside of the model. On a grander scale, it is equally easier to begin modelling by abstracting from the real dynamics of an economic system, and to treat the entire capitalist system of markets as self-regulating, and attuned to adjusting to external shocks, always returning to an equilibrium state. This is a state in which economic growth maintains a balance of supply and demand and therefore of prices between all the key variables. The proposition that the realism of the assumptions of economic models is less important than their outcomes has a long-established history among some sections of economists, but heavily criticised by others [5]. As explained below, this is a methodological choice which offers modelling benefits (primarily it takes less computational effort by economists) and would not find such widespread acceptance across the academic community of economists were it not for the fact that the most damaging effects of climate change are still to come, are still to be observed, are still to become inputted data.

The insurance industry is based upon the truism that things yet to come are difficult to estimate. Often times, the way people make an assessment is upon their perceptions of risk-probability of something happening versus the probability of how bad the outcome might be. This is illustrated in Table 3.1 where risk levels and outcome levels are set as either high or low. In this table, risk refers to something actually happening, and the probability refers to the level of severity of something happening.

It is worth pointing out here that the role of scepticism plays an obvious part in where an assessment places the risk/probability. For example, a prominent conservative critic of climate change politics, Roger Pielke [6], while not denying climate change is real and is man-made, has long argued that the findings of the IPCC do not support the contention that GHGs are responsible for more frequent, more violent weather events [7].

Discount Rates

Whether or not true, many of the risk levels, and the probabilities of high or low impacts of climate change, may seem too far ahead for people to experience in their lifetime. Although millions of people around the globe are already suffering difficult-to-manage temperature rises, droughts, violent storms, rising sea levels, and environmental degradation, many are not, at least not yet. Even in the richest country in the world, the USA, where tornadoes and other extreme weather events occur on a frequent basis, it remains a small minority who lose their homes, sometimes their lives, or call upon their insurance. And yet, concern for the future is widespread. A Lloyds Registry Foundation *World Risk Poll* report in 2024 covering 142 countries found that '72% of the world's adult population now say they feel "very" or "somewhat" threatened by climate change in the next 20 years' [8].

A concern for future generations is now a very real one. The way economists, and insurance brokers, manage expectations of risk is to apply an interest rate that discounts future values into today's values. If what is at stake is an investment, then a high rate of interest implies money today is highly valued over money tomorrow, and the converse holds with a low rate of discount. The choice of the discount rate in economic models of climate change therefore makes a big difference to the estimated value today of the costs and benefits that climate change may bring in the future. A high discount rate will downplay the implied longer-term concerns, reducing the present-day value of future benefits. This became a major point of criticism of the 2018 Nobel-Prize-winning economic model by Prof. William Nordhaus, who used a significantly higher rate than that used in 2006 by Prof. Nicolas Stern of the Grantham Research Institute on Climate Change, the London School of Economics (LSE), in his report *The Economics of Climate Change: The Stern Review* [9]. An LSE commentary in 2019 refers rather scathingly to the 2018 award to Prof. Nordhaus as 'A Nobel Prize for the creator of an economic model that underestimates the risks of climate change' [10]. Thus, economists disagree with one another perhaps even more than climatologists do, and with good reason if their models are equally poles apart.

The Concept of Equilibrium Versus *Tipping Points*

Equilibrium Climate Sensitivity (ECS) models are basic to climatology—see Chap. 1. The proposition is that the Earth emits back into Outer Space an energy equal to what it receives from the Sun. If due to GHS, less heat is naturally radiated back then average global temperatures rise and additional heat is then forced from the Earth's atmosphere to rebalance the energy budget the Earth has with the Sun. If this did not happen, the Earth would simply heat up until the atmosphere disappears, as is the case with Mercury, or becomes mostly CO_2, as is the case with Mars and Venus. Global warming, involving the heating of oceans and the melting of ice, releases methane, carbon dioxide, nitrous dioxide, and other GHGs (including most importantly water vapour) trapped in clathrates, which are a substrate of molecules of one substance contained within the molecular latticework of another. These GHGs interfere with the normal outgoing of infrared radiation and unless there is a way to bring the heat exchange between the Sun and the Earth back into equilibrium the long-term effect can be runaway warming. An equilibrium, albeit a changing one, can only be sustained long term if climate change avoids a tipping point, a runaway phenomenon.

Equilibria are also used in economic models, but in two very distinct ways. To explain this, it is necessary to acknowledge very broadly two schools of thought, although within each school there are many variants. From ancient times there has been what the Greeks called 'oikonomia' or economics, which was basically the good management of household budgets, what today corresponds to what is called microeconomics; and 'political arithmetick' or political economy, which was about how to run national or state finances, and would today be referred to as macroeconomics. In this sense, economics was all about equilibria or balanced budgets, a household should not spend more than it receives in income. In political economy, it was more complex. For example, in the Middle Ages the school known as the Mercantilists argued in favour of net surpluses in trade with other countries and for a country to accumulate as much wealth in gold as possible. These ideas were challenged by what have become known as the classical economists, widely regarded as the originators of modern economic thought.

Classical Economics

In 1776, Adam Smith published what is regarded as the foundation of Classical economics, *The Nature and Causes of the Wealth of Nations* to refute mercantilism, and to show how national prosperity would be increased if both imports and exports were allowed to grow in balance. The other most influential Classical theorist was David Ricardo, who in 1817 published *On the Principles of Political Economy and Taxation*, and argued that if each country concentrated its trade upon exporting what it does best compared with anything else it could export, and importing from others what they do best compared with anything else they could export, overall trade and welfare would be maximised. The Classical school in many ways broke down the separation of economics and political economy, and focused instead upon issues of land, labour, capital, and property ownership and incomes such as wage-labour, profits, rents, and interest.

In the mid-nineteenth century, Karl Marx developed a systematic critique of classical economics, emphasising capital accumulation as the source of economic power and growth in capitalist societies. He reasoned that capital extracted 'surplus value' from wage-labour (he termed it 'exploitation') which formed the basis of all property incomes such as rents, but especially of profits, and that as capital replaced labour, the source of profitability would grow slower, subject to the rate of productivity growth, than overall capital, resulting in a series of capitalist crises. Although this was a model of interrupted growth, one which Marx thought would ultimately destroy capitalism, in his posthumously published Volumes 2 and 3 of his book *Capital* central importance is given to what are termed the reproduction schemes and the circulation of capital in the system as a whole. These are modelled as an equilibrium between Department 1 which produces capital goods, and Department 2 which produces consumer goods. A falling *general rate of profit* would periodically upset this equilibrium resulting in an overcapacity to produce capital goods. It would take crises, such as widespread bankruptcies, to restore the equilibrium.

The important methodological point is that even in Marx, to explain the system of capitalist markets, there is a *hypothesised* equilibrium at its epicentre. The real story is not so much about this equilibrium, but more about how and why capitalist economies are constantly in disequilibria and having to adjust. The equilibrium conditions hold only *ceteris paribus* ('all other things being equal') which is a common feature of economic modelling. What differentiates the work of Marx from most others is that he viewed as the key variables the class interests that made up the competitive aspects of capitalist society.

Climate does not directly feature in Marx's model of capitalism, but nature does, albeit tangentially. In some ways like Adam Smith who sees land as being fenced off from access by common labour to become private property that yields a rent for the landlord, Marx sees labour as alienated from nature and from pre-capitalist society, an alienation that drives otherwise landless and unemployed populations into capitalist markets and changes social values. At the same time, the capacity of mankind to transform nature to the needs of society was an exemplifier of human capabilities, but the direction of those capabilities, although enabled by nature, was determined by the nature of society.

Today, most radical critiques of the systematic commercialisation and destruction of the biosphere, the loss of species, and of climate change see it as driven by the greed of capitalism, notably of the fossil-fuel industry. Actually, Marx did not place greed as the driver of the system, because he envisaged capitalist greed being limited by the 'walls of the stomach' of any given capitalist, a view that he might reconsider today given the obscene gaps in social wealth, but he saw the driver as the need by capitalists to accumulate capital to stay competitive and survive in markets. However, the focus of this chapter is not that, but rather upon the methods by which economists identify and calculate the social costs of climate change, and the policies that could help mitigate climate change. Supplanting capitalism with an alternative democratic socio-economic system may be the ultimate radical answer, but on the assumption that is not going to happen anytime soon, the planet Earth, *ceteris paribus*, still needs to be saved. Politically, that is known as reformism.

Neoclassical Economics

The term *ceteris paribus*, meaning 'all other things being equal' or simply let's hold all other things constant for the purpose of analysis, is fundamental to the concept of equilibria in economic modelling. It is like a car mechanic needing to keep a car stationary while working on it. Once the static equilibrium model has been formulated, the moving parts of an economy then need to be studied and replicated as the model is being built. However, and this is a big however, the way the model is designed, which is how it is conceptualised based upon the assumptions the economist makes about the society to be modelled, can cut off, or distort, certain directions in which the model could otherwise develop. Further, it should be noted that the fundamental assumptions can be either realistic or unrealistic, depending upon the approach to model building. In the case of Marx, for example, the labour theory of value was constructed upon what is arguably an observable, namely, the rate of exploitation or the value of hours worked relative to the value of hours paid. In mainstream Neoclassical economics, there is no such claim to realism. On the contrary, its most prominent exponent, Prof. Milton Freidman, argued that realism was totally unnecessary as long as the results could be verified [5].

Neoclassical economics pioneered by several mid-late nineteenth-century economists was a redirection away from Classical economics. Different Neoclassical economists made their own contributions, such as Alfred Marshall [11], who focused upon the *marginal* costs of production of firms and how they brought supply and demand *at their margins* into equilibrium—that is to say, markets were perfectly balanced when the costs of any *additional* (marginal or incremental) supply would be equal in value to any *additional* (marginal or incremental) demand. But Neoclassical economics then took two divergent paths: a mathematical path pioneered by Swiss economist, Léon Walras [12], who through the universal application of calculus generalised the model of individual markets into a mathematical model of general equilibrium for the entire capitalist system, and Austrian economist Carl Menger [13], who simultaneously with Walrus and William Stanley Jevons [14] developed the idea of marginal *utility*, a concept to explain consumer behaviour or demand. But whereas Walrus conceived that market prices could be determined by integrating marginal supply and marginal demand to determine prices, Menger and the Austrian school maintain that prices are only relevant to individual consumers, and deny the meaning and purpose of systematically mathematicising economics. This remains one of the most important differences between the Austrian school and the others—a useful YouTube video outlines some of the differences [15]. Nevertheless, both schools of thought maintain that a system of capitalist markets, if left unregulated, will be self-adjusting towards balanced equilibrated economic growth.

These are huge methodological divisions within Neoclassical economics, but not the only ones. An early leader of what became mainstream economics was Frank Knight [16], a founder of the Chicago school at the University of Chicago, and widely recognised for his distinction between risk as probabilistic and uncertainty as indeterminate, a key issue for climate change policy analysis today—see Table 3.1 and the discussion given below. One of the most prominent advocates of the Chicago school was Milton Friedman, a proponent of monetarism, which is the proposition that the money supply should remain in equilibrium with real output, and could be weaponised to maintain equilibrium—also, see above his defence of non-realistic assumptions in modelling. In contrast, but also working from a similar Neoclassical microeconomics foundation was the work of John Maynard Keynes [17], who argued instead that macroeconomic stability relied upon demand management, and that government deficit spending could bring an economy from slump to a full-employment equilibrium. Despite his break from conventional Neoclassical economics in terms of his advocacy of active government interventions, for example, pro-active government policies to combat climate change would be consistent with Keynesian thinking, the equilibria of his models remain mathematical, while his understanding of what drove the marginal utility of consumers was in terms of their 'animal spirits'. Like his contemporaries, issues, such as climate change, or even environmental issues, in general, did not enter his scope of economic thought, but his famous aphorism that 'in the long run we are all dead' suggests a dismissal of the inter-generational economic consequences for the future and a lack of time-perspective which is common to all Neoclassical thinking, other than the Austrian school.

Perhaps the eschewing of the use of mathematics by the Austrians freed them up in this respect, as Neoclassical mathematical models typically use quadratic equations to simultaneously determine the value of all variables in a static (timeless) equilibrium or in a linear growth equilibrium undisturbed by changes in proportionality between variables that a more realistic time-relevant growth (conceptual) model would permit. In reality, weather conditions and climate change are non-linear in their nature, and this renders predictions as no more than a game of extrapolation of possible scenarios with very uncertain probabilities—see the insightful comments on a YouTube video of Prof. David Stainforth [18].

Economics and Nature

In neither the Classical nor the Neoclassical economic systems did the biosphere or climate enter consideration, other than through their impact upon land values and rent. Biosphere and climate issues were external factors to these systems. Conceptually, nature does enter the Marxist system, but primarily as a critique in terms of the exploitation of nature (extractive farming, fishing, mining) by capital in what he terms 'primitive accumulation' in contradistinction to capital accumulation through technological reproduction. The exploitation of nature is therefore no longer for survival as in pre-capitalist modes of production, a process Marx refers to as a 'free appropriation from nature' [19], but for profits and capital accumulation. It is interesting to note that in his *Preface* to *A Contribution to the Critique of Political Economy* in 1859, Marx, trained in law as a subset of philosophy, states he was motivated to turn his attention to economic issues when confronted with Rhineland laws that prosecuted landless peasants for theft of forest wood for fuel [20]. What motivates writers is usually a good way into understanding their work, even in its most abstract form.

The Neoclassical economist who came closest to Marx was Joseph Schumpeter [21], a member of the Austrian school. Like all Neoclassicists, Schumpeter saw capitalism as a self-adjusting system, but his focus was upon the innovative role of entrepreneurs and upon what was subsequently referenced as the 'creative destruction' of innovation and of business cycles—a term it seems better attributed to Werner Sombart [22]. In this sense he shared Marx's view of capitalism as the most productive and dynamic economic system in human history, but subject to recurring disequilibria. Marx and Schumpeter therefore saw different causes of similar repetitive outcomes. Beyond this point, it would be a speculative counter-factual to argue that as both viewed innovation as a central explanation of capitalism's dynamic, driving capital accumulation, issues such as biodiversity and climate change insofar as they impact upon productivity would more easily dovetail into their respective world views. The basis of this thinking is that both systems of thought could easily regard biodiversity degradation and climate *change* issues as endogenous variables.

Despite the swings and arrows of misfortune, such as business cycles and slumps, much of Neoclassical economics, often now referred to as Neo-Liberalism, is focused on how markets work, on how 'economic agents', such as workers, landlords, business owners, seek to maximise their individual *marginal utility* by adjusting the supply of, and demand for, their 'factors of production' such as land, labour, and capital. As the economist Lionel Robbins, an opponent of Keynes for the most part, put it in the 1930s, economics was the study of the allocation of scarce resources, or 'of human behaviour as a relationship between ends and scarce means which have alternative uses' [23]. This is a classic Chicagoan formulation (although Robbins was based that the LSE) of what the study of economics is about, narrow and precisely focused. Keynes, on the contrary, was more concerned with social instability should capitalism fail. For Classical economics, a much wider concept of social welfare is implied, such as class inequalities, but not necessarily the welfare of the biosphere.

If humanity's impact upon nature and climate change is to be taken seriously, it needs to be brought inside the modelling of systems, and issues, such as innovative energy systems and the time-preferences of citizens as consumers, moved from being exogenous to endogenous variables.

Exogenous Versus Endogenous

A shock to the system at the macroeconomic level, such as a sudden unanticipated influx or printing of money (Chicago school), or a sudden rapid expansion of credit (Austrian school), or the announcement of a major government programme funded by deficit financing (Keynesian school), is treated by conventional economics as an exogenous shock to the economic system, a system that is otherwise a self-correcting (Neoclassical) capitalist market economy. To be counted as endogenous, the economic model would not only need to set such dramatic changes in the money supply, or in the availability of credit, or in deficit financing, as being directly related to some internal event, such as a government response to inflation, recessions, or unemployment, but would also need to construct the model such that its mathematical functioning replicated the same response whenever a similar event arose in the future. Without that condition, the model would only ever be a one-off, of use only as a retrospective, but not for scenario planning or for pathway forecasting.

There are reasons why this exercise in endogenous modelling rarely happens. First, in practical terms, the responses to internal events, such as high inflation, recessions, or unemployment, usually differ over time. Second, and more controversially, the Chicago school economist Robert Lucas [24] tried to show that policy shocks could only work once, because as soon as economic agents (consumers, investors) experienced the consequences of such policy changes, they would form 'rational expectations' allowing them (according to the model) to adopt strategies that nullify similar policy changes in the future. This is perhaps a classic example of unrealistic assumptions (the ultra-rational nature of human behaviour) producing results where the real is supposed to mirror the mathematical world, but the empirical evidence turns out to be insubstantial. Others, such as Keynesians, agreed that expectations play a role in the actions of economic agents, but suggest in most markets—financial markets may be an exception—reactions are rarely instantaneous, but respond rather hesitantly as 'adaptive expectations' over time.

Extending this line of reasoning, reactions to an increased frequency and impact of severe weather events could also be subject to adaptive expectations, as public perceptions of the costs involved and the need for mitigation and adjustment undergo changes. The causes of inflation or unemployment vary over time, but human activities that cause and accelerate climate change are pretty constant which, over time, should, if there is learning, make policy responses more consistent, as reviewed in Chap. 4.

Exogenous Models

Among the first economists to work on modelling the benefits and costs to society of climate change, or the 'damage function', was Nobel-Prize-winning American economist William Nordhaus [25]. But climate change shocks were treated in the normal way as exogenous to the model. Two implications follow: first, by not making the climate variable endogenous, the model ignores the role of human activity such as the burning of fossil fuels, as a source of the problem; second, the model ignores global warming as creating climate tipping points or runaway global heating. In terms of modelling, they have to be excluded because if and when they occur they lead to instability, and possibly to runaway global warming, and as such they undermine the inherent equilibrium of the Neoclassical model. They can only be exogenous to the model: nature impacting upon economics, and not economics impacting upon nature.

The economic model of William Nordhaus is known as an *Integrated Assessment Model* (IAM) which, as he says in a paper for the National Bureau of Economic Research (NBER) in 2017, 'can be defined as approaches that integrate knowledge from two or more domains into a single framework' [26]. This refers to both economic and physical data from numerous different sources regarding climate, such as the science models referred to in Chap. 1. The IAM model is a more general form of what, since the 1990s, have been named within IPCC circles as the (globally integrated) *Dynamic Integrated Climate-Economy* (DICE) model and the (similar model for 12 separate regions) *Regional Integrated model of Climate and the Economy* (RICE) model. As he explains, Nordhaus built upon previous work done on energy models that stretched back to the 1970s, but none of which included CO_2 among their variables [27]. Today the IPCC references dozens of IAM/DICE/RICE in use or under development, reviewed by the Integrated Assessment Modelling Consortium (IAMC) which, as of July 2023, listed 34 contemporary models on its website [28]. A typical IAM model (see Fig. 3.1) will have the following structure and consists of the following elements (Fig. 3.2).

Data on national income is traditionally broken down into data on returns (property incomes) to capital (K) and (wages) to labour (L). This is known in Neoclassical economics as the Cobb–Douglas production function [30], which is a simple way to express national output as a function of many different inputs, but in its reduced functional form (f) it is the proportion (1-β and β, where $\beta < 1$) of national output (Y) that accrues to K, ($K^{1-\beta}$) and to L, (L^{β}), or more simply put as $Y = f(K, L)$. At any given time (t), there can be external factors represented by A(t) where A could represent technological innovation or a change in climatic conditions, or anything else the modeler thinks should be included. In principle $A = A_1 + A_2 + \ldots.A_N$ according to how many externalities are to be included using relevant data: so $Y = fA(t)K^{1-\beta}L^{\beta}$, or simply $Y = fA(K, L)$ with A falling outside the parenthesis. The most famous example of this growth model was proposed by American economist Robert Solow [31].

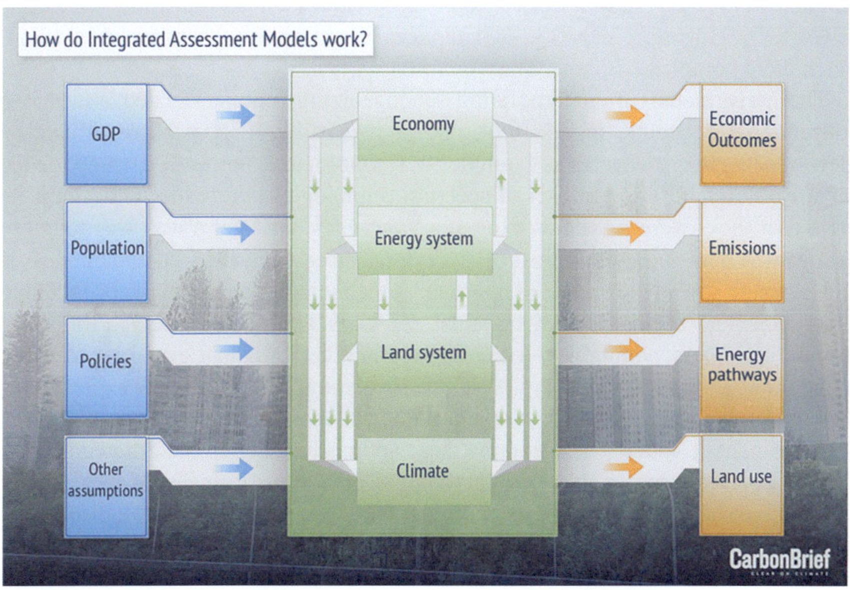

Fig. 3.2 How Integrated Assessment Models (IAM) work. *Source* Carbon Brief [29]

The mathematical beauty of this approach is that by formulating the model in terms of quadratic equations, the values of all the variables are simultaneously determined and revert to an equilibrium. The system therefore generates a *determinate* equilibrium, either static or dynamic. It does not pretend that the real world works in such a clockwork way, because, for example, monopolies and state regulations can interfere; rather it offers an abstraction of its key market elements. But its failure to account for endogenous sources of shock were challenged by other economists. For example, technological innovation was completely isolated in these equations and referred to as total factor productivity (TFP) as if it were a given that influenced the system's overall productivity.

Endogenous Models

Professor Paul Romer was among others who challenged and served a *coup de grace* to the exogeneity of the conventional Cobb–Douglas growth function [32]. In 2018, he shared the Nobel Prize with William Nordhaus whose IAM model treated climate change as an exogenous variable [33]—see above. Paul Romer changed the maths to show how shocks such as the impact of technological innovations can be generated within the system, and spread throughout the system 'if the right incentives are in place' [34]. Previously, technological change had only appeared in the models as an external 'given' at a static level of productivity, but Romer provided a dynamic model of growing productivity, a mathematical way to endogenize A, placing it inside the parenthesis as $Y = f(A, K, L)$. For a technical, yet accessible, explanation, see Romer (1994) [35]. However, $Y = f(A, K, L)$ makes the relationship between A and K, L much more complex.

The early models of incorporating (endogenizing) climate-related issues into conventional economic models took one of two paths. First, was the ground-up view of new energy systems as examples of technological innovation, which looked for ways to quantify the impact of specific technologies that reduce emissions. At the outset these were treated as exogenous variables, but later a way was found to endogenize new energy systems, by adding *learning curves* (as coefficients) to K and L, which spread the impact of new energy technologies across markets, reducing costs and enhancing productivity. Second, top-down approaches emerged following Romer and others, adding new technologies, and the R&D that gives rise to them, to the production function $Y = f(A, K, L)$. For example, this could be expressed as $A_l = f(R)$ where R stands for research and development. In a paper in 2002, Van der Zwaan, et al. found a way to unite these approaches, noting the following: 'The inclusion of endogenous technological progress leads to earlier investments in energy technologies, a different mix of technologies and a lower level of overall discounted investments, as compared to the case of exogenous technological progress' [36]. In other words, treating technological innovation, including new energy systems, as endogenous opens up a better understanding for their faster adoption.

Van der Zwaan, et al. further note that Nordhaus, when exploring climate-related technologies, uses a single abatement function—here we shall call this M for climate change mitigation—and concludes that innovation adoption is less important to abatement (mitigation) than the substitution of different energy systems, such as coal versus wind power, according to their relative prices. So empirically, this comes down to the falling costs of renewable energy versus the falling costs of new technologies such as carbon capture, utlisation and storage (CCUS), a debate of growing importance today, in part because policy-makers are not sure that renewables on their own will prove sufficient, and partly because the highly influential fossil-fuel lobby sees CCUS as a way of prolonging the use of fossil fuels. By contrast, Van der Zwaan, et al. split what we are here calling M, into M_1 and M_2 to represent two separate energy technologies, which, in their words, allow them: 'to explicitly derive conclusions on the feasibility of certain policies to make a transition between various energy options. Furthermore, these technologies are allowed to realistically develop along learning curves'. This discussion points to the possibilities of modelling in more complex detail than traditional models, ways to allow economists to analyse the dynamic functional effects of variables such as human activity, learning and of policy choices in the study of climate change. Given inputs from science models of climate change, endogenous economic models could make for a better understanding of the damage function with respect to these variables. Put otherwise, what are the implications of doing the wrong things, of doing nothing (which *is* the wrong thing), or of doing a medley of the right things.

Innovative Modelling

The focus of the above discussion has been upon methodology, or how to model the economic implications of climate change in terms of costs and benefits, in terms of technological innovation and the adoption of new energy systems, and in terms of policies that will encourage the shift to mitigation, such as carbon taxes. But using the production function $Y = f(A,K,L)$ adds greatly to the complexity of the model. It means that technology or innovation is no longer simply the residual of Y when the effects of K and L have been taken into account. For example, in the case of innovation, A_1 could be innovation correlated with L_H as highly skilled labour, and with K_H as start-up capital. Broadening innovation to specifically include new energy systems and associated learning curves are innovations in the modelling itself. It can go further if carbon emissions are made the key dependent variable. A_2 could be carbon dioxide (CO_2) emissions associated with L_{FF} (where FF represents fossil fuel) as households use petrol-driven vehicles, and K_{FF} as manufacturers use oil and gas. The strong reason for including climate change as an endogenous variable is the evidence that the majority of additional CO_2 emissions into the atmosphere is the result of human activity. Prior to the industrial era, for 800,000 years carbon molecules in the atmosphere never rose above 300 ppm (parts per million), but since the industrial era they have risen considerably, well above 420 by 2024 [37]; and more compelling evidence is that a rising proportion of that carbon is isotope carbon-12 [38], which comes from plants, that is, from burning fossil fuels [39].

Climate Models

While William Nordhaus's pioneering Nobel-Prize-winning *Integrated Assessment Model* (IAM), published in 1991, has been the most influential economic assessment of the damage function (costs and benefits) of rising GHGs in the atmosphere and of the estimated rate of rise of average global temperatures, it has also been heavily criticised by ecological economists on several grounds. These include treating climate change as an exogenous variable, to the functional forms chosen by Nordhaus which assume a self-adjusting equilibrium of the system, to his choice of a high discount rate which privileges today over tomorrow, to his assertion that all industrial and commercial activities taking place indoors are immune from the damages threatened by climate change.

One critic in 2020, Prof. Steve Keen, went so far as to call it *The Appallingly Bad Neoclassical Economics of Climate Change* [40]. An earlier critic, Prof. Martin Weitzman, arguing that the distribution function used by Nordhaus excluded the possibility of a fat tail (see Fig. 3.3), started his paper in 2011 entitled *Fat-Tailed Uncertainty in the Economics of Catastrophic Climate Change* [41], with the statement 'I believe that the most striking feature of the economics of climate change is that its extreme downside is nonnegligible. Deep structural uncertainty about the unknown unknowns of what might go very wrong is coupled with essentially unlimited downside liability on possible planetary damages. This implies that fat tails are more than likely in the extremes of critical probability distributions'. What follows is a review of these criticisms, starting with a paper by Profs. Nicholas Stern and Joseph Stiglitz, then Martin Weitzman's critique, then the general approach of ecological economists, followed by one of them, Steve Keen.

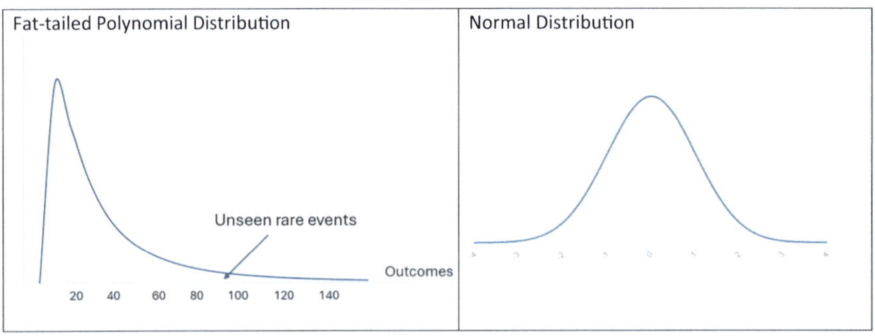

Fig. 3.3 Polynomial and normal distribution functions

Stern and Stiglitz Versus Neoclassical IAM Models

Among the most systematic critiques of Neoclassical models that attempt to estimate the social cost of carbon (SCC) and damage functions arising from climate change is that of Profs. Nicholas Stern and Joseph Stiglitz and their paper *The economics of immense risk, urgent action and radical change: towards new approaches to the economics of climate change* (2022) [42]. They note that 'The IAMs have been particularly influential in the calculation of carbon prices for use in appraisal of public programmes. They have been used to suggest relatively low levels for carbon taxes. And they have been used to argue that policy on climate change should be overwhelmingly dominated by carbon pricing'. (In fairness to Nordhaus, in his 2021 book *The Spirit of Green*, he does argue for much higher carbon taxes than those chosen by policy-makers influenced by his work, see Chap. 6.) One problematic aspect of these models is their sensitivity to the assumptions, parameters and functional forms they adopt. 'Some of the more plausible specifications of parameters and functional forms concerning damages, technologies and distribution can generate results more akin to those of the international consensus than to those of Nordhaus. But such sensitivity make the results of very limited usefulness'. And in refutation of Milton Friedman's view that assumptions are not important, only the outcomes to which they give rise, they cite from Dani Rodrik's *Ten Commandments for Economists*: 'unrealistic assumptions are ok; unrealistic *critical* assumptions are not ok' [43]. In other words, garbage in, garbage out.

But most important is that these IAMs make unwarranted assumptions about the working of markets, of treating climate change as an exogenous variable ('ignoring the endogeneity leads to an overestimate of the costs of tackling climate change'), and of asserting the tendency of the economic system to optimise or move towards equilibria, rather than to capture the deep uncertainty and extreme risk posed by climate change. 'The damages from a global temperature increase are reflected, in most IAMs, by a proportionate reduction in overall output. For example, in Nordhaus' DICE models, losses from a 3C temperature increase are around 2.1% of GDP, and for a 6C temperature increase, 8.5% of GDP … There is no damage to capital stocks in most IAMs, nor any reduction in the underlying growth rate, which is assumed to be exogenously determined'.

Stern and Stiglitz also take Neoclassical IAMs to task for their being 'focused disproportionately on intertemporal trade-offs, on how to evaluate the environmental benefits at some time in the future versus the sacrifices we make today'. The point here is that issues of a medium-to-long-term existential crisis cannot be treated as an investment decision to compare tomorrow's returns with today's returns. In the most extreme case there may be no returns. The criteria therefore should not be to discount rates on future GDPs, but on our social responsibility towards survival and the environment, which could easily justify zero or even negative rates of discount. Hence, 'the issues around discounting have received misleading treatment in much of the climate literature'. This issue is closely related to the distribution function used by the IAM models which frame the future in terms of a standard aggregated social utility function (more today, less tomorrow) which ignores the reality that it is the poorest who suffer the hardships, and those hardships need to be reflected in the weightings given to valuations of damage. 'Nor can it be assumed that government can or will simply introduce actions and policies to 'undo' these distributional effects'. They note that 'even mild modifications … more realistic damage functions or better taking into account distributive effects, yield markedly different results' which again raises the sensitivity issue.

Linear versus non-linear projections of the costs of climate change are basic issues in modelling. 'Nordhaus and others supporting the IAM approach are focused on the non-linearity in costs—if one attempts to reduce carbon emissions to achieve 2 degrees, the costs rise so high *as to be, for all practical purposes, infeasible*... As we note below, the international community is focused on the non-linearity in risks—if one doesn't reduce carbon emissions to achieve at least 2 degrees, the risks rise so high as to be intolerable. At the same time, the costs of limiting net emissions (including, if necessary, carbon capture and storage and carbon sequestration) are sufficiently limited that such trajectories are indeed feasible. The fact that there is such disagreement reinforces our emphasis on the importance of risk and uncertainty'.

The issues of risk and uncertainty lie at the centre of the critique. Among the examples cited by the authors are the regional effects of climate change such as: 'One of the most densely populated regions in the world, the North China Plain, would likely experience deadly heatwaves later this century with "wet-bulb" temperature exceeding the threshold defining what people can tolerate while working outdoors' where 'wet bulb' refers to temperatures at maximum humidity. Because 'climate change is moving us into unchartered territory: we do not know how an increase in greenhouse gas concentrations will affect weather and how changes in climate will affect the economy, simply because our economy has never experienced in the lifetime of homo sapiens anything like what we are likely to face'.

According to Stern and Stiglitz, the problems with the standard models are not limited to their assumptions, parameters, and functional forms, but notably include their assumption that markets consist of perfect substitutes (what economists call the convexity issue, the more of one, proportionally the less of the other) and that the variables of interest, such as climate change, technologies aimed to reducing CO_2, and technological change that drives GDP growth, are treated as externalities. 'The more fundamental problem is that these models are not well suited to analysing problems of deep uncertainty, extreme risk, endogenous preferences, intergenerational equity (where future generations are not discriminated against), and increasing returns to scale and rapid technological change'. Added to this list is the singular focus these models have upon prices as policy instruments, such as carbon taxes, however important these may be. Carbon credits could be added as another example.

To improve modelling, Stern and Stiglitz suggest a more integrated approach to economics, for example, giving institutional economics a central role to examine the work of regulators to incentivize carbon storage and flexibility in the use of power systems. As an aside, it should be noted that Prof. George Stigler of the ultra-orthodox Chicago school of Neoclassical economists pointed out the problem of 'regulatory capture' due in part to the regulator relying heavily upon the regulated for the information necessary to do the job [44]. There is truth in the observation that there can be 'moral hazard' in such a close relationship, but that is not a sufficient argument to discount regulation, or leave it out of models. Also needed are models that tackle head-on 'pervasive market failures in the financial sector, e.g. imperfections in information, through disclosure requirements and central banks analysing carbon risk in all of its dimensions; imperfections in risk markets, e.g. by governments taking a role in de-risking climate investments; and imperfections in financial institutions, e.g. by the creation of green development banks'.

This is casting the net widely. Whether single economic models could incorporate all of these issues is questionable. But the key seems to be in making the 'variables of interest' endogenous in whatever models are developed, and beginning, not ending, with the mitigation mechanisms that science suggests are necessary to keep temperature rises down. Starting with a damage function is therefore likely the wrong approach because apart from the methodological problems involved, the most it can do is to alert policy-makers of the challenge, something the science models do already, but without any firm foundation to assess the risks and uncertainty involved.

Weitzman Versus Nordhaus

It is worth noting that in 2011 when Weitzman's paper was published, CO_2 in the atmosphere was recorded as 390 ppm, a jump from 280 in 1850, and from 227 in 1750. By 2024 it was up to 417 ppm, but as Weitzman points out, what he calls the 'standard' benefit–cost analysis (BCA) of economic models of climate change foresee stabilisation (equilibrium) at 700 ppm, 'two and a half times above their highest level over the past 800,000 years'. Such unprecedented levels should send the alarm bells ringing, and these are only the hypothesised levels. Weitzman also highlights the insurance aspect of uncertainty, a 'better safe than sorry' calculus as suggested in Fig. 3.2.

An additional concern arising from standard modelling concerns equilibrium climate sensitivity, which in the words of the IPCC is a measure of the climate system response to sustained radiative forcing—see also Chap. 1. Radiative forcing is the process whereby infrared radiation from the Earth back into Outer Space, known as the albedo or the proportion of energy reflected back, falls below the solar energy level being received from the Sun due to it being absorbed by GHGs, causing the global average temperature on Earth to rise to a new equilibrium [45]. Weitzman notes the IPCC used, up to that time, 22 probability density functions (PDFs) each of which was used to estimate how high the Earth's average global temperature would need to rise to compensate for radiative forcing caused by an accumulation of GHGs in the atmosphere. By docking the outliers, the IPCC concluded the most likely temperature rises would be in the range 2–4.5 °C, on average 3 °C and not less than 1.5 °C. These are statistical outcomes, and technically they reflect the nature of the distribution curve that is used to assess them.

A perfectly standard normal distribution curve is bell shaped with steeply falling sides which would imply very high probabilities of the temperature ranges suggested by the IPCC. In a thin-tailed normal distribution curve, the sides tail out, but rapidly and evenly. In the case of the use of polynomial distribution curves, one or both of the sides may tail out for long distances. It is the tails that create the uncertainties because they imply a range of uncharted probabilities (the distribution curve on the left) and therefore possibilities. Some models of climate change, for example, suggest future possibilities of up to 10 °C taking into account the melting of artic ice and the warming of the oceans.

The probabilities of global average warming greatly exceeding the IPCC's range 2C–4.5C lie in the tail, and the proportion of tail to non-tail generates those probabilities, which Weitzman suggests are between 5 and 17%. More recent models cited by Weitzman range from 23 to 15%. However, there is no agreed way to aggregate and average all of the models. The method used by the IPCC suggests an average global temperature rise of 6.4 °C without discounting the outliers. The methodological question is how far is it the data that is responsible for these forecasts, noting the incompleteness of the data, and how far is it the mathematics of the models generating the outcomes. Either way, it is unsurprising that Weitzman finds the results nonnegligible.

The fourth major criticism of the standard economic models is that conventionally they do not take into account the 'fast feedback' mechanism of global warming releasing GHGs from the substrates of the Earth and from the clathrates of the oceans. Weitzman notes, even if the process is centuries away, that 'the possibility of a climate meltdown is not just the outcome of a mathematical theory but has a real physical basis'. The fifth criticism goes to the heart of the modelling issue. He takes aim at an IAM model by Nordhaus in 2008 which uses a quadratic-polynomial multiplier as a damage function. The difference between linear equations and quadratics is that in the former the variable is raised at most to the exponent of 1 producing a potentially limitless straight-line graph, and in the other to the exponent of 2 producing a more constrained parabolic graph, offering the option of substitution between variables, such as a lessening in general consumption as the price of food and energy goes up due to global warming and a growing consumption of air conditioning.

But, if climate change damages take the functional form of being additive, and separable, meaning not all things are capable of being substituted with rising temperatures, such as biodiversity or health, then the welfare consequences can be much greater than the conventional models suggest. Greater in the sense that not only must consumption levels fall, which in itself may not be a terminal problem, but irremediable damage to human and planetary welfare may result. Weitzman's cautious conclusion is that no one really knows what appropriate functional forms the models should take, which leaves us again in the risk-probability dilemma.

Weitzman's final argument relates to the appropriate discount rate that models should use to estimate the damage function. Once the various models have forecast the likely future global average temperature impacts upon the commercial world, they use a discount rate to estimate what that future damage equates to in today's money or current GDP. There are really two issues here. First, if the discount rate is high, then the implication is citizens are less concerned with the future than they are with the present, and vice versa if the discount rate is low. The arguments at this point revolve around perceived risk-probability, or how seriously citizens take the threat, and issues of inter-generational commitments, in other words, do citizens want their progeny to enjoy an Earth in no worse a condition than it is today. These are all subjective matters. Second, and more fundamental to modelling, do the time periods ahead provide any basis at all for even choosing a discount rate. Many effects of climate changes may manifest themselves in the twenty-first century, but the models suggest they will stretch well into the twenty-second century and probably far beyond. Apart from not knowing with any precision where these climate changes impact regionally and what those changes might be, less is also known about what changes will take place in technologies, including carbon abatement technologies and renewable sources of energy such as nuclear fusion. Other uncertainties include movements in the Earth's tectonic plates, the migration of peoples and birth rates, changes in food science, and so forth. It is simply not possible to use discount rates over such timelines. Usually, discount rates are used in financial markets, and even then timelines are limited to decades, especially where fat tail distributions are involved which maximise the unknowns. As Weitzman concludes on discount rates, 'the appropriate interest rate for discounting damages is not the rate of return on capital but the rate of pure time preference, which for a normative evaluation of climate change is arguably near zero'. This was also the choice of Nicolas Stern in *The Stern Review*, for the UK Treasury, see above [46].

Ecology Economists

Rather than a focus on BCA (using Weitzman's expression for cost–benefit analysis) the economics of ecology works with a system-based analysis (SBA) that can embrace multiple criteria either to maximise or to optimise. Whereas maximise implies a maximum possible value, optimise implies a value per variable that is consistent with other target variables also being at an optimum. For example, employment achieving its highest level consistent with stable prices, or with a more equitable distribution of wealth and income. These represent a series of 'second-best' options since not all of them can be simultaneously at their maximum levels. In other words, policy target trade-offs. A helpful summary is provided by Yannis Dafermos and Maria Nikolaidi [47], who emphasise that SBAs follow 'path dependency, disequilibrium phenomena and the endogeneity of risks'. Clearly, very distinct from Neoclassical models. They illustrate the role of such models by way of a *DEFINE* (Dynamic-Ecosystem-Finance-Economy) model.

Modern Monetary Theory (MMT)

The introduction of finance is significant because it draws upon Post-Keynesian Modern Monetary Theory (MMT) which offers a view of fiscal and monetary management that can, in principle, deliver the multiple policy targets mentioned above. This requires a short explanation. Traditional Neoclassical monetary theory sees money creation being in the hands of a Central Bank which prints paper money ('fiat' money or 'promises to pay') and provides the rules by which private banks can relend a certain percentage of customer funds deposited with them. This enables the Central Bank through monetary easing or squeezing to control monetary growth. By contrast, MMT sees governments *in practice* creating money by having the Central Bank honouring the real-world purchases the government and public sector agencies make, thereby creating a bookkeeping asset in their own book and a bookkeeping liability in the book of the government's Treasury. This is called double-entry book-keeping. The new money used to support these public-sector purchases circulates among business corporations and employees, finding its way into private banks as deposits, and non-bank financial institutions (NBFI) where private debts and credits are created. When the Treasury wishes to borrow money back from the economy to reduce its liabilities (public debt) those private banks and private corporations buy government bonds at a discount to their promissory monetary value at their maturity date.

These market operations are an ongoing feature of the central financial management of a country, but what gives money its value to hold and to accept as a payment is that it is legal tender, which means all taxes, such as income tax, value-added tax, stamp duties, and road tax, need to be paid in that currency. If governments wish to slow down the economic rate of growth because, for example, supply chains are unable to grow as fast as demand, then some taxes (fiscal policy) can be raised and public sector growth cut back. Fiscal policies in this way curb the growth in the money supply, assuming governments do not simply inflate the currency. Many of these ideas were pioneered by economist Prof. Hyman Minsky, and a helpful summary is available by Richard Murphy, Prof of Accounting Practice, on YouTube, including what MMT is not [48]. Steve Keen (see below) devotes a chapter to the influence of Minsky and model building in *The New Economics* [49], in which he takes to task William Nordhaus' model of climate economics.

The main difference between New Keynesian MMT and traditional Keynesian theory is that New Keynesians, unlike traditional Neoclassical economists who treat government expenditure as if it were a family budget such that taxation must *precede* public sector expenditures to avoid running up inflationary debt, instead recognise that a growing economy pays off or reduces the value of debt as taxation *follows* public sector expenditures. Nevertheless, the nature of monetary policy remains, in effect, separated out from fiscal policy and the real economy, and acts as a kind of scaffolding that supports the real economy as it expands or contracts. The caveat for MMT theorists is that the real rate of growth of public expenditure is constrained not by the supply of money but by the full employment of available productive resources, otherwise inflation will result. A good explanation is provided by Richard Murphy on YouTube. [insert fn: Richard C. Murphy 'Keynes vs MMT: which economic theory fits our world?' www.youtube.com/watch?v=ZGFpBU37riM] Where the Keynesian and New Keynesians come closer together is their focus upon the leading role of demand, for example, in increasing or decreasing public expenditure to help regulate the economy, and in seeing debt as paying for itself through economic expansion resulting in higher tax revenues.

Within this context, ecology economists believe that, if adopted, a well-orchestrated and balanced public policy approach could achieve the environmental goals of the Paris Agreement and the UN Sustainable Development Goals (SDGs). The point is well made in a YouTube video of a post-COVID-19 TED Talk by MMT economist Stephane Kelton [50], who makes the point about investment in 'things worth doing' such as housing, health, and environmental issues when she says that the question is not about the money and finances but whether the real resources are available to realise them, 'how to resource them'. However, if there is a shortage of resources, inflation could result—inflation may be seen as essentially supply-sided rather than demand-sided—so increasing the supply of resources has to be managed over time. Government reactions to COVID-19 showed, in principle albeit often in poor practice, how policies can react to emergencies when needed, but old ways of economic thinking ('how can we afford them') have reasserted themselves. A good example would be the failure in Western economies of public investment in a nationwide recharging infrastructure for electric vehicles (EVs); by contrast China has combined exactly that with battery innovation and 'new energy vehicle' innovation, with EV and battery manufacturer BYD ('Buy Your Dreams') becoming a world leader [51].

System-Based Analysis (SBA)

Climate change targets are incorporated in SBA models along with their contribution to other targets, such as green job growth, emission levels, inflation levels, and social justice. From multiple real data inputs in the benchmark year, e.g. 2024, multiple path-dependent scenarios are extrapolated for up to decades ahead through positive and negative feedback mechanisms. For a simple example, $Ag = f(\pm Z)$ where Z could be a regional average temperature and Ag could be regional agricultural output. Short-run gains/losses could end up as longer-run losses/gains, so as ever, risks are involved in the scenarios, such as resource depletion or severe inflation, but so also are opportunities, which could be policy changes or advances in technology. The obvious problem is weighing the risks and opportunities, as these are all hypotheticals, but clearly there is a high risk associated with the possibility (not necessarily a probability) of temperature rises above, say, 5% of having a higher negative environmental feedback effect than, say, a 10% rate of inflation. But the relative level of risk depends also upon the priority placed on the target of interest. These constitute uncertainties in the models which call for additional qualitative assessments, such as surveys of experts as also used by Nordhaus (see below).

The models of ecology economists submitted to the IPCC as are referred to as 'shared socioeconomic pathways' (SSPs) [52], and these are combined with the scientific models referred to as 'representation concentration pathways' (RCPs) [53] to create benchmark baselines. As of 2024, five SSPs/RCPs were incorporated into IPCC AR6 [54].

Steve Keen versus Neoclassical Economics and Climate Change Modelling

In his book *The New Economics: A Manifesto* (2022) [55], Prof. Steve Keen expounds an alternative approach to *all* economics, from Classical to Neoclassical to Post-Keynesian, including a Chap. 4 where he confronts the climate change models of William Nordhaus and others. Keen's starting point is the framework of the pre-Classical Physiocrats [56], who were influential in revolutionary France in the eighteenth century. They saw land as the source of all value, only for the Classicals to replace land with labour, and the Neoclassicals to add capital as jointly responsible for output, value, or utility. Keen's innovation is to translate the 'free gifts' of nature into the free sources of energy, and to integrate his analysis with the Laws of Thermodynamics: the First Law being that energy cannot be created nor destroyed, and the Second Law being the law of entropy, namely, that as heat flows from hot to cold objects, usable energy is lost. That implies that in every productive system there is unavoidable loss.

Keen's contention is that the productivity of labour and of machines (capital) is a direct result of the energy they use, but according to the Second Law there is always entropy or wasted energy, whether trapped in waste materials, or in lost heat, evaporation, etc. This changes the physiology of Earth, as the energy that constitutes natural phenomena, such as vegetation and rocks, is transformed leading to loss of flora and fauna and an increase in molecular disorder in the Earth's environment, including its atmosphere. Thus accumulated 'development' on the scale of capitalist production and consumption inevitably heads towards ecological and, by implication, social collapse, over an as yet unspecified timescale. This trajectory is cited as independent of global warming which is an additional factor.

Keen on Cobb–Douglas Versus Leontief Functional Forms

Modelling such processes requires changes to the functional form, similar to making GHG emissions which appear as exogenous variables in the Cobb–Douglas function, $Y = f\text{A(KL)}$ (see above) as endogenous $Y = f(\text{A, K, L})$. Keen replaces Y with Q to represent output in terms of work, so that $Q = f[(K(E), L(E)]$ where (E) represents the energy levels used by K and L, respectively, in producing work.

Consistent with the driver of output being technological progress, Keen also argues that while *energy consumption per worker* (E) has risen over time (with real wages, for example), *energy per unit of output of product 'i'* by labour given as $xe_{L,}$ has not, hence $E_L xe_L$ is treated as a constant C_L, while *energy consumption and output per K or per machine*, $E_K xe_K$, have both risen. Thus, the Cobb–Douglas equivalent becomes $Q = C_L[(E_K xe_K)^{1-\beta}(K^{1-\beta})(L^{\beta})]$ to be read as Output (Q) from work is a function of a constant (C_L), the energy-labour contribution to work, multiplied by a variable, the energy-machine contribution to work (K_{xk}), each weighted by their respective shares of total work outputs. The weights 1-β, β (which is a tautology) as assigned by Keen have simply been reversed in this text to be consistent with the example of the Cobb–Douglas function given earlier.

Keen also illustrates what is a called Leontief production function [57] as an alternative to the Cobb-Douglass. Named after Russian-American economist Wassily Leontief [58], it measures changes in the quantity of inputs (L, K) keeping their proportions unchanged. An example is the production of four wheels per car chassis, always the same ratio however many cars are produced. The equivalent Leontief function would become $Q = \mu(K)(E_K)(e_K)$ where the symbol μ represents the level of productivity of the work process, which hinges upon technological progress or how efficiently the machines use energy per unit of output. Keen adds that using the standard Leontief function shows the capital-output ratio, or level of work, which is K/Q, or $K/\mu(K)(E_K)(e_K)$, as the inverse of the energy efficiency of capital (e_K) which is intuitively correct as a reduction in the ratio of machines to output implies an increase in productivity. Keen concludes 'In this sense, the Leontief function has implicitly incorporated the role of energy in production'.

Non-linear Damage Functions and Statistical Modelling

The most significant policy-related conclusions from the economic models are those of the damage function, by how much will GDP slow down or actually fall (or nature harmed in socioeconomic models) as a result of global average temperatures rising to any given level. The problems with this are, first, the data points in the models are not genuine data points because they are not observable, but are rather extrapolations from various climatology models as referenced in Chap. 1; second, their functional forms, determined by their assumptions, largely determine their outcomes, and in that sense, what goes in, comes out. In a lengthy interview on Lex Fridman's *YouTube podcast* in 2023 [59], well worth watching, Steve Keen argues that, from a mathematical perspective, and despite climate change as a contiguous process over time, these models are using inappropriate equations: difference equations, which measure data at discreet intervals, rather than differential equations, which measure continuous changes over time. To make matters even more complex, as climatologist David Stainforth has pointed out (see above) the equations need to be non-linear because the pattern of atmospheric chemistry does not evolve in a straight line [18]. Nordhaus uses quadratic equations which can capture small changes in one variable triggering major responses in another, and if he had inserted tipping points, such as the melting of icecaps, the resulting damage function could have been substantially greater. But he does not, for reasons Keen rejects as unwarranted, adding also that there is 'no empirical or scientific evidence for choosing a quadratic equation to represent damages from climate change'. As a result, Keen notes and objects to the fact that Nordhaus, from his early work in 1992 through to his work in 2018, has systematically reduced the damage to GDP from 7% to 3.6%, as arising from an imaginary rise in temperature of 4 °C, which Nordhaus considered the 'optimum' as it minimised the joint costs of damage and abatement in his model.

Keen further berates all climate change models that take the simplistic statistical route of taking the spatial spread of GDP, for example, across US states, and relating that spread to average temperatures across those states, so that, for example, agricultural productivity will vary from low-to-high temperature states, and then extrapolating them over time as if global warming would not affect the ratios in different ways. The supposition that, say, a 3C rise in average temperature would over time be uniform across all states or regions, and that relative productivities would remain unchanged, is, as Keen puts it, a case where 'the assumption is the conclusion'. The simple truth of the matter is that it is an unknown, the outcome of which could have serious consequences for overall GDP. Keen notes that Nordhaus uses an 'Experts Opinion' survey to lend credibility to his low estimates of the damage function. Among economists in the survey the average estimate of the impact of a 3C temperature rise upon GDP was 0.4%; by contrast, the impact estimated by natural scientists in the survey was 12.3%. Nothing sensible can be concluded from these estimates other than the biases of the models, which include the assumed in-built adaptability of human societies in tandem with market self-adjusting equilibria.

Climate Versus Weather Versus GDP Growth

Besides the statistical and mathematical problems, Keen also takes Nordhaus to task for simply excluding 87% of US industry from consideration, on the grounds that most of it takes place in sheltered environments. The model assumes that whatever climatic changes take place, it is really only open-air activities that will be seriously affected, mostly agriculture, aquaculture, and horticulture. Impacts upon the biosphere which fall beyond the market economy are marginalised out in these models. As Keen points out, this assumption conflates the impacts of climate change with those of weather. In other words, if global temperatures rise significantly, they will not necessarily rise in equal measures in all regions, which will bring major disruptions to regional production, costs and global trade, rising costs to manufacturing supply chains, and to the costs of keeping industrial plants cool, especially high-tech plants such as data centres. The irony is that GDP could actually rise in some of these cases, as investments in air conditioning or in re-engineering a water supply, for example, are recorded as financial benefits outweighing the costs involved despite them being opportunity costs of climate change.

Energy and Growth

To support his energy model, Keen produces a regression chart showing a correlation of 0.977, that is almost one-for-one, between world energy production and world GDP growth between 1971 and 2017—hardly surprising as energy consumption and production must be closely correlated—followed by one showing a correlation between atmospheric growth of CO_2 and GDP growth of 0.988, 1960–2019 [40]. He argues that given fossil fuels by 2020 were roughly 80% of energy production, any reduction in CO_2 through mitigation efforts would inevitably significantly lower GDP growth rates, contrary to the predictions on Nordhaus. Keen is implying that the rate of substitution with renewables may be too low and too slow to make up the difference, but more recent data from the International Energy Authority (IEA) shows that GDP rates of growth have over recent decades diverged from, and outstripped CO_2 emissions, in all regions of the world, including China, and may have started to do so after 2020 in the economies of Southeast Asia, and after 2022 in the oil-rich Middle East [60]—see Fig. 3.4. But this may not apply to GDP per capita—see Chap. 7. This is a remarkable turnaround, and COVID-19 notwithstanding, attributed by the IEA principally to four factors: the growing use of renewable energy, the spread of electrification, increased technological efficiencies, and the shift away from coal. By 'growing use' is implied the growing availability of these sources of transformation on the supply side, and the growing adoption and use of them on the demand side. They are the hidden variables.

These charts seem to shift the argument away from suggesting, as Nordhaus implies, that activities that cause or contribute to climate change only impact upon a few of the activities that make up GDP, to its opposite, that by promoting the adoption of these four factors the impact upon GDP growth across the board is potentially substantial. In the case of renewables, the correlations between energy supplies/usage and GDP growth undoubtedly substitute for those between fossil-fuel supplies/usage, but not between GDP growth and global temperatures. Average global temperatures continue to rise, with 2024 being the hottest since records began in around 1880 [61]. The implications are that, assuming carbon dioxide emissions are a major contributing factor, *net additions* of CO_2 continue to drive up global average temperatures even though they are declining for many regions as GDP continues to rise. In other words, CO_2 loses its correlation with GDP growth, but remains on track to exceed the Paris limits. This de-linking of GDP growth and CO_2 implies GDP growth should not be the primary variable of interest if climate change is taken to be a threat to civilization and life on Earth. The switch to renewables and the conservation of biodiversity is more crucial. However, this does not rule out the long-term impacts of climate change on GDP as global flooding, firestorms, migrations, and deaths driven by unbearable temperatures (high or low) would have catastrophic effects

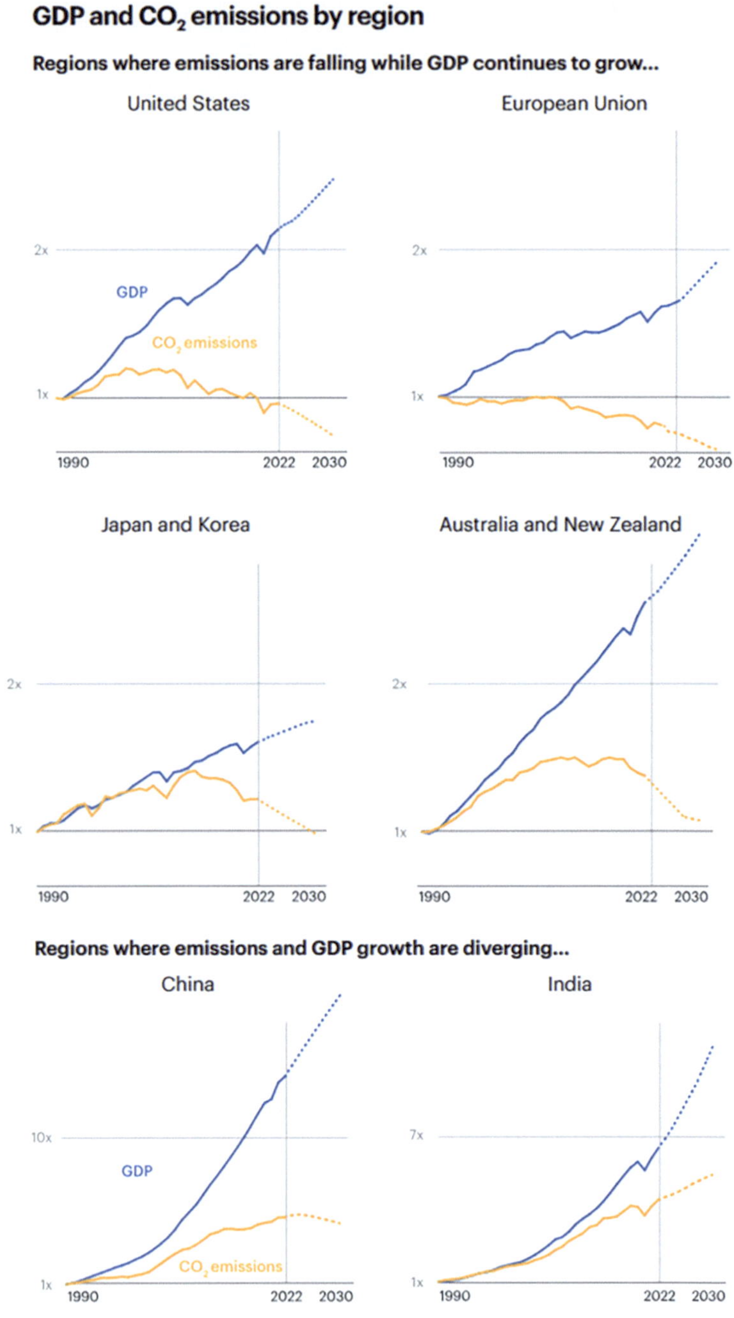

Fig. 3.4 Regions where emissions are falling while GDP continues to rise, and those where they have grown in tandem. *Source* IEA (2024) https://www.iea.org/commentaries/the-relationship-between-growth-in-gdp-and-co2-has-loosened-it-needs-to-be-cut-completely

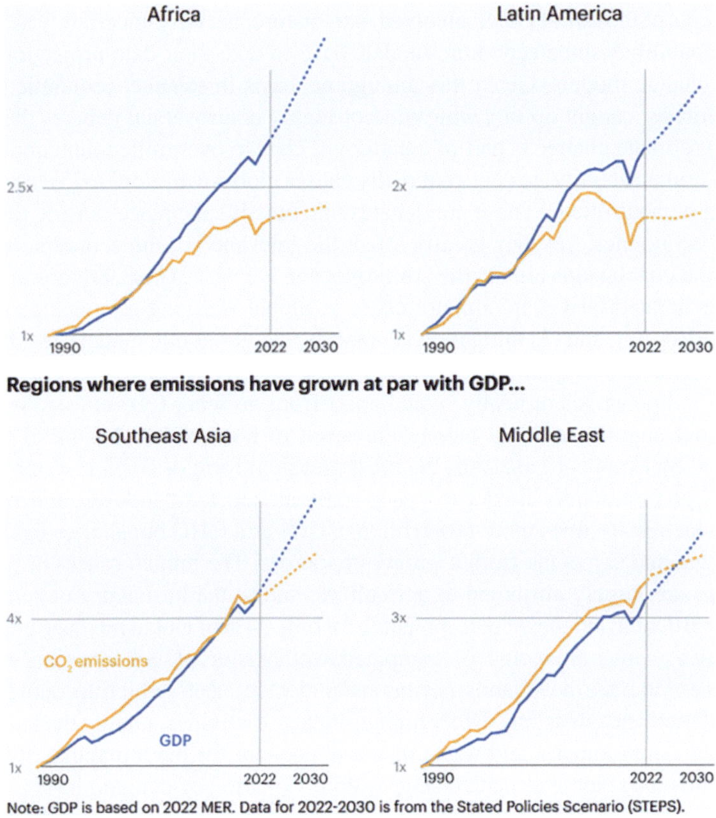

Fig. 3.4 (continued)

upon economic well-being as well as medical health. What is required are models that take these factors into account and compare them with the counterfactuals of failing to do so.

Science Versus Economics

Even when economics adorns itself with the mantle of mathematics and the pretentions of a science, it is incapable of generating more than a variety of climate scenarios based upon the choice of assumptions which are subjective and then hard-wired into computerized function forms; mathematical functional forms that give rise to the appearance of a science. This is not to say that the scenarios are unimportant. On the contrary, the risk-probability matrix (see Table 3.1) is an important way to conceptualize and frame uncertainty to guide policy choices such as those between mitigation and adaptation. Feeding into this framework is the damage function, which can offer assessments of the risks to GDP involved, or to nature, but only uncertainty as regards their probabilities since economic models have to take their data projections from climate change models. Given this ambiguous basis in science, economic models will be forever caught up in a whirlwind of highly controversial debates that range from (1) 'climate change is part of a pattern of change over millennium and not the result of human activity' to (2) 'what really causes global warming' to (3) 'can global warming be controlled'. These are debates that mostly take place outside the discipline of economics, yet they greatly affect the relevance of the economic models. Only if the conclusions are (1) the Anthropocene is real, (2) besides natural events, human-generated GHGs are a major cause of global warming as concluded by the IPCC's AR5 [62], and (3) mitigation is possible, can economic models ever be more than just guides to adaptation.

What is known scientifically is that apart from growing CO_2 emissions, which are without significant doubt mostly the result of humans burning fossil fuels as evidenced by isotope-12 analysis, the most prevalent form of GHG is water vapour, which is predominantly linked to ocean currents and temperatures, and methane (CH4) which is 40 times more powerful than CO_2 as a GHG but shorter-lived in the atmosphere and can come from a variety of sources. The human causes of methane emissions are largely attributed to agriculture, and to the husbandry of ruminants, cows in particular, but these are responsible for only 24% of global methane emissions (40% if rice paddies are included), compared to other sources such as natural environments like clathrates in wetlands, oceans and under ice sheets, which account for 40% of global methane emissions. Other human-induced activities, mainly the burning of fossil fuels and biomass, and waste disposal, account for the remaining 36% [63]. Despite this data, some climate change sceptics (not to say deniers) have proposed the *Clathrate Gun Hypothesis* [64], which suggests that in times past clathrate emissions have been responsible for major and rapid increases in global warming, and the same could be true in the twenty-first century. For example, the geologist and combative political conservative, Prof. Ian Plimer, argues that clathrates disturbed by eruptions in the Earth's surface giving rise to volcanoes and seabed fissures have been historically responsible for global warming and cooling, including crucial fluctuations in sea temperatures promoting or reducing water vapour, which is the single most important GHG. In his opinion, there is no scientific evidence to show that human activities carry any responsibilities, and was a strong advocate of the USA leaving the Paris Agreement. His hopes were fulfilled (again) on 20 January 2025. Although he appears to accept the role of GHGs in global warming, in a video interview he seems oddly equivocal about the evidence of carbon isotope-12 analysis that clearly associates it with human activity [65].

As discussed in Chap. 1, ideology, preconceptions and passions seem to be overrepresented in scientific debates around climate change, even when they can be drivers of the research. To be clear about this, there is nothing inherently wrong with designing a model in such a way that it privileges certain assumptions over others, and even produces results that are more or less preordained by those assumptions. Such models can be useful for showing where they lead, but the real problem is that they are blind to alternative explanations, which may be simpler. For example, the Hungarian mathematician and philosopher Imre Lakatos showed how researchers tend to hang on to the veracity of their models by continually adding to their complexity to accommodate new data or evidence, what he called a regressive research programme [66]. Alternatives are excluded by *de fault*, when in science they deserve to be shown to be wrong or misguided.

An example of the latter is a paper in 2019 purporting to show that because the position of the Sun's rotation around its axis moves between the centre and what is termed the 'barycentre' of the solar system as it is pulled by the gravitational waves of the planets, especially by Jupiter, its oscillations give rise to a magnetic field and solar irradiance that interferes with the Earth's climate [67]. If true, this could be alternative explanation of climate change, adding to the geological explanation of Plimer. However, the paper was retracted by the publishing journal after peer reviews rejected the claim as untrue [68]. It would seem that the Earth retains its regular orbit around the Sun whatever its position. The lead author was quickly involved in a heated protest, questioning the credentials of one reviewer, accusing him of unprofessionalism and of cheating the data—the video, which is educational, is worth watching as indicative of the emotional tensions involved in so-called scientific discourse [69].

Conclusions

What the American historian Melvin Kranzberg once said about technology applies to economic models, they are neither good, nor bad, nor neutral [70]. They all serve a purpose, unwittingly or not, to promote or reconfirm a particular paradigm, such as Neoclassical economics, because contrary to the way they are often presented, they are not simply tools of analysis. The work of William Nordhaus has been recognised as seminal in the economics of climate change, largely due to its pioneering nature, but is based entirely upon Neoclassical principles. Martin Weitzman's criticisms imply he is stepping outside the Neoclassical paradigm, but actually he remains within it and focuses upon the functional forms of the Nordhaus model, specifically upon (1) a potentially fat-tailed distribution curve that would suggest many more extreme weather events and temperatures than a normal thin-tailed distribution curve, and (2) challenging the use of a high discount rate in the damage function to price-in the current cost of future climate change damages.

Sometimes the use of conventional Neoclassical modeling is insisted upon by funding bodies, by patrons, and by journals. As an anecdotal aside, the author of this book once asked a highly respected professor of the economics of technology, who was a leader in his field, why he used general equilibrium models when it was evident that technologies drove disequilibria. He replied simply, it was the only way to get published in leading journals. Unfortunately, it is not the papers that represent alternative viewpoints, but the journals. Nevertheless, some papers bridge the gulf between different schools of thought. The work of Paul Romer has been cited for bringing technology in from the cold as an endogenous variable. Following that, ways to analyse economic growth directly linking GHG emissions to the work of labour and machines have been suggested.

More radically, Steve Keen has repurposed the paradigm of the Physiocrats, substituting energy for land as the 'free gift of nature'. By introducing functional forms for the use of energy in work by labour and by machines (in his model Keen confines capital to machines and by implication to anything besides labour that produces output) he argues two conclusions. First, due to the Second Law of Thermodynamics, entropy means energy is inevitably lost, which, given the unremitting pace of capitalist development, is transforming the physiology of planet Earth at the expense of the environment. This process stretches into the timeless future, which is why Prof. Keen somewhat bizarrely promotes the Elon Musk view of a future colonisation of Mars where waste can be disposed of. More importantly, the entropy process is quite distinct from the threat of global warming, meaning environmental damage and its impact upon the sustainability of GDP, is built-into the development process. So with respect to the damage function, cutting CO_2 by reducing the use of fossil fuels will impact upon GDP growth much more profoundly than Nordhaus assumes. However, it is pointed out above that as renewables are replacing fossil fuels at pace, the data now shows that GDP growth can and has departed from its direct correlation with CO_2. Yet the continued rise in global warming is still related to the use of CO_2, so GDP growth and energy use are closely correlated. There also remain hidden variables as part of the 'development' process, such as the systematic destruction of the biosphere (carbon sinks), over-fishing, and the over-use of water resources that lowers water tables, decreasing returns in the use of fertilizers and microbe-resistant drugs.

Since neither passion nor ideology can be the basis of truth even if they are sometimes correct, something more is required. The physical complexity of climate change calls for a more collaborative multi-disciplinary approach with teams that can bring together researchers across various disciplines, such as astrophysics, solar-physics, meteorology, climatology, seismology, geology, oceanography, plus for the damage function economists, actuaries, anthropologists, demographers, and sociologists. Further, such multidisciplinary teams would need to combine tools of analysis, such as image processing and AI, mathematical modelling, statistical analysis, statistical mechanics, system dynamics, and econometrics. Lastly, but genuinely not least, there should be a role for communities, not just to assess damages at first hand, but to chart out the priorities of mitigation and adjustment. This is really too much to hope for, partly because such teams would be expensive to assembly, even at a distance, and because climate change complexity is only matched by the complexity and the politics of research grant funding, plus the egos and professional rivalries between scholars. But it should be the path of direction. The best guide is that of the French Enlightenment philosopher/mathematician René Descartes 'I think, therefore I am' and do the best we can with that.

References

1. WMO (2024) Latest news https://wmo.int/
2. NOAA (2025) About our agency https://www.noaa.gov/
3. NOAA What is the Atlantic Meridional Overturning Circulation (AMOC)? https://oceanservice.noaa.gov/facts/amoc.html
4. CarbonBrief (11 February 2020) Guest post: Could the Atlantic Overturning Circulation 'shut down'? https://www.carbonbrief.org/guest-post-could-the-atlantic-overturning-circulation-shut-down/
5. Kevin D. Hoover (2004) Milton Friedman's Stance: The Methodology of Causal Realism https://papers.ssrn.com/sol3/papers.cfm?abstract_id=902062
6. DeSmog (2025) Roger Pielke Jr. https://www.desmog.com/roger-pielke-jr/
7. THB (19 July 2023) What the IPCC Actually Says About Extreme Weather https://rogerpielkejr.substack.com/p/what-the-ipcc-actually-says-about
8. Lloyds Registry Foundation (2024) World Risk Poll 2024 Report https://wrp.lrfoundation.org.uk/sites/default/files/2024-11/World%20Risk%20Poll%202024%20Report%20%20-%20What%20the%20world%20worries%20about%20.pdf
9. H.M.Treasury (2010) Stern Review final report https://webarchive.nationalarchives.gov.uk/ukgwa/20100407172811/https:/www.hm-treasury.gov.uk/stern_review_report.htm
10. LSE (January 2019) A Commentary: A Nobel Prize for the creator of an economic model that underestimates the risks of climate change https://www.lse.ac.uk/granthaminstitute/news/a-nobel-prize-for-the-creator-of-an-economic-model-that-underestimates-the-risks-of-climate-change/
11. Wikipedia (2024) Alfred Marshall https://en.wikipedia.org/wiki/Alfred_Marshall
12. Wikipedia (2024) Léon Walras https://en.wikipedia.org/wiki/L%C3%A9on_Walras
13. Wikipedia (2024) Carl Menger https://en.wikipedia.org/wiki/Carl_Menger
14. Wikipedia (2024) William Stanley Jevons https://en.wikipedia.org/wiki/William_Stanley_Jevons
15. Academic Agent (2019) Economics: The Austrian School vs. The Chicago School YouTube video https://www.youtube.com/watch?v=s5Kx6-nyVkk
16. Wikipedia (2024) Frank Knight https://en.wikipedia.org/wiki/Frank_Knight
17. Wikipedia (2024) John Maynard Keynes https://en.wikipedia.org/wiki/John_Maynard_Keynes
18. David Stainforth (2021) 'Uncertainty: a view from Climate Modelling, with David Stainforth' YouTube Video https://www.youtube.com/watch?v=6I5Q_IHCvoQ
19. John Bellamy Foster (2018) 'Marx, Value, and Nature' Monthly Review (July 2028) https://monthlyreview.org/2018/07/01/marx-value-and-nature/
20. Karl Marx (1859) Preface: A Contribution to the Critique of Political Economy https://www.marxists.org/archive/marx/works/1859/critique-pol-economy/preface.htm
21. Wikipedia (2024) Joseph Schumpeter https://en.wikipedia.org/wiki/Joseph_Schumpeter
22. Wikipedia (2024) Werner Sombart https://en.wikipedia.org/wiki/Werner_Sombart
23. Wikipedia (2024) Lionel Robbins https://en.wikipedia.org/wiki/Lionel_Robbins
24. Wikipedia (2024) Robert Lucas https://en.wikipedia.org/wiki/Robert_Lucas_Jr.
25. Wikipedia (2024) William Nordhaus https://en.wikipedia.org/wiki/William_Nordhaus
26. NBER (2017) William Nordhaus Integrated Assessment Models of Climate Change https://www.nber.org/reporter/2017number3/integrated-assessment-models-climate-change
27. William Nordhaus (2013) 'Chapter 16 - Integrated Economic and Climate Modeling' Science Direct V.1, 2013, Pages 1069–1131 https://www.sciencedirect.com/science/article/abs/pii/B978044459568300016X
28. IAMC Wiki (July 2024) Participating models https://www.iamcdocumentation.eu/index.php/IAMC_wiki
29. CarbonBrief (2025) How 'integrated assessment models' are used to study climate change https://www.carbonbrief.org/qa-how-integrated-assessment-models-are-used-to-study-climate-change/

30. Mark Bounthavong (2019) Cobb-Douglas production function and costs minimization problem https://mbounthavong.com/blog/2019/2/19/cobb-douglas-production-function-and-total-costs#:~:text=The%20Cobb-Douglas%20(CD),product%20of%20capital%20(K)
31. Wikipedia (2024) Robert Solow https://en.wikipedia.org/wiki/Robert_Solow
32. Wikipedia (2024) Paul Romer https://en.wikipedia.org/wiki/Paul_Romer
33. Royal Swedish Acadmy of Science (2018) The Prize in Economic Sciences 2018 https://www.nobelprize.org/uploads/2018/10/popular-economicsciencesprize2018.pdf
34. Paul Romer (2024) Exploring New Growth Strategy with Paul Romer YouTube video https://www.youtube.com/watch?v=qslTHxP_CEc
35. Paul Romer (1994) 'The Origins of Endogenous Growth' Journal of Economic Perspectives, Volume 8.1 (Winter 1994), pp. 3–22 https://pubs.aeaweb.org/doi/pdfplus/ https://doi.org/10.1257/jep.8.1.3
36. B.C.C van der Zwaan, et al. (2002) 'Endogenous technological change in climate change modelling' Energy Economics, V.24.1 (January 2002, pp.1–19) https://www.sciencedirect.com/science/article/abs/pii/S0140988301000731
37. Metoffice.gov.uk (2024) Mauna Loa carbon dioxide forecast for 2024) https://www.metoffice.gov.uk/binaries/content/assets/metofficegovuk/pdf/research/climate-science/seasonal-to-decadal/co2-forecast-2024.pdf
38. Climate.Gov (October 2022) How do we know the build-up of carbon dioxide in the atmosphere is caused by humans? https://www.climate.gov/news-features/climate-qa/how-do-we-know-build-carbon-dioxide-atmosphere-caused-humans
39. Arvin Ash (2023) Is Climate Change Man Made? The Surprising Data! YouTube Video https://www.youtube.com/watch?v=-skE4jCuf-w
40. Steve Keen (2020) The Appallingly Bad Neoclassical Economics of Climate Change https://mahb.stanford.edu/wp-content/uploads/2021/07/The_appallingly_bad_neoclassical_economics_of_clim-1.pdf
41. Martin L. Weitzman (2011) Fat-Tailed Uncertainty in the Economics of Catastrophic Climate Change https://scholar.harvard.edu/files/weitzman/files/fattaileduncertaintyeconomics.pdf
42. 43. Stern, N., Stiglitz, J., & Taylor, C. (2022). 'The economics of immense risk, urgent action and radical change: towards new approaches to the economics of climate change' Journal of Economic Methodology, 29(3), 181–216. https://doi.org/10.1080/1350178X.2022.2040740
43. Dani Rodrik's Weblog (1017) The Economics Debate, again and again https://rodrik.typepad.com/dani_rodriks_weblog/2017/12/the-economics-debate-again-and-again.html
44. Sam Peltzman (1993) How George Stigler Changed the Analysis of Regulation https://www.chicagobooth.edu/review/how-george-stigler-changed-analysis-regulation
45. EPA (2024) Climate Change Indicators: Climate Forcing https://www.epa.gov/climate-indicators/climate-change-indicators-climate-forcing#:~:text=by%20each%20gas.-,Radiative%20forcing%20is%20calculated%20in%20watts%20per%20square%20meter%2C%20which,change%2C%20that%20affect%20radiative%20forcing
46. HM Treasury (2010) Stern Review final report https://webarchive.nationalarchives.gov.uk/ukgwa/20100407172811/https:/www.hm-treasury.gov.uk/stern_review_report.htm
47. Yannis Dafermos and Maria Nikolaidi (2022) 'Assessing climate policies: an ecological stock–flow consistent perspective' ElgarOnline Research Article V.9.3 (2022) pp.338–356 https://www.elgaronline.com/view/journals/ejeep/19/3/article-p338.xml
48. Richard Murphy (2024) 'What is modern monetary theory' YouTube video (https://www.youtube.com/watch?v=1_vNAY2Nrm0
49. Wikipedia (2024) Hyman Minsky https://en.wikipedia.org/wiki/Hyman_Minsky
50. Stephanie Kelton (2022) The big myth of government deficits | TED https://www.youtube.com/watch?v=FATQ0Yf0Fhc
51. BBC.Com (2 January 2024) China's BYD overtakes Tesla's electric car sales in last quarter of 2023 https://www.bbc.com/news/business-67860232
52. Climatedata.ca (2025) Understanding Shared Socio-economic Pathways (SSPs) https://climatedata.ca/resource/understanding-shared-socio-economic-pathways-ssps/

53. Australian Government What are the RCPs? https://coastadapt.com.au/sites/default/files/infographics/15-117-NCCARFINFOGRAPHICS-01-UPLOADED-WEB%2827Feb%29.pdf
54. Ministry of Environment, New Zealand (May 2024) Intergovernmental Panel on Climate Change SSP-RCP scenarios https://environment.govt.nz/what-you-can-do/climate-scenarios-toolkit/climate-scenarios-list/ipccs-ssp-rcp-scenarios/
55. Steve Keen (2022) The New Economics: A Manifesto Polity Press
56. Wikipedia (2024) Physiocracy https://en.wikipedia.org/wiki/Physiocracy
57. Wikipedia (202 4) Wassily Leontief https://en.wikipedia.org/wiki/Wassily_Leontief
58. Wikipedia (2024) Leontief production function https://en.wikipedia.org/wiki/Leontief_production_function
59. Steve Keen (2023) Steve Keen: Marxism, Capitalism, and Economics Lex Fridman Podcast #303, YouTube https://www.youtube.com/watch?v=1XGiTDWfdpM
60. IEA (January 2024) The relationship between growth in GDP and CO2 has loosened; it needs to be cut completely https://www.iea.org/commentaries/the-relationship-between-growth-in-gdp-and-co2-has-loosened-it-needs-to-be-cut-completely
61. Nasa.Gov (January 2025) Temperatures Rising: NASA Confirms 2024 Warmest Year on Record https://www.nasa.gov/news-release/temperatures-rising-nasa-confirms-2024-warmest-year-on-record/
62. IPCC (2014) 'IPCC AR5: Process, Projections, and Predictions' YouTube video https://www.youtube.com/watch?v=Xpz8H8jRbdg
63. IEA.Org (2023) Global Methane Tracker 2023 https://www.iea.org/reports/global-methane-tracker-2023/understanding-methane-emissions
64. Lewin Day (2023) The Clathrate Gun Hypothesis: Unearthing Puzzles Of Warming Events Past https://hackaday.com/2023/08/21/the-clathrate-gun-hypothesis-unearthing-puzzles-of-warming-events-past/
65. Ian Plimer (January 2025) 'Climate Change: The Facts 2025' YouTube: Tom Nelson Pod #270 https://www.youtube.com/watch?v=a9NX9a2vyQE
66. Stanford Encyclopedia of Philosophy Imre Lakatos https://plato.stanford.edu/entries/lakatos/#FalsMethScieReseProg1970
67. Nature.Com/Scientific Reports (June 2019) RETRACTED ARTICLE: Oscillations of the baseline of solar magnetic field and solar irradiance on a millennial timescale https://www.nature.com/articles/s41598-019-45584-3
68. Science Alert (March 2020) Paper That Blames The Sun For Climate Change Was Just Retracted From Major Journal https://www.sciencealert.com/a-paper-that-blames-the-sun-for-climate-change-has-been-retracted#:~:text=In%20remarks%20to%20retraction%20website%20Retraction%20Watch%2C,the%20paper%2C%20and%20the%20authors%27%20scientific%20standing.
69. Dr Valentine Zharkova (2024) 'Solar Wander, Climate & the Surface of the Sun - Dr. Valentina Zharkova' YouTube DSPod #200 https://www.youtube.com/watch?v=OqLQyQtCNGg
70. Technologystories.org (2018) Kranzberg's First and Second Laws https://www.technologystories.org/first-and-second-laws/

Chapter 4
State Policies Towards Greenhouse Gases

Local Versus Global Reach

At the individual level, pessimism and resignation are understandable responses to a feeling of futility, that individuals cannot on their own make a difference, whatever their impact or influence may be on others. At societal levels, the differences that *can* be accomplished will be essentially local. Citizens who rail against the single use of plastic or the use of fossil fuels can achieve changes locally, even nationally, but global warming is what it says, it is global irrespective of regional differences. It is extremely difficult for citizens, even well-organised citizens, to be citizens-without-borders, to reach out to their peers in other countries, although international movements of mass protests have indeed raised public awareness and unsettled otherwise complacent governments. A *Climate Protest Tracker* [3] is one of many sources covering these movements, and in parallel there is also a *Climate Action Tracker* [4]. It would be interesting research to examine any correlations between the two.

What citizens can and should do is to engage with their own states and governments who *are* more able to achieve a global reach. However, that will be an exercise in frustration, because inter-governmental relations between independent states are always issues of negotiations and trade-offs. The point is well illustrated by the European Union's *Cross Border Adjustment Mechanism* (CBAM) which imposes a tax on the carbon content of goods entering the EU [5]. Discussing its introduction in 2024, one of the political supporters of the CBAM in a YouTube video [6] makes clear that, first, no mention of tax arises in its title as it is passed off as an environmental issue that avoids politically sensitive local tax issues within EU states. Second, the choice of imported goods selected for the not-to-be-mentioned tax deliberately does *not* include any coming *directly* from the USA in order to avoid a trans-Atlantic economic and political dispute. Since January 2025, that might not prove sufficient to placate the US Government. Pragmatism requires that the CBAM first establishes itself as a working mechanism before widening its scope in terms of goods and countries. There is some hope of that. In an interesting parallel development,

J. Ure, *Climate Change and Carbon Markets: Late or Too Late?*,
https://doi.org/10.1007/978-3-032-07619-9_4

John Podesta, who succeeded John Kerry as US climate envoy in 2024, hinted at a possible US cross-border carbon tax in an interview with the *Financial Times*: 'the US would fight against "freeriding" by foreign producers [read China—author] of carbon-intensive industrial imports' [7]. However, under the new US administration, the policy focus has initially turned towards tariffs, not towards regulatory alignment, still less towards carbon controls.

Two major issues follow from the above. First, that society as-a-whole needs to be engaged in debates and decisions, from woke (awoken) individuals, as in 'taking responsibility', to civil society at all levels, to governments and state decision-making and implementing bodies. Second, no progress is possible without the commitment of governments and states not only to make the right decisions, but also to pursue UN-agreed international conventions. The withdrawal in 2025 of the USA from the Paris Agreement and its targets and the UN Sustainable Development Goals (SDGs) obviously work against this. The remainder of this chapter is devoted to government policies, and to reviewing the policies of the countries that are heavy polluters and/or who are taking important initiatives: namely China, India, Singapore, the EU, the UK, and the USA.

Nationally Determined Contributions (NDCs)

Under the Paris Agreement, countries set targets as Nationally Determined Contributions (NDCs) for mitigating greenhouse gas emissions to achieve their climate goals. During 2025 these are being revised upwards as agreed every 5 years: from 2015 to 2020–21 to 2025 [8]. In November 2024, COP 29 (Conference of Parties) took place in Baku, Azerbaijan to increase countries' green funding contributions—see Chap. 7 for details. These are all government targets, but in achieving them the role of the market economy is left up to each country. Examples include, in addition to commitments to Net Zero by 2050, commitments in the USA to reduce GHG emission by 2030 over 1990 levels by 50–52%, until the new administration rescinded the commitment [9], in the UK by 68% [10], and across the EU by 'at least' 55% [11]. In industrialising economies, China committed to a peak in its GHG emissions before 2030, and carbon neutrality by 2060 [12], and India to 45% by 2030 and carbon neutrality by 2070 [13]. The next 5 year revisions will be a testimony to the levels of progress countries have made since the 2015 Paris Agreement.

Government Policies

Broadly put, there are three ways in which governments can meet their NDCs to combat climate change and global warming. The most direct is to stop issuing licences to fossil fuel enterprises, and to provide fiscal encouragement to renewable energy enterprises. The most effective way may be to create a carbon tax which polluters have to pay, together with a tax incentive where carbon emissions are being reduced. A good example of this comes from Singapore [14]. A third method is the Compliance or Mandatory Carbon Market (MCM)—also known as Emission Trading Schemes (ETS)—where governments issue carbon credits to designated polluters who are required to buy according to their emission levels, but which they can trade with other polluters, such that if they reduce their emissions they end up with excess credits to be sold to others who are perhaps even worse polluters. There are, of course, other measures governments can take, such as grants to industry promoting carbon reduction strategies, and research grants to develop emerging carbon-saving technologies. Separately, there is aid, grant and investment money to poorer developing countries - see Chap. 6.

There is also an area beyond immediate government policy-making that nevertheless impinges upon state policies, namely how to encourage action by the private sector which is not under a compliance requirement to reduce its carbon footprint, and how to encourage the use of private sector funds to finance carbon reduction projects. This is the Voluntary Carbon Market (VCM) which makes its own rules of assessment and carbon credits, but a lack of regulation and standardisation has proved an ongoing problem. After being close to collapse following the economic downturn post-COVID, there is a trend within the VCM to generate standards that more closely conform to those within the MCM, and within the UN to extend compliance principles towards the voluntary markets. See especially the section on Article 6 below and Chap. 5.

Government, Governance, and Civil Society

One way to view how the public (government) sector can influence the much broader private sector of a country is illustrated in the Venn diagram Fig. 4.1. The work of government is normally seen as consisting of government policies and the instruments of state through which they are administered, namely the civil, judicial, and military services. The governance of a country will consist of organisations which are not part of the state apparatus but rather make up the constituent agencies of civil society, such as stock markets, chambers of commerce, trade unions, the commercial media, non-governmental organisations (NGOs) which represent different sets of stakeholders in the economy. Government policies will be filtered or 'mediated' through the reactions and initiatives of these bodies as they inform and influence civil society at large consisting of all citizens. This constitutes the governance of a country.

The next sections of this Chapter are examples of ways in which Government policies towards carbon and GHG reductions, at national and international levels, overlap with private sector actions which are only indirectly regulated, if at all, by official policy approaches.

Carbon Reporting: It has been encouraged by the *Sustainable Stock Exchanges Initiative* (SSE) as part of the 2015 UN *Sustainable Development Goals* that included

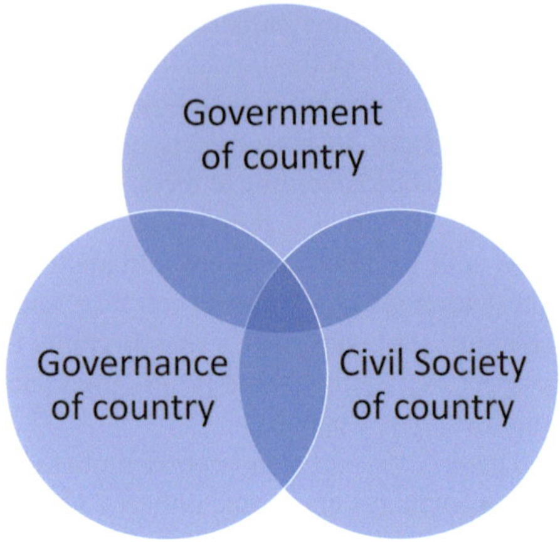

Fig. 4.1 Venn diagram of government, governance, and civil society

Goal 12, target 12.6: 'Encourage companies, especially large and trans-national companies, to adopt sustainable practices and to integrate sustainability information into their reporting cycle.' By 2024, 74 of the 122 stock exchanges (61%) tracked by the SSE had adopted the initiative [15], among them the London Stock Market [16]. Following a consultation in 2022, the London Stock Exchange decided to list funds focused upon carbon reduction projects [17].

Licences and Fiscal Incentives: Two types of licences are issued to fossil fuel companies, the first is an exploration licence to undertake drilling and survey work for oil, gas, or coal. The second is a production licence to extract and refine fossil fuels into a variety of derivative products, ranging from petroleum to plastic, and sell the stuff. Different companies may be involved along this supply chain, so the critical part is the initial production licence. However, in reality not all licences, which are typically but not necessarily allocated through a bidding process, are used. For example, in the UK, North Sea oil was first explored in the 1960s and since then over thirty rounds of licensing have been conducted by the North Sea Transition Authority (NSTA), often involving over 100 applications, in a rush to exhaust the supplies [18]. The announcement in 2023 that licensing would continue in an apparent contradiction to the UK's commitment to a green energy policy came less than a year ahead of a General Election, with the preservation of jobs being given priority. An additional factor was the drive towards 'energy security' as the UK and the EU turned away from Russian gas supplies following the invasion of Ukraine in 2022 [19].

The commitment to gas and oil exploration and extraction remains strong globally, supported by a powerful and combative lobby of 'oil majors' and often by their governments. For example, the USA since 2019 has been close to self-sufficient in oil, gas, and coal as shown in Fig. 4.2.

That might suggest less interest in international exploration, but poorer countries with crude oil reserves are dependent upon sector investments from richer countries. Venezuela is an example, which under the authoritarian rule of President Nicolas Maduro has been considered a pariah state by the USA. Nevertheless, the USA

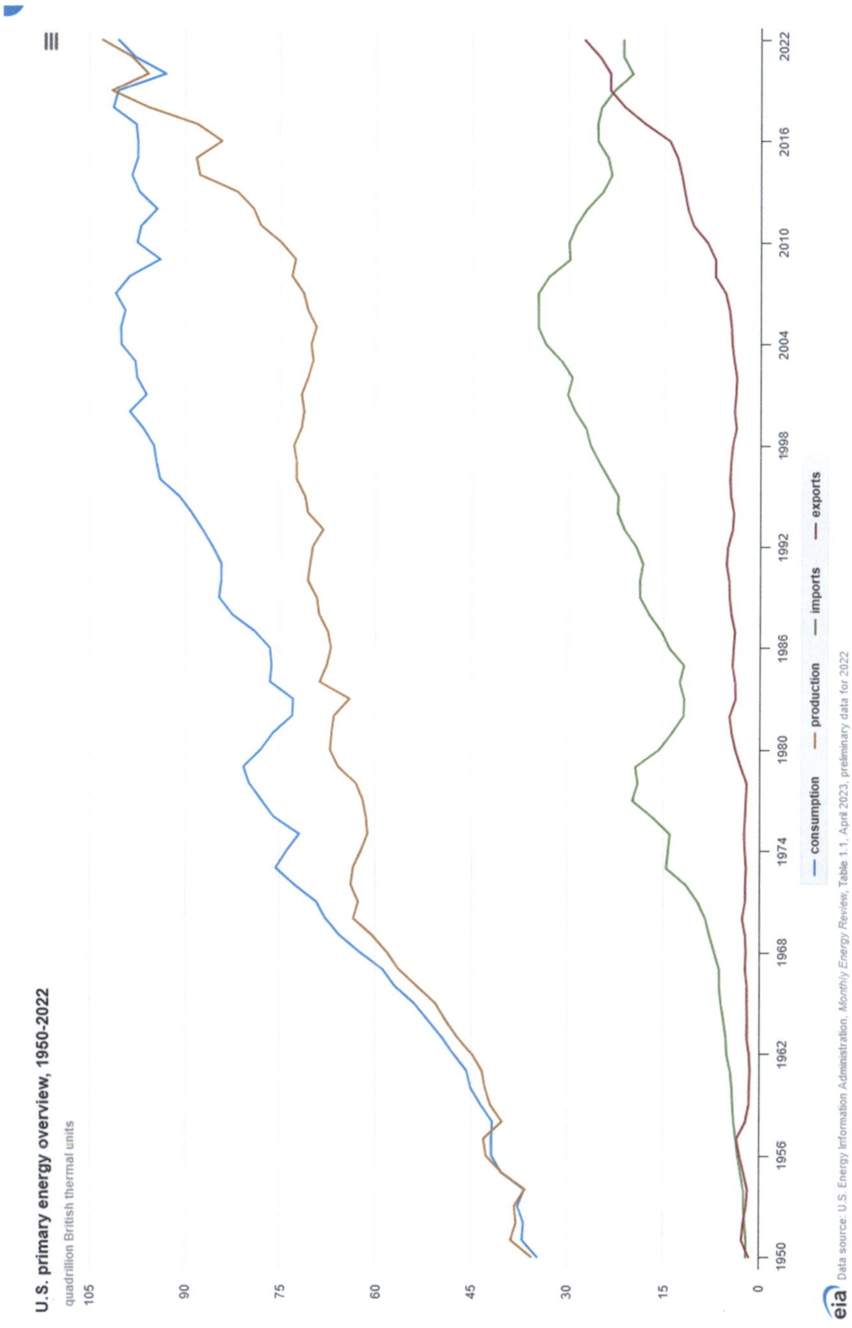

Fig. 4.2 US Coal, Oil and Gas Output and Consumption, Imports and Exports 1950-2022. *Source* US Energy Information Administration [20]

permitted US companies investing in Venezuela's oil industry, and after withdrawing a general authorisation of licences replaced it with a selective licensing system. These investments are profitable for the US companies, and if Venezuela allows genuinely free elections then restrictions will be lifted [21]. But the selectivity should also be noted, as the US does not impose similar restrictions upon countries with even worse human rights records. At this level, the priorities are politics and profits, not a 'green' agenda. Under President Donald Trump however, licences are being rescinded, although oil company lobbying may extend some of them [22].

The commitment of the USA to oil and to gas, especially to gas in its liquified state as Liquified Natural Gas (LNG), directly reflects the *status quo* in terms of energy dependency, where the *status quo* is in competition with more forward-looking government commitments as set out in their Nationally Determined Contributions (NDC) under the Paris Agreement, or before the USA withdrew from the Agreement in 2025. Commitments to an uncertain future are always at risk of being discounted to what actually exists today. To achieve the NDCs requires a political act of will. To move away from the *status quo* also involves a political decision.

Subsidies: They are still widely available to the fossil fuel industry to the contradiction of the goals of Paris Agreement. They are mostly justified in the grounds of 'energy security' or on the grounds that by 2050 renewable energy is likely to be insufficient to meet end-user demand. But the power of the fossil fuel lobby seeks to deflect the debate away from subsidies and shift it towards carbon capture technologies [23]. In a very real sense, these are last resort technologies.

Box 4.1 Subsidies

Explicit or 'direct' subsidies, designed to keep prices low, can either go to producers or to end-user consumers. In the case of fossil fuels, it has been estimated that 80% of subsidies go to consumers and 20% to producers, according to *Our World in Data* [24]. Ending subsidies to consumers is only realistic if there are cheaper alternatives, which is why renewables are so important. But even as 'free gifts of nature' solar, wave, wind, and hydro energy require capital-expensive infrastructures to transform the incoming power levels to lower grid-level storage and transit requirements, which is a strong argument for shifting the subsidies from fossil fuels to these infrastructure costs until renewable energy is both widely accessible and affordable. The oil emergency caused by Russia's invasion of Ukraine pushed global explicit subsidies in 2022 to over USD1 trillion, although they fell back to USD600 billion in 2023. As recorded in *Our World in Data*, the largest subsidies come from petrol-producing states such as such as Saudi Arabia, Turkmenistan, Algeria, and are often over USD500 per person and sometimes over USD1,000 per person. By contrast, across Europe, North and South America and East Asia subsidies per person are typically below USD100, and below USD20 per person in Africa and South Asia. Together these subsidies represent over 10% global GDP. These enormous sums could be redistributed towards renewables.

Additionally, there are 'indirect' subsidies, which the IMF (International Monetary Fund) has suggested could be over USD7 trillion, or 7% of global GDP [25]. These are the estimated social costs, or externalities, of burning fossil fuels, such as ill-health and environmental damage. By levying carbon taxes and other measures these social costs could be reduced, but subsidies contradict such policies. The good news is that, as cited in another *Our World in Data* blog, over recent years several countries, such as Indonesia, Italy, Thailand, and Ukraine, have reacted to the reality of global warming and have cut subsidies to fossil fuels [26]. A study of the US fossil fuel sector lists eleven distinct ways in which fossil fuels are subsidised, six explicitly including a tax discount (typically 15%) on gross revenues, and five indirectly [27], and references a study in 2017 in *Nature Energy* that found that nearly half of new U.S. oil fields would not be viable without this support [28].

Carbon Taxes: These are taxes levelled upon a given metric of carbon, or carbon-equivalent (CO_2e) GHG emissions, usually on each tonne of carbon emitted, or emitted above a baseline level, or upon the *intensity* of carbon per value of goods produced, or as a proportion of GDP, etc. A carbon tax is the most straightforward means of penalising GHG emissions and a direct way to encourage carbon abatement. Nobel Laureat economist William Nordhaus [29] has been a prominent advocate of 'green taxes' because the costs of GHG emissions mostly fall upon the public and not the polluter; they do not figure as costs of production because they are largely *externalities*. In that sense, and in the absence of a market that could put pressure upon polluters to invest in cleaner, greener fuels and production methods, a carbon tax creates a social price to be paid by the polluter. It internalises the externality. Nordhaus argues the green or carbon tax should be the marginal cost to the public of the pollution created. For example, road transport emissions cause health problems such as lung diseases, time-off-work, and in serious cases hospitalisation and lost productivity. A green tax on transport that incentivises a switch to electric battery or hydrogen-driven vehicles would pay for itself, even as tax revenues fall towards zero. Similarly, a tax set at the marginal cost to society of pollution and imposed upon industries that emitted high volumes of GHGs would be deemed successful as it brought in less and less revenue. The critical economic issue is how high the tax needs to be. The political issue is how acceptable are high carbon taxes if applied to all sectors if they become an electoral liability, and if estimating and enforcing taxable emission levels across all small and medium-sized enterprises becomes an administrative nightmare. The alternative in practice has been to confine carbon taxes to the largest and heaviest polluters.

Compliance or Mandatory Carbon Markets: Emission allowances, or the right to pollute, are sold by government to companies designated as high polluters, with a cap, which can be enlarged over the years, placed upon their permitted levels of pollution. If the companies pollute less they can sell their excess credits. If they pollute more they must either buy more credits or pay a penalty. The baselines against which carbon reductions are set are by convention established according to the principles developed by the UN Framework Convention on Climate Change (UNFCCC) [30]. The allowances are issued and purchased under Mandatory Carbon Market (MCM) rules, whereby carbon removals or reductions, and any credits based upon them, can only be counted as part of a company's emission reductions if they arise within the company's own sphere of operations, as Scope 1, Scope 2 or as part of their supply chain Scope 3. This is sometimes referred to as *in-setting*. If the reductions arise from carbon credits relating to projects outside of the company's operations, they are referred to as *off-setting* and do not qualify as MCM reductions. They become part of the Voluntary Carbon Market (VCM).

International evidence suggests in many cases the enforcement of compliance or MCM rules has resulted in significant CO_2 reductions. A research paper in 2024 [31], estimated that where reductions were identified, the average reduction effects of 37 carbon tax schemes ranged from −10 to −8.9%, with the highest recorded at −21% and the lowest at around −5%. For 36 cap-and-trade schemes the equivalent figures were for eight China ETS pilots an average of −10.4 to −5.4%, and with ETS across the EU and British Columbia (Canada) averaging −7.3 to −5.4%. Only three smaller reduction of 5% or less were reported for ETS in Australia, Korea, and Switzerland.

Voluntary Carbon Markets: They fall beyond government regulations, but this has led to problems of integrity—see Chap. 6. UN think tanks and governments alike are seeking ways to encourage greater private capital market participation in climate change finance, because public funds on their own will be insufficient. Direct financing by the corporate enterprise sector means investing in Scopes 1, 2, and 3 carbon emissions reductions. Indirect means buying carbon credits as offsets, either as a transition to cover the gap between the present and the day when they can reduce their own emissions, or as an ongoing compensation for emissions they are unable to reduce. In this way, companies buying credits hope to be seen by investors and governments, as 'good actors' in *Environmental, Social, and Governance* (ESG) parlance, and, if a listed company, through acknowledgements in their annual reports—the *financial* logic being that ESG compliance reduces future risks, including risks arising from climate change, and can thereby add to long-run business sustainability [32].

Environmental, Social, and Corporate Governance (ESG)

Whereas there is no conclusive evidence of a correlation between ESG and longer-term financial sustainability or profitability, there are several studies that do find positive correlations, such as one study of 1,000 papers published between 2015 and 2020 on sustainability and financial performance by the NYU Stern Center for Sustainable Business [33]. The study revealed that 58% of the papers found a positive relationship, 8% a negative relationship, 13% no relationship, and 21% mixed results. Reference to this paper can also be found in the literature review of a paper published in 2022 using a Bloomberg database of the 5,000 largest listed firms 2013–2021, finding that 'Individual Social and Governance scores have a positive and significant relationship while Environment score does not have a significant relationship with firm value. On the other hand, ESG combined scores… have positive and significant relationships with firm profitability' [34].

Research in 2023 by the International Federation of Accountants (IFAC) found that 66% of the 600 'largest stock exchange-listed companies (by market capitalization) from the 15 most industrialised jurisdictions' disclosed emission targets [35]. Smaller companies are less likely to do so. Another analysis in 2023 by S&P Global of its Corporate Sustainability Assessment (CSA) data compared over 13,000 companies reporting climate-related information using the *International Financial Reporting Standards* (IFRS) of the International Accounting Standards Board (IASB)—an acronym-heavy procedure, but see also under United Kingdom below—and found that SMEs (< 250 employees) typically made only half as many emission disclosures (14%) as larger companies (33%). Not surprisingly, it also found that in Europe where such disclosures are a requirement, reporting was higher than in all other regions [36].

The SME finding underscores the fact that internationally the voluntary market has more width than depth, and the EU requirement finding underscores the importance of demand in an over-supplied carbon credits market. Bloomberg has noted, due to the market being 'heavily oversupplied, to the tune of nearly 50% in 2023', the total value of credits traded in 2023, after a fall in 2022, was just USD163.6 million, up from USD161 million in 2021. The compelling reason for this is a hesitancy of demand, due to quality issues fuelled by years of 'greenwashing' and associated financial scandals, driving many international corporations to prefer their own direct investment in carbon reduction projects. However, if quality assessment and project auditing improve to the point where demand, under one of three BNEF's scenarios, becomes price-inelastic, 'companies could purchase 1 billion offsets annually in 2030, levelling off to 2.5 billion in 2050' [37]. But this forecast, although often cited, is heavily dependent upon the integrity of credits improving and is very much at the higher end of the scale, as illustrated by Fig. 4.3 of various other estimates, taken from a posting on Linkedin by Tea Bano.

Given the poor history of the voluntary market, voluntary credits are usually excluded from qualifying as offsets in the compliance market, including eligibility for tax credits. But given the growing concerns with aiding the financing of private sector carbon reduction projects, as long as they are genuine, governments are beginning to look at ways to build bridges between the MCMs and the VCMs by, for example, promoting more rigorous standards and codes of conduct, and integrating private sector GHG reductions into the Nationally Determined Contributions (NDCs).

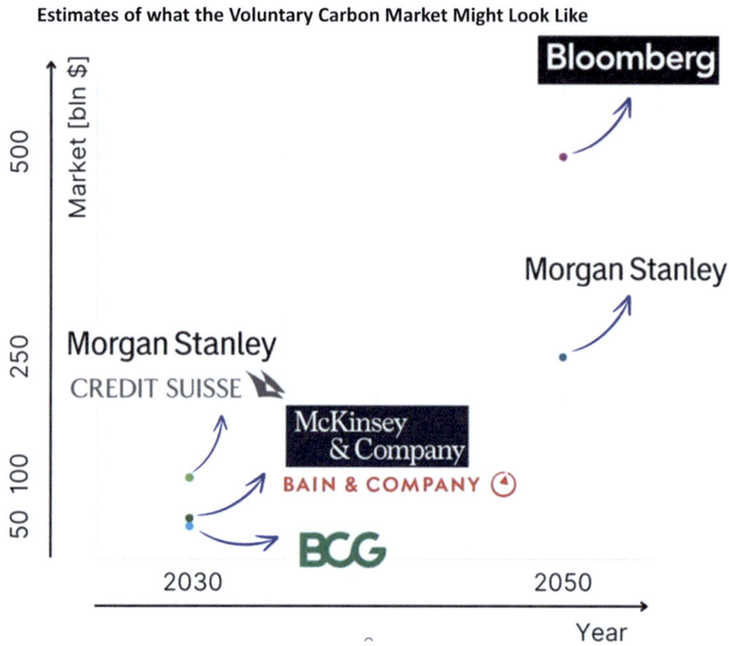

Fig. 4.3 Forecasts of the value of voluntary carbon market credit transactions up to 2050. *Source* Tea Bano@teo-bano, Linkedin, Carbon Compass (2024)

Article 6 of the Paris Agreement

Article 6 of the *Paris Agreement* is seen by many as a way for States to breathe new life into the voluntary market after the failure of the *Kyoto Protocol* under which *Certified Emissions Reductions* (CERs) credits were issued [38], see Box 4.2.

Box 4.2 A Brief History of CERs

The history of CERs is not a hopeful one—poorly administered, over-issued and open to fraudulent projects that gamed the system. One especially egregious example was gaming the *Montreal Protocol* [39], designed to protect the Ozone Layer by banning the use of hydrochlorofluorocarbons (HCFCs) in refrigerators and cooling systems. Rogue manufacturers, mostly located in China and India, managed to secure carbon credits (CERs) to dismantle such systems that had been created simply for the purpose of claiming the credits. In other words, they dodged the 'additionality' audit that is the need to show the GHG reduction would be *additional to* GHG reduction without these systems, and they actually increased GHG emissions [40].

As the UN's *Clean Development Mechanism* (CDM) did not offer incentives to reduce emissions but only credits to finance them, it was in reality a subsidy scheme waiting to be abused. Because the CER scheme was not an *Emissions Trading System* (ETS) 'cap-and-trade' scheme—see below for CDM and ETS schemes—some developing countries even suggested that the baseline from which the emissions reductions were estimated should not have to conform to the additionality requirement. In other words, they should be allowed to raise as much finance from the selling of CERs irrespective of how much they continued to pollute. But this would imply a loss of project integrity, exactly as in the case of the hydrochlorofluorocarbons. Scandals and incompetencies such as these, together with a slowdown in European economic growth, reduced demand as buyers turned away, and the price of CERs fell from a high point of USD20 per tonne to below USD1 per tonne until the system finally collapsed in 2010 (see below).

Although the Kyoto Protocol was agreed in 1997, it did not come into effect until 2005 due to the lengthy process of ratification by individual states [41]. Such are the unavoidable delays in securing international agreements by consensus, even after differences of opinions and interests have been resolved. Under the auspices of the *UN Framework Convention on Climate Change* (UNFCCC) every year a COP ('Conference of the Parties') is organised, and after each one frustration is expressed due to the urgency of climate change mitigation and the failure of the COPs to reach consensus on the outstanding issues. For example, COP27 (Sharm El-Sheikh, Egypt) in 2022 was criticised for failing to call unequivocally for the phasing out of fossil fuels [42]. A partial resolution was made using diplomatic language the following year in COP28 (Dubai, United Arab Emirates), but measures to help developing countries finance their carbon footprint reductions fell short [43].

Article 6 offers three possible ways forward, as outlined in Fig. 4.3 from the must-read *The Nature Conservancy* (TNC) report in 2024 [44]. What follows draws heavily from that report. Articles 6.2, 6.4, and 6.8 offer three different routes towards carbon and GHG mitigation.

Article 6.2: It is labelled by TNC as 'market' but a better hybrid term might be 'traded' as the main actors are states even if the projects can be private. Developing countries have an option to sell the financing of projects within their borders to wealthier countries, either bilaterally or multilaterally, as *Internationally Transferred Mitigation Outcomes* (ITMOs) and by agreement of the parties involved in the ITMO can also count towards the NDC of the buyer—see Fig. 4.5.

This is where an ITMO differs from the earlier CERs. A 'corresponding adjustment' a register in each country will record the transfer to prevent double-counting. Figure 4.4 from TNC illustrates the process.

The contract or treaty arrangement between the two parties has to cover many issues such as transparency, where each country adds the transaction to a national register, a process for *Monitoring, Reporting and Assurance* (MRA) [45] and in the

Fig. 4.4 Article 6 of the Paris Agreement. *Source* TNC (2024)

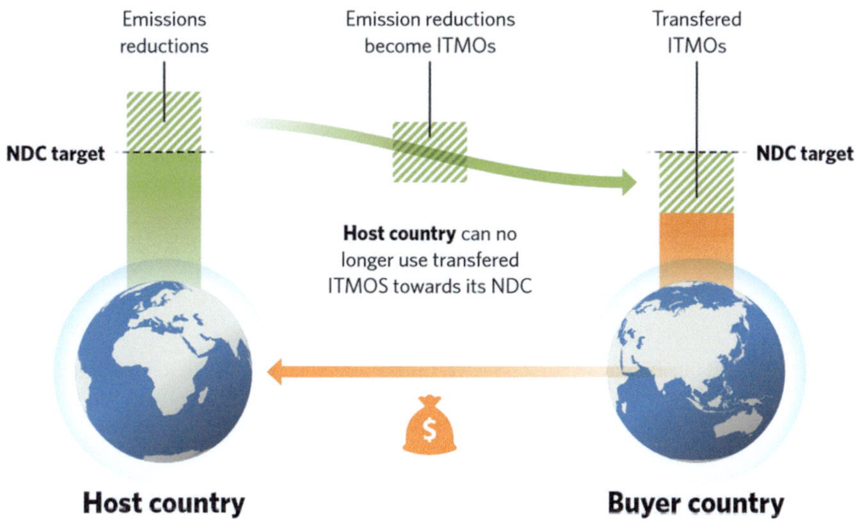

Fig. 4.5 The role of ITMOS under Article 6.2 of the Paris Agreement. *Source* TNC (2024)

case of an agreement termination, for example, should a change of government in a developing country want to withdraw the project to boost its own NDC, or the buyer revokes funding, issues such as arbitration and compensation become necessary. The legal, technical, financial, and political details are many, which explains why up to mid-2024 only one ITMO had been created, between Switzerland and Thailand [46], involving emissions from Bangkok Transport system, but many more are under negotiation as listed in Table 4.1.

Despite the contractual tangles, Article 6.2 offers the flexibility of no restrictions on the types of units (1 unit being 1 tonne of carbon which can be sub-divided between investors) to be traded, or the sectors of the economy or the project assessment methodologies used, as long as they conform to Article 6.2 requirements. For example, an ITMO agreement could be linked to an Emissions Trading System (ETS) cap-and-trade or MCM allowing trading and tax breaks. In this way the compliance and voluntary markets in some countries could interconnect, and if sufficient quality assurances were forthcoming, this could open the door to a wider range of financing opportunities [47]. But given the VCM's checkered history, this process is inevitably a gradual one, and a learning process, but driven uppermost by the need for finance.

Article 6.4: Unlike the decentralised trading mechanism permitted under Article 6.2, Article 6.4 creates a centralised *Paris Agreement Crediting Mechanism* with a *Supervisory Body* [48] which reports to the *Parties to the Agreement* [49], with responsibility for 'approving methodologies, registering activities, accrediting third-party verification bodies, and managing the Article 6.4 Registry' [50], see Box 4.3.

Table 4.1 Internationally transferred mitigation outcomes (ITMOs) *in Progress* (2024)

Buyer country	Host country
Switzerland	Chile, Dominican Republic, Georgia, Ghana, Kenya, Malawi, Morocco, Peru, Senegal, Thailand, Tunisia, Ukraine, Uruguay, Vanuatu, Zambia
Japan	Azerbaijan, Bangladesh, Cambodia, Chile, Costa Rica, Georgia, Ethiopia, Indonesia, Kazakhstan, Kenya, Kyrgyzstan, Lao PDR, Maldives, Mexico, Moldova, Mongolia, Myanmar, Palau, Papua New Guinea, Philippines, Saudi Arabia, Senegal, Sri Lanka, Thailand, Tunisia, Vietnam, United Arab Emirates, Ukraine, Uzbekistan
South Korea	Gabon, Fiji, Lao PDR, Mongolia, Vietnam, Uzbekistan
Singapore	Bhutan, Cambodia, Costa Rica, Chile, Columbia, Dominican Republic, Fiji, Indonesia, Kenya, Lao PDR, Mexico, Mongolia, Morocco, Peru, Papua New Guinea, Paraguay, Rwanda, Senegal, Sri Lanka, Thailand
Norway	Indonesia, Morocco, Senegal
Sweden	Nepal, Dominican Republic, Ghana

Source TNC (2024) [44]—updated by the author

Box 4.3 High-Quality Carbon Offset Guide as used by UNFCCC, States and VCM Standards Bodies

- **Additional**: an additional project is one that would not have occurred without the incentive provided by carbon credit revenues.
- **Not Overestimated**: means the GHG emissions avoided or the removals that are enhanced by a mitigation project are quantified conservatively relative to a realistic baseline.
- **Permanent:** offset credits must be associated with GHG reductions that are permanent.
- **No Leakage:** must not be replaced by emissions elsewhere.
- **Not Claimed by Another Entity**: carbon credits must be exclusive and not be counted or used more than once.
- **Not Associated with Significant Social or Environmental Harms:** must avoid causing any (new) social and environmental harms.

Source Adapted from Carbon Offset Guide [51]

When ITMOs are issued a levy on the transactions accrues to the UNFCCC account at the World Bank. The arrangement is similar to the previous UN's *Clean Development Mechanism* (CDM) under the *Kyoto Protocol*, but ITMOs can be used to offset the buyer's NDC. Carbon credits known as *Certified Emissions Reductions* (CERs) issued under the CDM and registered before 2013 can only be used to offset a county's very first NDC. This is a concession to allow double-counting as in these cases both buyer and host countries can offset their NDCs. It is also a retrospective arrangement to avoid overwhelming NDCs with a plethora of over-issued pre-Paris Agreement CERs. One estimate by the New Climate Institute (cited by TNC 2024) suggests that the difference between the 2013 and 2021 cut-off dates is the difference between upwards of 340 million CERs versus nearly 4 billion. ITMOs on the other hand can be used as offsets to future revisions of NDCs [38]. But trading in ITMOs is initially off-limits as they are regarded as a 'mitigation contribution' from the buyer. For example, Switzerland retired the ITMO from Thailand immediately upon reception to avoid any chance of it being traded twice and double-counted [52].

Besides offsetting a buyer country's NDC, under some circumstances the buyer country could also offset their commitment to reducing the carbon footprint of their airlines using the *Carbon Offsetting and Reduction Scheme for International Aviation* (CORSIA). An ongoing debate is whether host countries could likewise divert some of their ITMO funding to CORSIA. At stake in these debates is the trade-off between host countries being offered greater flexibility in the use of the funds coming from richer buyer countries, against fund risk in poorly regulated and assessed projects in host countries.

Article 6.8: It remains a work-in-progress, creating a framework for a UNFCCC centralised website to record mitigation projects that are planned, pending or receiving public funding, and which conceivably could also attract private sector funding. TNC makes the point that most historical payments for forest-based REDD+ projects [53] (*Reducing emissions from deforestation and forest degradation in developing countries*) came from bilateral deals and multilateral funds, such as the *Green Climate Fund* created by the UNFCCC, [54] and the World Bank under the *Warsaw Framework for REDD+* [55]. TNC adds: 'These non-market payments helped countries to improve their REDD + programs and now many REDD + countries can apply for market-based, or hybrid public/private funding through programs like the Carbon Fund of the *Forest Carbon Partnership Facility (FCPF),* [56] and *Lowering Emissions by Accelerating Forest Finance* (LEAF)' [44, 57]. An alignment of standards is required if, for example, some funds require a 'buffer' pool or stock of stored credits in case of project reversals or *de faults*. This is therefore another route towards integrating the compliance and voluntary markets, again driven by the need to mobilise finance from all available sources.

Chapter 5 notes that since the COP-29 decision on Article 6, countries including the EU, UK, China, India, Japan, and Indonesia are evolving their policies towards recognition of carbon credits but subject to very strict quality requirements and buffer stocks to insure against defaults [58].

United States

To achieve its NDC, in 2022 Congress passed a bi-partisan bill called the Inflation Reduction Act (IRA). In 2025, on his second day in office, 'Trump ordered all federal agencies to immediately pause the disbursement of funds under the IRA and through his executive orders, the president issued a directive to cease financial support for the development of electric vehicle charging infrastructure and other policies that favour electric vehicles' [59].

As its name implies, the broad objectives of the IRA were to address domestic supply-side influences on output, productivity, and prices. Included were many measures to boost the electrification of America's infrastructure and to provide loans to community environmental projects [60]. Subsidies and tax breaks were available for foreign direct investments (FDI) in cleantech, with USD278 billion pledged to attract investment mostly from Asian economies such as South Korea, Japan, and China [61], but one assessment suggests the CHIPs + Act, also passed in 2022, has been more influential in attracting FDI [62]. The Act provides USD39 billion for chip manufacturing, 25% investment credits for the costs of manufacturing equipment, and USD13 billion for semiconductor R&D and workforce training, seen as a means to strengthen the USA in its competition with China [63]. By March 2025, President Trump was describing the bi-partisan Act as 'a horrible, horrible thing' and declared he wanted to get rid of it [64]. For good measure, he also signed the Congressional Review Act blocking the pricing of methane emissions which were scheduled to rise from USD900 per tonne (USD36 per tonne of CO_2e) to USD1,200 in 2025, and USD1,500 from 2026 [65].

Unlike the EU, the UK, and several other countries, the USA does not have at the *federal* level an Emission Trading Scheme (ETS). In what has become a decades-long toxicity in American politics, in 2009 the Republican Party, having gained control of the US Senate, adopted an unremitting stance to oppose virtually anything and everything President Obama's Administration proposed, including the *Waxman-Markey bill* that would have brought the *1970 Clean Air Act* up to date [66]. The funding of Republican candidates, Representatives, and Senators by fossil fuel companies is well documented [67]. Despite that, California (2013), Washington D.C. (2017), and eleven Democrat-leaning North Eastern states which are party to a Regional Greenhouse Gas Initiative (RGGI) have put in place cap-and-trade programmes [68].

Carbon Trading

The failure of the US nationally to adopt an ETS not only makes a potential alignment difficult with the Cross-Border Adjustment Mechanisms (CBAMs) of the EU and the UK, but it is somewhat ironic given that the first *national* effort to create a carbon cap-and-trade market was the initiative of the Chicago Climate Exchange

(CCX) in 2003. This offered a more centralised, organised and regulated market in contrast to the chaotic and irregular 'over-the-counter' trading by which traders are responsible to no-one [69]. Each member company joining the CCX committed to a 1% annual reduction in emissions, to be reported annually for transparency, and those exceeding 1% could either bank or trade their reductions as credits to others. Although this scheme was for members only, CCX did allow offsets coming in from third parties, mostly Mid-Western farming projects. But the CCX failed, and closed in 2010, according to one assessment, because the demand for credits was too low, the penalties for failing to meet reduction targets were negligible, new members were very few, and because the supply of incoming credits was too high and their quality often not well substantiated. By 2010, when credits were trading at just 5 cents per tonne of carbon, the trading volume ground to a halt [70].

California

The *California Air Resources Board* (CARB) [71] was created in 2006 to administer a compliance 'Cap and Trade' Emissions Trading Scheme (ETS) under which allowances are issued, each equivalent to one tonne of carbon emitted, and in total (the cap) equaling California's emissions target level [72]. The targets include driving emissions 40% below 1990 levels by 2030, and 80% below by 2050. Allowances can be allocated or auctioned in which case the revenues go to a *Greenhouse Gas Reduction Fund* (GGRF). Electricity and natural gas utilities are allocated allowances according to their caps. If they overshoot, they either buy additional allowances by trading, or pay a fine of around USD100 per tonne of carbon. Allowances are auctioned to major polluters in industry, transport, and construction, but at reserve-prices low-enough to discourage leakage of companies moving their operations to States that do not operate carbon pricing schemes. Figure 4.6 shows the annual reduction in available allowances as the cap is lowered each year.

In the absence of a Federal ETS—which might be considered unconstitutional in the USA unless pollution was deemed as crossing state borders—the problem of arbitrage is ever present. Around 450 companies are covered by the scheme [73]. Remaining allowances either go to state-owned agencies or to a price stabilisation fund to adjust the availability of allowances according to fluctuations in demand.

Besides a compliance market, in California the CARB ETS supports a voluntary market. Unlike allowances which relate to emissions already taking place and with payments to CARB (the government), carbon credits, which can be bought and sold as offsets, relate to future anticipated emissions and the payments move horizontally between traders. By 2023 the average price of credits in the ETS was around USD30 per tonne of carbon. For compliance obligations, a limit of 8% is placed upon CARB offset credits, and CARB sets a strict verification process for projects, mostly nature-related or CCUS (carbon capture, utilization, storage).

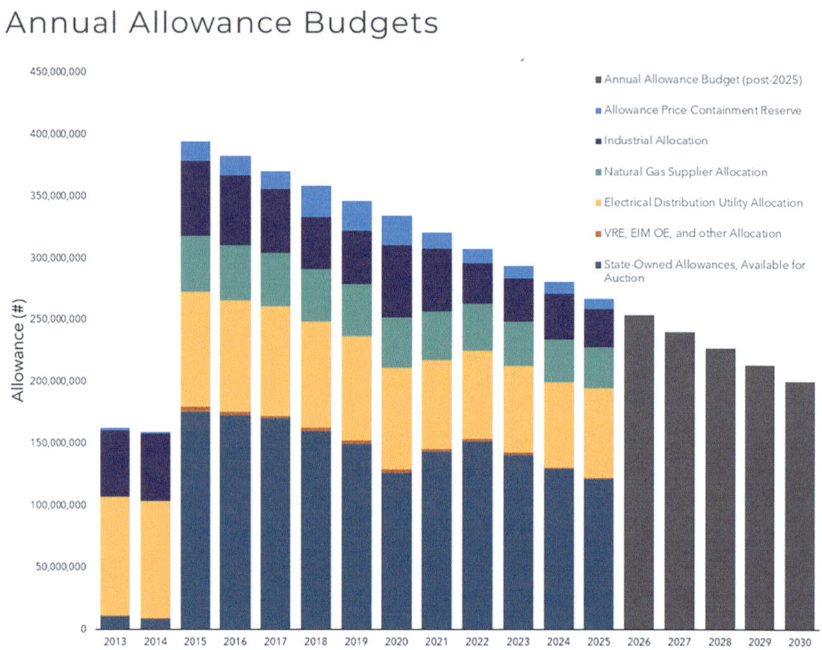

Fig. 4.6 Progressively declining allowances under the California 'Cap-and-Trade' Emissions Trading Scheme. *Source* CARB Cap-and-Trade Program Data Dashboard [https://ww2.arb.ca.gov/our-work/program/cap-and-trade-program/program-data/cap-and-trade-program-data-dashboard#Figure1]

United Kingdom

In the UK the Adaptation Committee of the Climate Change Committee (CCC) [74], which arose from the Climate Change Act (2008), has a statutory duty to advise Government on climate change policies. It delivers a *National Adaption Programme* (NAP) every two years, and conducts annual assessments [75]. Until 2023, when *Climate Action Tracker* [76], an independent scientific research consortium, downgraded the NAP performance to 'insufficient' the UK seemed to be doing well. A government news release in May 2024 asserted that '[s]urplus from the UK's third carbon budget will not be carried forward—proving the UK's pragmatic approach to net zero is compatible with its climate targets' [77]. However, the CCC 2022 annual report to Parliament suggested otherwise [78].

> In targets, the UK is indeed a world leader. However, this Progress Report reveals that, despite important achievements in renewable energy and electric vehicles, the Government is failing in much of its implementation… For that reason, the CCC will in future be concentrating even more centrally upon the delivery and implementation of the targets which are now enshrined in statute and international agreements. The Government has set the right course. It has now to deliver on the scale and urgency that is required.

In response the then Conservative UK government introduced a *National Adaption Programme* 2023–2028 (NAP3) [79] to.

- Protect the natural environment
- Support business in adapting to climate change
- Adapt the infrastructure (e.g. electricity and rail networks)
- Protect buildings and their surroundings from hotter temperatures
- Protect public health and communities
- Mitigate international impacts on the UK (e.g. on imports of food supplies)

However, in 2024 the rebuke from the CCC was even more damning [80]. While noting that 'NAP3 is an improvement on its predecessor' an assessment of the measures proposed and undertaken concluded that.

> NAP3 falls far short of what is needed. NAP3 lacks the pace and ambition to address growing climate risks, which we are already experiencing in the UK. It fails to set out a compelling vision for what the government's 'well adapted UK' entails, and only around 40% of the short-term actions to address urgent risks identified in the last Climate Change Risk Assessment are progressed. The lack of a measurable vision will prevent effective delivery of adaptation by public agencies, local authorities, and the private sector, as well as inhibiting a clear assessment of progress…. Slow progress across three NAPs shows that the current approach is not working, and that change is needed.

The CCC recommended an 'urgent refresh of NAP3 and adaptation governance should be undertaken in the new Parliament'. The words 'across three NAPs' in the report imply a more radical approach is required. Specifically, the areas where NAP3 fell short are summarised in Fig. 4.7.

All areas were judged to be under 40% addressed, none more so than 'business and industry' where GHG emission reductions are urgent for the Net Zero target to be reached by 2050. That raises the issue of how much importance a UK government attaches to mandatory measures, and how much reliance is placed upon the VCM.

Contracts of Difference (CfDs)

A Labour Government took over in July 2024, declaring to 'deliver our clean power mission, Labour will work with the private sector to double onshore wind, triple solar power, and quadruple offshore wind by 2030' [81]. One policy innovation of the UK carried over since 2014 is the 'contract of difference'(CfD) [82] available to operators of onshore and offshore wind, solar PV, geothermal parts, hydropower, ocean power (tidal and wave), and other renewable environmental projects. The CfD stabilises revenues to the renewables operator based upon the difference between the market price and an agreed 'strike price'. When the market price rises above or falls below the strike price, the renewables operator repays or receives the balance from the CfD. The Low Carbon Contracts Company (LCCC) was created to manage around 240 CfDs which last for 15 years, 'enabling generators to deliver 29.4GW of new low-carbon capacity by 2030'[83]. However, this pricing mechanism relies far

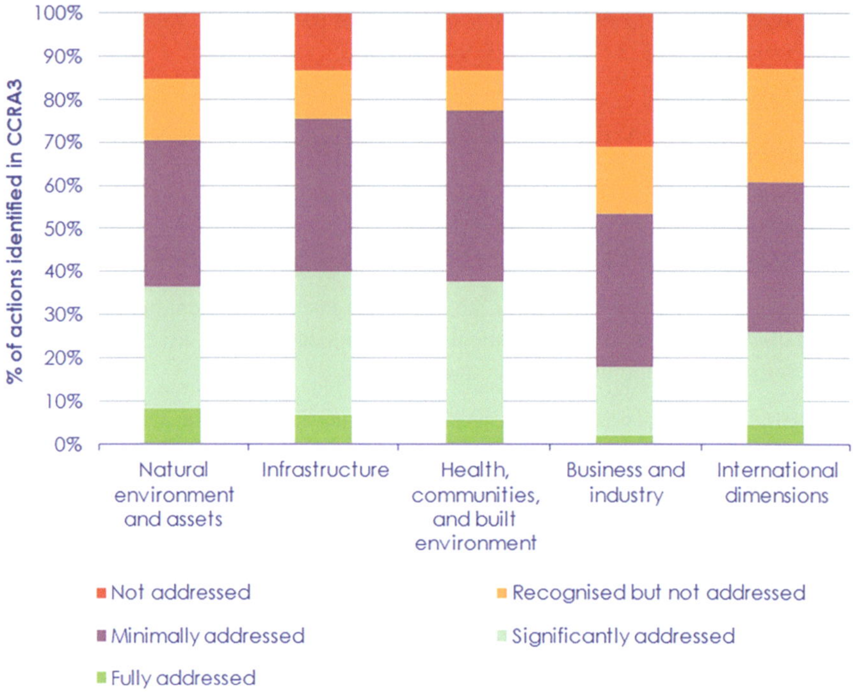

Fig. 4.7 A Critical Review of the National Adaptation Programme 2023-2028. *Source* CCC (2024)

too heavily upon underlying costs that ignore the falling average cost of renewables, see Chap. 7 and the *Cost of Energy Review, 2017* [84].

The first budget of the new Labour Government came in October 2024 against a background of a £28 billion revenue-expenditure gap inherited from the outgoing government, but with a strong election manifesto commitment to green policies. Under the circumstances, and with the Government's self-impose fiscal rules, the green budget was cut (see below) but there was new funding for green technology, including £3.9 billion for nascent CCUS technologies and a continued support for an additional nuclear power station, with opinions among climate activists divided between those recognising the need for trade-offs such as carbon removals and those exclusively advocating carbon reductions [85]. A £7.3bn flagship *National Wealth Fund* was announced, in addition to the existing £12bn of capital in the *UK Infrastructure Bank* that was launched in 2021 [86], to decarbonise Britain's heavy industry. The fund intends 'to drive a further £20bn of private sector money into low carbon investment in the economy' [87]. This 'catalysing' approach towards private funds is very much the philosophy of the UNFCCC, but with highly questionable success—see Chap. 6. A lifting of a de facto ban on onshore windfarms was also part of a *National Planning Policy Framework* (NPPF) [88], as a challenge to NIMBY ('Not in My Back Yard') resistance from local residents. This underscores a difficult and somewhat ironic trade-off between the need for renewables and for environmental protection. But similar reactions are being recorded in other countries, for example in Italy where the spread of solar farms is being criticised as endangering food security and despoiling the countryside [89]. A further step was confirmation that the UK would introduce a Carbon Border Adjustment Mechanism (CBAM) from 1 January 2027, covering aluminium, cement, fertiliser, hydrogen, iron, and steel.

Licences

For many years the UK took pride in being at the forefront of combatting climate change and other environmental issues, such as a ban from 2022 on single-use plastics [90], but the issuing of 27 licences in January 2024, in addition to 17 issued in 2023, for drilling for gas in the North Sea, was widely condemned by environmentalists and the opposition Labour Party as reneging upon national Net Zero targets, and abandoning global leadership [91]. For the Conservative Government it was a way to extend the declining potential of North Sea oil and gas. Following COVID and then Russia's invasion of Ukraine, 'energy security' suddenly loomed large as a way to revive a flagging economy [92].

Even before the 2024 General Election, the UK's Supreme Court ruled against a local authority issuing an oil drilling licence on the grounds that the environmental impact must also take into account the GHG emissions from the use of the oil [93]. In 2025, the International Court of Justice in a landmark decision went further and ruled climate-inaction by any government was contrary to international law [94]. However, one possible consequence of reduced licensing of oil and gas exploration in a home market is an increase in imported fossil fuels, and these may have a lower energy intensity than domestically-produced fuels, resulting in higher CO_2 emissions. Banning imports and restricting domestic licensing only adds to the urgency of renewable energy.

The UK trade unions have been somewhat split between those wanting the licences to go ahead until such time as there is alternative employment, and those supporting a ban on North Sea gas and oil excavation. In September 2024, Ed Miliband, the Secretary of State for Energy Security and Net Zero announced the new Labour Government would not issue new licences, and would set up 'Great British Energy, a publicly owned energy company headquartered in Scotland, which will invest in technologies… that use the expertise of North Sea workers such as floating offshore wind and carbon capture and storage' [95]. In early 2025 a consultation was announced to avoid a 'culture war' underscoring the sensitivity of balancing climate action against high energy prices [96]. This sensitivity was further exacerbated when the new government was forced to slash its overall Green Agenda budget from £28 billion to £15 billion due to a substantial budget deficit inherited from the previous government [97]. The implication being fewer subsidies and tax incentives, for example for wind or solar farms, or to insulate buildings better, or to invest in energy-efficient infrastructure, unless a way can be found to encourage further private investment in reducing GHGs, and/or raise more in tax revenues. As the government had pledged no rises in working class taxes, a carbon tax could be part of an answer.

Carbon Taxes

In several ways 'Carbon taxes already exist in the UK through things like the climate change levy and the UK Emissions Trading Scheme' [98]. Carbon taxes in the UK address three sectors, the consumers of energy, the most polluting industries, and by 2027 the carbon content of selected imports through the CBAM. The *Climate Change Levy* (CCL) [99], introduced in 2021, is an environmental tax charged on the energy that businesses use, paid at either the main rate, which applies to industrial, public services, commercial and agricultural sectors, or at the *Carbon Price Support* (CPS) rate for operators of electricity generation and power stations. The CPS varies according to commodity: kilowatt-hours for gas and electricity, and kilogrammes of carbon for all other taxable commodities [100].

The *UK Emissions Trading Scheme* (UK ETS), a mandatory 'cap and trade' scheme replaced the UK's participation in the European Union Emissions Trading Scheme (EU-ETS) [101] in 2021 owing to Brexit. It applies to energy intensive industries, the power generation sector, and aviation as outlined in Schedule 1 (aviation) and Schedule 2 (installations) of the *Greenhouse Gas Emissions Trading Scheme Order 2020* [102]. For practical administrative purposes, including the costs involved, a granulated approach is adopted. Emissions above a 'total rated thermal input exceeding 20 MW' are included, except when primarily used for the incineration of hazardous or municipal waste. Emission targets will substitute for rated emissions below a thermal capacity of 35 MW for hospitals and for smaller operators emitting below 25,000t CO_2e (where the 'e' stands for GHG carbon-equivalent), while those below 2,500t CO_2e are only required to monitor and report to the regulator when they exceed that level [103]. The trading of carbon 'future prices' under the UK ETS have fluctuated widely, from over £100 per metric tonne of CO_2 in 2022 to under £40 in 2024 [104], exacerbated by Russia's invasion of Ukraine which temporarily pushed up the price of gas.

A consultation on a proposed UK *Cross-border Adjustment Mechanism* (CBAM) [105] to come into effect by 2027 began in 2024 to bring a carbon levy on certain imported goods in line with the domestic *Climate Change Levy*—see above. But each sector will be attributed an individual CBAM rate 'determined by a set methodology and will be updated on a quarterly basis, reflecting the UK's moving carbon price under the UK Emissions Trading Scheme (ETS)' [106]. Discounted rates will be levied on goods that are already subject to carbon-taxes in their countries of origin. The scheme is similar to the EU CBAM, but with differences. A key factor for CBAM laws is to minimise 'carbon leakage' meaning carbon-emitting production processes relocating to carbon tax-havens. Whereas the EU CBAM covers all goods in the EU ETS, the proposed UK CBAM will only cover goods considered *at risk* of leakage under the UK ETS. Besides Scope 1 and 2 sectors, the EU CBAM includes Scope

3 material inputs, but as of 2025 only iron and steel inputs into 'complex goods.' Rather unnecessarily, the UK CBAM renames Scope 1 and 2 'direct' and 'indirect', and includes an input into a 'complex good' only if it already qualifies as a Scope 3 emission under the CBAM. The UK CBAM also includes glass and ceramics, not included in the EU CBAM, but excludes electricity [107].

UK Regulation Challenges

Following the Kyoto Protocol, the evolving guidelines for carbon trading took place against a history of growing global financial volatility and uncertainties following the 2007–2009 'Great Recession' [108], including in 2017 the default of Carillion, a major construction company leaving debts of around £7 billion, the largest default in UK financial history [108]. The response of the G20 [110] grouping of the world's leading economies was to create a Financial Stability Board (FSB) which under the chairing of Mark Carney, Governor of the Bank of England (Prime Minister of Canada since April 2025) turned its attention to the global challenge of the financing of climate change remediation—see Chap. 6. The Taskforce on Climate-related Financial Disclosures (TCFD) was created in 2015 and two years later the TCFD released climate-related financial disclosure recommendations designed to help companies provide better information to support market transparency and more informed capital allocation [111].

The UK ETS used these as guidelines for the enterprise sector, following the *International Financial Reporting Standards* (IFRS) [112] as adopted by the UK Financial Reporting Council (FRC) in 2021 [113] However, the FRC itself came in for serious criticisms from an independent auditor, noting it was founded in 1990: 'an institution constructed in a different era—a rather ramshackle house, cobbled together with all sorts of extensions over time' and highlighted its lack of a strong statutory powers. The government of the day accepted it should be replaced by a new Audit, Reporting and Governance Authority (ARGA), but that did not happen before the General Election in 2024 [114] In 2023, the *Sustainability Reporting Standards* (UK SRS) were established to bring corporate reporting in line with the IFRS *Sustainability Disclosure Standards* [115] 'creating future obligations under company law and the Financial Conduct Authority (FCA) requirements for listed companies' [116] In this way, standards appropriate for the MCM were beginning to be extended to the VCM.

'Natural' Capital Assessment and Investment Standards

A further important area of standards development in the UK concerns 'natural capital' markets to reduce GHG emissions on farmland and promote forests as carbon sinks. These policies have been led by the Department for Environment, Food &

Rural Affairs (Defra) and include *Nature Investment Standards,* efforts to harmonise leading carbon calculators, an *Agriculture Transition Plan*, and other initiatives to be found in the *Nature Markets Framework* update March 2024 [117].

European Union

The primary objective of the EU is to harmonise laws, regulations, and standards across the Union to create a sustainable *common* market. Policies are set by the Heads of State in the European Council, while laws are proposed by the European Commission, are passed by the Parliament, and issued by the Commission as Directives [118]. The laws come into effect only after being accepted by all Members States who then need to pass their own locally aligned versions. Member States in opposition can delay the process by years, but overall, in the area of climate change and environmental or 'green' policies, the EU does not face quite the same level of conflicting interests as do states in the USA, partly because in 2021, oil and gas made up only 3.3% of EU primary energy production. However, the rise of populism from the Right could change that.

EU Carbon Trading

To promote pro-environmental policies *and* an efficient market, the EU introduced a set of Directives governing online services, the *Digital Services Act* (DSA) and the *Digital Market Act* (DMA), which form a single set of safety and competition rules for citizens and competitors that apply across the whole EU [119]; and a *Green Claims Directive* to empower consumers through product labelling that meets EU standards and bans generic labelling such as 'green' or 'eco-friendly' [120] The only exception is for claims restricted to being business-wide and not product-specific, where companies are participating in the voluntary carbon markets. Green labelling 'greenwashing' concerns were fueled by the Commission's research that found, among other things, '40% of all environmental claims have no supporting evidence and half of all eco-labels are barely or not at all verifiable' [121]. Although this would suggest the need to make QR codes compulsory on most products, by 2025 this remains not the case [122].

By contrast, the EU takes a more proactive approach to GHG emissions. Before COP28 (Dubai, United Arab Emirates, 2023) the EU argued that CCUS (Carbon Capture, Usage, and Storage) should play no role in the abatement of fossil fuels, and the focus should be on reducing their use altogether [123]. This places the EU at odds with powerful fossil fuel lobbies around the world, who argue CCUS is absolutely necessary if Net Zero is to be reached by 2050, an argument that, while not necessarily being wrong in its forecast, obviously extends their business life. The public policy choice is clearly between *either* accepting a necessary redistribution of finances to renewables and to grid infrastructures as the first-best option—assuming the opportunity costs of such redistribution are ignored—and/*or* substantially reducing the use of energy as the second-best option—again ignoring any opportunity costs—*or* accepting a prolonged use of fossil fuels and offsets as the third-best option—and accepting the opportunity costs.

However, CCUS is not absent from EU policy. The EU is developing a *Carbon Removal Certification Framework* (CRCF) aimed at removing 300 million tonnes of carbon by 2040, using various carbon capture, removal, storage, re-use and sink methods [124]. As of early 2024 there were over 100 proposed projects, mostly focused on the North Sea [125]. In light of the quality problems hounding the voluntary markets, the immediate challenge for the implementation of CRCF is creating a set of standards for project assessment. As CRCF will allow polluters to use certificates as carbon credits to offset emissions, the selective assessment of projects becomes crucial to the success or failure of the scheme. As of 2024 quality control is to use '*QU.A.L.ITY*' criteria (QUantification, Additionality, Long-term storage and sustainabilITY) to avoid low quality certificates [126]. A critical note from the Institute for Agriculture & Trade Policy (IATP) makes two insightful comments: first, that when applied to agriculture the CRCF has to deal with carbon stored in forests and soil being returned to the atmosphere which will necessitate the effective cancellation of expired certificates in the registries, a process involving 'monitoring, reporting and verification' or just MRV (or MMRV if measurement is included in the acronym) and any failure to do so will undermine the entire process; second, the use of CRCF credits on voluntary carbon markets will be governed by the EU's *Green Claims Directive* legislation (see above) which will impact upon claims of offsetting. 'In fact, the CRCF might become the main standard for voluntary carbon markets for companies operating in the EU' [127]. This is another example of the emerging interlacing of concessionary and voluntary markets.

Licences

Under EU law, each Member State has the right to issue licences. Article 2 of the 1994 Directive reads: 'Member States retain the right to determine the areas within their territory to be made available for the exercise of the activities of prospecting, exploring for and producing hydrocarbons' [128]. Only ten countries in Europe have pledged to put an end to fossil fuel exploration [129], and to fossil fuel usage, by varying dates [130]. Although no longer an EU member, there is growing pressure in the UK to put an end to issuing further fossil fuel licences—see above. Licences for nuclear reactors are likewise decided by each Member State and are seen by advocates as a way to bridge the gap between the phasing out of fossil fuels and replacement by renewables. By mid-2023, France had 56 reactors in operation, by far the most within the EU [131]. Outside of Russia and Ukraine, the next highest was the UK with nine. By mid-2024, Italy [89] was considering to revive nuclear after an absence of over 20 years, by investing in a new generation of small modular reactors (SMRs) which typically have a capacity of less than 100 MW, but offer a modular structure which means they can be scaled-up using standardised designs in response to demand. This lowers the costs of construction [132].

Carbon Taxes

Finland first introduced a carbon tax in 1990, and 23 European countries had done so by 2024, ranging from €1 per tonne of carbon to €120 per tonne in Liechtenstein. The scope differs, covering just fluorinated gases in Spain to over 70% of all gas emissions in Albania, Liechtenstein, and Luxembourg. As a result, all EU states, plus Iceland, Liechtenstein, and Norway, are members of the 'cap and trade' EU Emissions Trading Scheme (EU ETS). Switzerland has its own ETS, but linked to the EU ETS, and since Brexit in 2020 when it left the EU, the UK also has its own [101].

A separate ETS2 scheme was added in 2023, auctioning emission allowances covering buildings, road transport and sectors not covered by the existing ETS. The aim is to reduce GHG emissions by 42% from 2005 levels by 2030. A stabilization fund will be maintained to balance an insufficient or excessive issuing of allowances. For the first three years the stable price will be kept at €45 per metric tonne of CO_2. Businesses will annually register their GHG estimates as part of an ongoing MRV process and verifiers shall be appointed in each Member State. The Devil lies in the verification details. This is part of the *Delivering the European Green Deal* [133] which includes a *Social Climate Fund* [134], to provide €86 billion of green transition support to the EU's most vulnerable citizens and small businesses, funded by ETS2 auction revenues and echoing the UN's resolve of 'leaving no one behind' [135]. But a key difference with the UK ETS is that carbon credits from the VCM are not accepted as part of the cap-and-trade [136]. Also announced was the *Carbon Border Adjustment Mechanism* (CBAM)—discussed above. In doing so, the EU proposed adding to its Paris Agreement commitments a reduction of GHG emissions by 90% over 1990 levels by 2040 [137].

China

The People's Republic of China (PRC) is both the world's leading polluter and the world's leader in the generation and use of renewable energy. China's economy is in transition from one dependent for its GDP growth upon the output of highly polluting industries, and upon supplies of cheap labour migrating from rural areas to the coastal cities, to GDP growth turning towards productivity, the promotion of a low-waste circular economy [138], and what are termed 'new quality productive forces' [14]. These include new energy (solar power, electric vehicles, batteries, etc.), new materials (construction, wearables, super-conductors, particle physics, etc.), AI, next-gen IT, robotics, green technologies, aerospace, bio-tech, and so on [140]. Box 4.4 provides a context.

Box 4.4 China Context for Environmental Control

China's early economic growth was highly dependent upon industries powered by fossil fuels and an abundant supply of cheap labour migrating from rural areas. That supply has dried up—China's working population peaked in 2015 [141]—and population growth has gone into reverse, in part a legacy of restrictions on family size, the one child policy. China's great challenge was feeding and housing its vast and growing population; China's subsequent challenge is to transition from a focus upon production to a focus on productivity, and from a polluting economy to a greener one, that is driven by 'new quality productive forces.'

But there is an additional challenge that has historically haunted China. It is summarised in a Chinese saying that '*the emperor lives over the hills and far away.*' Provinces, local state-owned-enterprises (SOEs) and banks, and semi-autonomous investors regularly ran up debts to fund capital investment in rapidly expanding businesses, such as coal mines, steel works and the construction industry, to support China's infrastructure development and industrialization. Difficult to confirm statistics showing annual growth rates of well over 7% were regularly reported from the periphery (provinces) to the centre (Beijing). Reining in provincial autonomy, along with associated high levels of corruption and falsification of data, and asserting financial discipline from Beijing, has been a continuing policy dilemma facing China's leaders, including for President Xi. See Anni Dai, et al. Berkeley Law [142].

However, not only has President Xi centralised the power of Beijing, he has centralised his own power within the Chinese Communist Party (CCP), to an extent unprecedented since Chairman Mao Tse Tung, as a way to weaken resistance by those who are '*over the hills and far away*'. One demonstration of this is 'Xi's focus on environmental issues [*that*] led to decisive measures in environmental regulation campaigns. The campaign aimed to ensure lower-level compliance by making environmental targets vital in the cadre evaluation' [143]. This became most evident during China's controversial use of a 'blunt force' lockdown policy to eliminate COVID-19. But when applied to climate change and pollution in general, it may be as one scholar argues: 'blunt force regulation is used not to scare polluters into respecting pollution standards, but to scare bureaucrats into respecting central orders' [144].

Following China's participation in the Paris Agreement and its subsequent Nationally Determined Contribution (NDC), the CCP and State Council issued two documents considered by the United Nations Conference on Trade and Development

(UNCTAD) [138] to be 'the most important top-level assessments of Chinese policy': an *Opinion* on China's commitments to climate change policies designed to meet the Paris goals, and an *Action* plan which covers 10 actions and 43 tasks in areas such as energy transition, industrial development, urban and rural construction, and others. The policy framework is referred to as '1 + N' where 1 stands for the single overarching policy aim of climate policy, and N stands for the numerous supporting action plans [145]. The policy is set as an integral part of China's *High Quality Development Framework* [146]. [Note: access to some search engines are unavailable in China.]

China's NDC is enshrined in the *14th Five Year Plan, 2035 Vision Outline* and the *1 + N Policy Framework for Carbon Neutrality*. The announcement in 2020 of a scaled-up NDC to peak carbon emissions by 2030 and carbon neutrality by 2060 has become known as the '30–60' goals, and a further commitment in 2021 was to tackle methane emissions and to 'phasing down coal' and protect forests [147]. But China also brought forward its carbon *intensity* targets to 2025 from 2026 (originally to 2030) the difference being total carbon emitted *versus* carbon emitted *per unit of output* [148]. However, by 2025 China was missing its carbon intensity target reduction by 0.5% [149], and the efforts by China to reduce carbon emissions were rated in 2024 by the Climate Action Tracker as 'highly insufficient' [150]. Ahead of COP 29 China also passed a new energy law to decarbonize the economy by 2060 [151]. Among the key targets are.

- Peaking carbon emission before 2030 (2026 was later specified; then 2025)
- Installed wind and solar capacity to 1.2 billion kW by 2030
- Carbon neutrality before 2060
- Lower CO_2 per unit of GDP (energy intensity) by 65% over 2005
- Increase renewables in primary energy consumption by about 25%
- Increase share of non-fossil fuels in primary energy consumption by over 80% by 2060
- Increase forests by 6 billion cubic metres above 2005 levels

Source Carbon Change Laws.

Coal

'Phasing down coal' is the most immediate challenge as coal has been the staple backup for China's industrial surge, the 'go-to' fossil fuel enabled by provincial authorities benefitting from decades of local development of energy supplies, industrial employment and revenues, and urban construction.

According to IEA data [141], over 70% of the rapid growth in the demand for electricity since 2015 has arisen from China's transition to electricity, for example in transport and heating, and more than 60% of the use of coal arose in the generation of that electricity. China added close to 40GW of coal capacity in each of the five years 2018–2013, more than the rest of the world, and has even approved an additional

100GW capacity for the remainder of the 2020s. The use of coal was scheduled to peak by 2025, but in 2024 China decided to add 94.5 GW capacity of coal-generated energy, casting doubt over its announced target [152]. The strategy was to be that production will be used more to add flexibility to electricity generation during demand peaks or when renewables cannot cope, and used less as the main generative fuel. This will require a major shift from the current energy-mix, of which by 2020, coal made up 56.8%, oil 18.9%, natural gas 8.4% and others, including renewables, 15.9%. UNCTAD notes this policy is known as 'dual control of carbon emissions' wherein coal emissions are lowered alongside GDP carbon intensity, and each ministry is charged with reducing their GDP carbon intensity, a target metric [138].

Gas and Oil

Seventy-five percent of oil in China, mostly imported, is used for energy and 25% for non-energy derivatives such as plastics and chemicals, otherwise known as 'feedstock'. According to the IEA, China was on course to add as much petrochemical feedstock 2019–2024 as the combined capacity of all the OECD countries in Europe and Asia [141]. The chemical industry, it is estimated, accounted for 10% to 14% of China's total CO_2 emissions in 2017, and 4% globally [153]. Natural gas emissions rose from 350 $MtCO_2$ in 2015 to just under 700 $MtCO_2$ by 2021, accounting for 7% of China's total emissions of CO_2 [154].

Heading up China's major national oil and gas companies, known as NOCs, are: the China National Petroleum Corporation (CNPC), the China Petrochemical Corporation (Sinopec) and for offshore exploration the China National Offshore Oil Corporation (CNOOC). A 50% rise in emissions from oil from 2010 to 1,500 $MtCO_2$ in 2021 has triggered a shift, as one research paper puts it [156], in China's NOCs from 'oil and gas' to 'gas and oil' with a growing focus upon relatively cleaner Liquified Natural Gas (LNG). As transport accounts for 49% of oil in energy consumption, it is indicative of the reset now underway that Sinopec is seeking to 'leverage its network of more than 30,000 retail stations to become a leading supplier of hydrogen fuel across China' as part of China's *Hydrogen Industry Development Plan (2021–2035)* which targets bringing 50,000 hydrogen fuel-cell vehicles on the road by 2025, supported by refuelling stations [157]. The shift to hydrogen, which is a clean yet dangerously volatile fuel, parallels China's ongoing shift to electric battery-driven cars, buses and trains as 'new energy vehicles', and to hybrids that combine electricity and H2 or diesel.

However, in 2023 only 1% of China's hydrogen production was 'green hydrogen', that is electrolysis using renewable energy [158]. Electrolysis uses electricity to separate oxygen and hydrogen from water. If the electricity has been generated from renewable sources, then the hydrogen is declared 'green', otherwise is it 'blue'—or 'black' if coal was used. It is an inefficient process, using more electricity input than can be generated as a hydrogen fuel output, but hydrogen, if carefully handled, can replace gasolene and diesel in vehicles, and refuelling a vehicle is much faster than recharging its battery.

Renewables

A joint statement from China and the USA in November 2023 backed a global goal to triple renewable energy capacity by 2030 to further abate the use of fossil fuels [159]. China's NDC included a commitment to have 1.2 billion kW wind and solar capacity installed by 2030, and by 2023 China already accounted for 65% of the world's wind power and 60% of the world's solar power capacity [160]. This was initially driven by subsidies but they have been phased out for solar and onshore wind, and remain in place only for offshore wind [161]. In 2024, China launched the world's largest 3.5 GW, 33,000 acre solar farm, in a desert area of Xinjiang Province with an annual generating capacity of over 6 billion kWh of electricity, 'enough to power the country of Papua New Guinea for a year' [162].

Grid Infrastructure

Electricity generated in remote regions needs a transmission grid to carry it to where it is needed, and this is proving a challenge within China. China aims to expand its grid capacity from 540 TW-km in 2022 to 1,200 TW-km by mid-century, owing to the vast amount of renewables coming online [163] The two national legacy grids, North and South, need a supplementary grid running West–East, and to be upgraded to transmit energy at '800,000 V and above, double the voltage of conventional high-voltage lines, allowing them to transmit up to five times more electricity at minimal energy loss' [164]. Engineering these changes has been fraught with design discrepancies while competing commercial interests have greatly complicated the project. In addition, 'the current technology cannot handle the instability of load from renewable sources'. Hence energy supplies have to resort to more stable resources such as coal to iron-out the inconsistences and fluctuations of renewable energy being uploaded onto the grid. The vastness of China and many competing jurisdictions are issues here, but when the technology and the engineering are completed, many smaller countries in the Global South could be beneficiaries of the infrastructure when the technology involved becomes available for export.

Carbon Taxes

The world's largest ETS came into being in China in 2021 [165], run by the Ministry of Ecology and Environment (MEE) and covering in its initial phase annual gas and coal-fired power sector emissions of around 4.5GtCO_2 [166]. Limits are determined by their respective emission intensities, not volumes, and allowances are issued, allocated not paid for, against these limits by the power stations. Where emissions are reduced below set limits the allowances can be traded on the Shanghai Environment Energy Exchange as credits. However, no cap has yet been placed on emission volumes, mainly as a precaution not to disrupt energy supplies post-COVID, and probably for the same reason, fines on exceeding the emission limits are very low, at a maximum of 30,000 RMB (little over USD4,000). There is, in effect, an incentive to be profitable by exceeding the emission limits with little to pay in fines, rather than create tradeable credits.

One report notes trading volumes are very low, tend to be seasonal, and are beset by a lack of data integrity and effective legislation. Consequently, China's ETS 'hovers at around 60 yuan (USD8) per tonne, while its EU equivalent is more than 10 times higher' [167]. As a result, the MEE sees the initial aim to be creating a carbon emissions awareness rather than significant carbon reductions. Close to 2,000 companies are involved, and are responsible for about 40% of China's emissions [168], so getting the ETS to work more effectively, imposing a 'cap' as well as the 'trade', introducing legislation that clarifies how the system should operate rather than leaving it to a regulatory trial-and-error process, introducing data integrity and MRV standards, and extending the scheme to other sources of carbon emissions will all be necessary steps in the right direction. Whether China would ever open up the trading to international markets is an interesting question, but that would need much more rigorous data and MRV standards. There are ongoing talks between China, the EU and Singapore to create a Multi-Jurisdiction Common Ground Taxonomy (M-CGT) that could provide a basis for coordinated efforts between them.

The Voluntary Market

Due to the lack of standards, China's issuing of *China Certified Emissions Reductions* (CCERs) under the Kyoto Protocol was halted in 2017. But reports suggest that China is again preparing to allow trading in this voluntary market as CCERS can be offset against 5% of annual emissions [169]. One report cites a conservative estimate that 'the CCER spot market will reach about 20 billion yuan (USD2.8 billion) by 2025' [170]. China's Supreme Court has already ruled that anyone damaging forests, while they can still be prosecuted, can nevertheless buy forest carbon credits as offsets to fines [171], and in 2024 a court found that illegal fishermen had paid for blue carbon credits as an offset [172]. These represent a perverse concept of an offset, and suggest standards and controls in the voluntary markets remain weak. The response of the courts may be no more than to recognise and alert others rather than to condone such a reality.

India

India's GHG emissions have doubled since the turn of the century, reaching around 4 billion metric tonnes equivalent of CO_2 in 2024 [173], placing India (7%) as the world's third highest emitter of GHGs after China (29%) and the USA (11%) [174], but with the world's lowest *per capita* emissions at 2.5 tonnes CO_2 per person [175], as India's population expansion overtook China's [176]. The principal sources of GHGs in India are coal power plants, extensive rice paddies [177], and cattle [178]. The principal causes include coming late to the climate change challenge, a lack of resources to combat the problem, and an apparent lack of determination by some parties to phase out coal. According to a Power Ministry spokesperson, there are sufficient coal reserves to power India for another 50 years [179], and to focus upon coal when there are other sources of emissions is, according to the spokesperson, to be regarded as a diversion because India lacks the technology to store sufficient energy for its needs [180].

In some cases, such as rice paddies, changes in field management can help. The Asian Development Bank (ADB) reports that instead of continuously flooding rice fields, '*Alternate Wetting and Drying* (AWD) water management allows rice fields to be periodically dried. Studies found that AWD reduces methane emissions by 30–50% and water use by 10–20% without reducing rice yield in a season' [181]. Reducing methane emissions from cows is also possible if feeding techniques developed in the USA can be used in India, using less fibrous content and more fats and grains [182]. Indoor and outdoor pollution from cookstoves is another major issue for India. Poorer families typically use wood, bio-waste or coal to heat stoves. A study in 2015 estimated up to 50% of outdoor $PM_{2.5}$ particulates resulted from cookstoves in heavily urbanised areas and villages [183]. The Ministry of New and Renewable Energy (MNRE) is doing its best to introduce programmes to better manage cookstoves, their fuels, and to promote alternatives such as solar energy [184]. Among other major sources of GHGs is transport, in over-crowded cities, using fossil fuel and responsible for 12% of India's energy-related emissions. An IEA report warns that emissions could increase by 50% by 2050, but with corrective policies the levels could be reduced by one-third [185].

Despite these challenges, India has claimed sufficient progress to update its NDC [186]. In 2023, the Ministry for the Environment, Forest and Climate Change (MoEFFCC) informed the UNFCCC that its emissions intensity (tonnes per GDP) had been reduced by 33% between 2005 and 2019 [187]. The new commitment is to reduce the emissions intensity by 45% by 2030 over its 2005 level, and to achieve around 50% of cumulative electric power installed capacity from non-fossil fuels by 2030, with the help of transferred technology and low-cost international finance, including from the South Korean-based global *Green Climate Fund* (GCF) [188]. The planting of up to 3 billion trees in a programme of reforestation as a carbon sink is also planned. One issue that is not raised in the NDC is CCUS, although India in 2022 did announce the establishment of two National Centres of Excellence in Carbon Capture and Utilization (NCoE-CCU) under the Ministry of Science and Technology [189].

Carbon Trading in India

A national mandatory *Carbon Credit Trading Scheme* (CCTS) was proposed in 2023 by the Bureau of Energy Efficiency to start operating in 2025–26 [190], to be followed by a voluntary carbon market [191]. Trading in CCTS could easily become the world's largest market and is to be modelled similarly to California's ETS and China's ETS, allowing some trading between the compliance and voluntary sectors. It follows upon an earlier attempt in 2012 to promote a cap-and-trade system for 13 energy-intensive sectors, such as power-plants, iron and steel, aluminium, cement and chemical fertilisers under what became known as the *Perform, Achieve and Trade* (PAT) scheme [192]. Tradeable *Energy Savings Certificates* (ESCerts) were issued. A parallel but separate renewable electricity generation scheme was launched by the Power Ministry in 2010 requiring the purchase of tradeable *Renewable Energy Certificates (*RECs). These it seems will be available for trading within the CCTS when it becomes operational once sectoral methodologies for project assessments are available [2002].

Singapore

Despite its small size, approximately 50 by 26 km, and lack of mineral resources, Singapore has expertly taken advantage of its geographical position, sandwiched between Indonesia and Malaysia, as a trade and transport hub for the region. It could be described as a 'Ricardian' economy—sometimes referred to as Ricardian socialism [194]—in which most of the land is state-owned, with the government as the rentier reaping the benefits, enabling Singapore to keep taxes low while state spending high, a model that was proposed to societies by American economist Henry George towards the end of the nineteenth century [195]. The wealth of Singapore has been used to accumulate a national wealth fund, to fund innovation turning the country into a highly-developed digital economy, and to transform the country into a regional, and global, wealth-hub, sometimes referred to as the Switzerland of Asia.

Carbon Tax

Singapore instituted a carbon tax in January 2024. The tax applies to all industrial activities which emit annual direct GHG emissions of 25,000 tonne of carbon dioxide equivalent (tCO_2e) in the following categories: manufacturing and manufacturing related services; supply of electricity, gas, steam, compressed air and chilled water for air-conditioning; and water supply and sewage and waste management. The National Environment Agency (NEA) provides an excel spreadsheet for calculating various GHG emissions [196]. The carbon prices will rise progressively from SGD25/USD19 per tonne 2024 and 2025, to between SGD50/USD38 and SDG80/USD60 per tonne by 2030. Under the scheme, international carbon credits (ICCs) can be used by taxable entities to reduce their local carbon tax by 5%, although these prices are well below the social costs, estimated to be USD185 per tonne of CO_2e emitted [197].

Singapore has fast-tracked its regional advantage as a transactions centre for climate change finance by taking a lead in negotiating ITMOs (Internationally Transferred Mitigation Outcomes). These offer Article 6.2 arrangements under the Paris Agreement for richer countries to buy carbon credits from poorer countries, which can then be retired to count towards the buyer's NDC—see above. Singapore is one of four countries in 2024 to offer ITMOs to poor and middle-income countries, which include: Bhutan, Cambodia, Costa Rica, Chile, Columbia, Dominican Republic, Fiji, Indonesia, Kenya, Lao PDR, Mexico, Mongolia, Morocco, Peru, Papua New Guinea, Paraguay, Rwanda, Senegal, Sri Lanka, and Thailand. In support of ITMOs, Singapore hosts the Climate Action Data (CAD) Trust, a decentralised metadata platform 'that links, aggregates and harmonises all major carbon credit registry data to enhance transparent accounting in line with Article 6 of the Paris Agreement'.

Conclusions

In addition to government commitments to their Nationally Determined Contributions (NDCs), direct government actions such as the issuing of licences and the funding of renewable energy projects, and the imposition of cap-and-trade ETS schemes to create compliance or mandatory carbon markets, have been relied upon to decarbonise the atmosphere, however insufficient they have so far proved to be. The reality is that since mid-2023, average global temperatures have consistently exceeded the Paris goal of no greater than 1.5C above pre-industrial levels by 2050.

All six of the jurisdictions discussed above, China, the EU, India, Singapore, the UK and the USA, have been following similar the paths of mandatory (complaint) carbon trading schemes covering the major polluting industries, in some cases allowing carbon credits in the voluntary market to be traded in these schemes. The compliance rules are mostly confined to the major polluting sectors, but where they have proved successful in bringing down GHG emissions (see above) there is no reason in principle not to broaden their scope. The EU and the UK are experimenting with Cross-Border Adjustment Mechanisms (CBAM) which threaten some friction with China and the USA as their exporters may be subject to these carbon duties. Brexit aside, the EU and the UK are mostly in alignment, but the USA is not at the federal level, but at the local level some Democratic states have also adopted Emission Trading Schemes (ETS). Singapore has linked its ETS with a carbon tax scheme, and is also exploring Article 6.2 ITMO trades as a way to promote the island

State as a regional hub for carbon trading, while China has pioneered an ETS similar in nature to the scheme run in California, but it is as yet only partially functional. India is following suit.

What mostly differentiates these jurisdictions is the nature of their economies. Because the EU economies, the UK and the USA have long surpassed industrialization and are now more focused upon a mix of commercial and digital enterprises, they have accelerated their commitments to the Paris goals, even if their carbon abatement performances remain more slow-paced. The USA however differs in two major senses. First, it is energy rich, especially in fossil fuels which empowers the sector in its lobbying to demote the urgency of the climate change challenge. Second, since January 2025, under the Presidency of Donald Trump the USA has rescinded its commitments to the Paris Agreement and to the UN Sustainable Development Goals (SDGs) and to helping finance the UNFCCC, in addition to halting domestic climate change-related payments, subsidies and tax breaks under the Inflation Reduction Act (IRA). Also halted have been payments under USAID to developing countries. This constitutes a huge blow to the efforts of the COPs to mobilise finances for poorer countries to sustain climate mitigation and climate adaptation policies. See also Chap. 7.

By contrast, industrialization is an ongoing process in both China and India, and both are reluctant to sacrifice the productive power of their domestic development which remains principally driven by fossil fuels. China is best positioned to shift the balance of investments into renewable energy and into green(ish) expansion, boosting exports while doing so in areas such as 'new energy' vehicles and AI products. In India the challenges are far greater in terms of percentage of the population (over 60%) that lives outside the cities in rural areas often poorly served by electricity and families reliant upon highly polluting cook stoves using biomass as fuel [198].

One lesson the rest of the world could learn from China is a focus upon national infrastructures, especially the electricity grid infrastructure. The grid is the platform for green growth. Grid development and its upgrade is essential if renewable energy from sources such as oceanic wave technologies, solar farms and wind farms, and hydroelectricity are to be converted, where necessary, by transformers from DC to AC, then lowering the renewable voltages to enter the grid, then upping the voltage from the generators (typically 110 kV or more) [199] for efficient long-distance transmission (anywhere up to 1,200 kV depending upon distance) [200] and then lowering the voltage again for distribution to consumers. Many existing grids do not have the capacity to handle surges in renewables that are dependent upon the weather conditions, which is one reason why wave technology is useful. Power cables need to connect to the generators, otherwise battery storage needs construction. Infrastructure investment can create 'green' jobs on a large scale, from the new energy technology manufacture to the generators and transformers, to the construction of storage facilities, and the distribution network, to the 'last mile' connectivity including the installation of smart metres to ration usage to off-peak hours whenever possible, and home and building battery storage systems and heat pumps. There are however doubts over the extent to which green jobs are being created. While Google searches produce many reports claiming they are growing rapidly, the Tony Blair

Institute (TBI) for Global Change reports 'In many economies, the promise of green jobs has not materialised at the scale expected' [201].

There are inevitably GHG emissions along a green industry supply chain, notably across Scopes 2 and 3, but over time the more renewable energy that can be generated and applied to each segment of the chain the lower emissions will become. New AI-driven technologies can also serve to reduce the energy intensity of these processes. Unit operating costs will also fall as a consequence of scale. These are some of the central components a green agenda and growth strategy that can only be led by governments. They also lay the foundations for the private sector, with government encouragement, to switch to greener production processes as they substitute renewables for fossil fuels. But the longer governments hold back from such long-term investments in the grid and its associated production activities, the longer will countries get caught up in the vicious circle of unusable renewables and the continued use of fossil fuels.

References

1. NASA Responding to Climate Change https://science.nasa.gov/climate-change/adaptation-mitigation/
2. IEA (2023) World Energy Outlook 2023: Executive Summary https://www.iea.org/reports/world-energy-outlook-2023/executive-summary
3. Carnegie Endowerment.Org (2023) Climate Protest Tracker https://carnegieendowment.org/features/climate-protest-tracker?lang=en
4. Climate Action Tracker Find Your Country https://climateactiontracker.org/
5. European Commission (2024) Carbon Border Adjustment Mechanism https://taxation-customs.ec.europa.eu/carbon-border-adjustment-mechanism_en
6. CPER and VideoVox Economics (2024) Symposium: First Steps of the EU CBSM: An Evaluation YouTube video https://www.youtube.com/watch?v=OPB3JLhc4_4
7. Financial Times (2nd July 2024) US examines carbon pricing on imports
8. UN.Org All About the NDCs https://www.un.org/en/climatechange/all-about-ndcs#:~:text=After%20the%20original%20NDCs%20in,intended%20climate%20actions%20through%202035.
9. Whitehouse.Gov (November 2021) The Long-Term Strategy of the United States https://www.whitehouse.gov/wp-content/uploads/2021/10/us-long-term-strategy.pdf
10. OGL (September 2022) UK's Nationally Determined Contribution – updated September 2022 https://unfccc.int/sites/default/files/NDC/2022-09/UK%20NDC%20ICTU%202022.pdf
11. UNFCCC (October 2023) Update of the NDC of the European Union and its Member States https://unfccc.int/sites/default/files/NDC/2023-10/ES-2023-10-17%20EU%20submission%20NDC%20update.pdf
12. Mee.Gov. Cn (2022) Progress on the Implementation of China's Nationally Determined Contributions https://www.mee.gov.cn/ywgz/ydqhbh/qhbhlf/202211/W020221111760730462299.pdf
13. UNFCCC (2022) India's Updated First Nationally Determined Contribution Under Paris Agreement https://unfccc.int/sites/default/files/NDC/2022-08/India%20Updated%20First%20Nationally%20Determined%20Contrib.pdf
14. NCCS https://www.nccs.gov.sg/singapores-climate-action/mitigation-efforts/carbontax/
15. SSEI (2024) ESG Disclosure https://sseinitiative.org/esg-disclosure

16. LSE (May 2022) Voluntary Carbon Market: Supporting Direct Investment on The Path to Net Zero https://www.londo nstockexchange.com/discover/news-and-insights/voluntary-carbon-market-climate-effect
17. LSE (May 2022) Market Notice https://docs.londonstockexchange.com/sites/default/files/doc uments/n1222_2.pdf
18. BBC (July 2023) What is an oil and gas licence? https://www.bbc.com/news/business-663 58380
19. https://www.theguardian.com/business/2023/nov/07/north-sea-oil-gas-licensing-scheme-rishi-sunak-kings-speech-bills
20. U.S. Energy Administration (2023) U.S. energy facts explained https://www.eia.gov/energy explained/us-energy-facts/imports-and-exports.php
21. Reuters (May 2024) Up to 50 firms seek US oil licenses in Venezuela, official says https://www.reuters.com/business/energy/up-50-firms-seek-us-oil-licenses-venezu ela-official-says-2024-05-23/
22. Reuters (21 March 2025) Trump weighs extending Chevron's license for Venezuela operations, sources say https://www.reuters.com/markets/commodities/chevron-ceo-wirth-seeks-more-time-wind-down-operations-venezuela-wsj-reports-2025-03-19/
23. CIEL (16 November 29024) Growing Fossil Fuel Influence Exposed: 480 Carbon Capture Advocates at COP29 https://www.ciel.org/news/ccs-presence-cop29/
24. Hannah Ritchie (2025) How much in subsidies do fossil fuels receive? https://ourworldindata.org/how-much-subsidies-fossil-fuels
25. IMF (2025) Fossil Fuel Subsidies https://www.imf.org/en/Topics/climate-change/energy-sub sidies
26. Max Roser (2021) Fossil fuel subsidies: If we want to reduce greenhouse gas emissions we should not pay people to burn fossil fuels https://ourworldindata.org/fossil-fuel-subsidies
27. Sarah Carballo (14 March 2025) 'Fossil Fuel Subsidies: The $760 Billion Lie About 'Free Market' Energy' Fractracker Alliance https://www.fractracker.org/2025/03/fossil-fuel-subsid ies-free-market-myth/
28. Peter Erickson et al. (2017) Effect of subsidies to fossil fuel companies on United States crude oil production' Nature Energy 2, pp. 891–898 (2017) https://www.nature.com/articles/s41560-017-0009-8
29. William D. Nordhaus (2021) The Spirit of Green Princeton University Press
30. UNFCCC (2024) News https://unfccc.int/
31. Niklas Döbbeling-Hildebrandt, et al. (2024) 'Systematic review and meta-analysis of ex-post evaluations on the effectiveness of carbon pricing' Nature Communications v.15.4147, 2024 https://www.nature.com/articles/s41467-024-48512-w#:~:text=The%20estimated%20redu ction%20effects%20range,substantial%20and%20highly%20statistically%20significant.
32. Stanford Business (May 2024) Big Investors Say They Use ESG to Reduce Risk (But Mostly Focus on the E and G) https://www.gsb.stanford.edu/insights/big-investors-say-they-use-esg-reduce-risk-mostly-focus-e-g#:~:text=Nearly%20half%20of%20all%20respondent s,could%20tank%20a%20company%27s%20performance.
33. Stern.nyu.edu (2024) ESG and Financial Performance https://www.stern.nyu.edu/experi ence-stern/about/departments-centers-initiatives/centers-of-research/center-sustainable-bus iness/research/research-initiatives/esg-and-financial-performance
34. Mahmut Aydoğmuş, Güzhan Gülay, Korkmaz Ergun (2022) 'Impact of ESG performance on firm value and profitability' Borsa Istanbul Review, V.22.2 December 2022, Pages S119-S127 https://www.sciencedirect.com/science/article/pii/S221484502200103X#bib37
35. IFAC (November 2023) Understanding the Voluntary Carbon Markets https://ifacweb.blob.core.windows.net/publicfiles/2023-11/23-3284%20-%20Carbon%20Credit%20Teaser_EN_final.pdf
36. S&P Global (September 2023) As companies prepare to implement new international climate standard, disclosure varies widely around the world https://www.spglobal.com/esg/insights/featured/special-editorial/as-companies-prepare-to-implement-new-international-climate-standard-disclosure-varies-widely-around-the-world

37. Bloomberg NEF (February 2024) Carbon Credits Face Biggest Test Yet, Could Reach $238/Ton in 2050, According to BloombergNEF Report https://about.bnef.com/blog/carbon-credits-face-biggest-test-yet-could-reach-238-ton-in-2050-according-to-bloombergnef-report/#:~:text=Under%20the%20Voluntary%20market%20scenario%2C%20prices%20reach%20just%20%2413%2Fton,up%20from%20%242%20billion%20today.
38. UNFCCC The Clean Development Mechanism https://unfccc.int/process-and-meetings/the-kyoto-protocol/mechanisms-under-the-kyoto-protocol/the-clean-development-mechanism
39. UNEP The Montreal Protocol https://www.unep.org/ozonaction/who-we-are/about-montreal-protocol
40. Inside Climate News (2020) China, India Emissions Pledges May Not Be Reducing Potent Pollutants, Study Shows https://insideclimatenews.org/news/22012020/china-india-hydrofluorocarbon-hfc-23-greenhouse-gas/
41. UNFCCC What is the Kyoto Protocol? https://unfccc.int/kyoto_protocol
42. UCL News (November 2022) Opinion: COP27 will be remembered as a failure – here's what went wrong https://www.ucl.ac.uk/news/2022/nov/opinion-cop27-will-be-remembered-failure-heres-what-went-wrong
43. Bond.Org. UK (December 2023) What were the real successes and failures of COP28, and what does it mean moving forward? https://www.bond.org.uk/news/2023/12/what-were-the-real-successes-and-failures-of-cop28-and-what-does-it-mean-moving-forward/
44. The Nature Conservancy (2024) Article 6 Explainer https://www.nature.org/content/dam/tnc/nature/en/documents/TNC_Article_6_Explainer.pdf
45. VCMI-MRA Monitoring, Reporting & Assurance (MRA) Framework. https://vcmintegrity.org/wp-content/uploads/2023/11/VCMI-MRA-Framework.pdf
46. Klik (2023) We buy ITMOs in Thailand https://www.klik.ch/en/international/partner-countries/thailand/
47. IMTA (2019) Current Status of Article 6 of the Paris Agreement: Internationally Transferred Mitigation Outcomes (ITMOs) https://www.iisd.org/system/files/publications/status-article-6-paris-agreement.pdf
48. UNFCCC Article 6.4 Supervisory Body https://unfccc.int/process-and-meetings/bodies/constituted-bodies/article-64-supervisory-body
49. UNFCC Paris Agreement—Status of Ratification https://unfccc.int/process/the-paris-agreement/status-of-ratification
50. UNFCCC What is the Supervisory Body of the Article 6.4 mechanism? https://unfccc.int/process-and-meetings/bodies/constituted-bodies/article-64-supervisory-body
51. Carbon Offset Guides What Makes a High-Quality Carbon Offset? https://offsetguide.org/high-quality-offsets/
52. In a communication with the author.
53. UNFCCC What is REDD+? https://unfccc.int/topics/land-use/workstreams/redd/what-is-redd
54. Green Climate Fund Areas of work https://www.greenclimate.fund/areas
55. UNFCCC Warsaw Framework for REDD+ https://redd.unfccc.int/fact-sheets/warsaw-framework-for-redd.html
56. Forrest Carbon Partnership About https://www.forestcarbonpartnership.org/
57. LEAF Coalition (2024) The LEAF Theory of Change https://www.leafcoalition.org/about
58. EUI (January 2025) Policy Developments in Carbon Markets Rewarding High-Quality Carbon Removals STG Policy Brief https://cadmus.eui.eu/entities/publication/b26e21a7-cdf6-5029-bdf5-faf5cac29a85
59. Thomson-Reuters (27th February 2025) IRA's uncertain future: How the Trump administration's approach could impact corporate tax functions https://www.thomsonreuters.com/en-us/posts/corporates/ira-uncertain-future/
60. USDA (August 2023) FACT SHEET: Marking One Year of the Inflation Reduction Act https://www.usda.gov/media/press-releases/2023/08/16/fact-sheet-marking-one-year-inflation-reduction-act

61. Electrek.co (September 2023) Chinese solar giant Trina is opening a 5 GW factory in Texas https://electrek.co/2023/09/13/trina-solar-factory-texas/
62. E3G (October 2023) One Year Inflation Reduction Act Initial outcomes and impacts for EU-US trade and investment https://www.e3g.org/publications/one-year-inflation-reduction-act/
63. Wikipedia CHIPS and Science Act https://en.wikipedia.org/wiki/CHIPS_and_Science_Act
64. Reuters (5 March 2025) Trump wants to kill $52.7 billion semiconductor chips subsidy law https://www.reuters.com/technology/trump-wants-kill-527-billion-semiconductor-chips-subsidy-law-2025-03-05/
65. Carbon Risk (27 March 2025) Step on the Gas by Peter Sainsbury, https://carbonrisk.substack.com/p/step-on-the-gas
66. EPA Summary of the Clean Air Act (1970) https://www.epa.gov/laws-regulations/summary-clean-air-act
67. Cap 20 (2024) Anatomy of a Senate Climate Bill Death https://www.americanprogress.org/article/anatomy-of-a-senate-climate-bill-death/
68. C2ES (2024) Market-Based Strategies https://www.c2es.org/content/market-based-strategies/
69. Merriam-Webster over-the-counter https://www.merriam-webster.com/dictionary/over-the-counter
70. CSIS (May 2023) Voluntary Carbon Markets: A Review of Global Initiatives and Evolving Models https://www.csis.org/analysis/voluntary-carbon-markets-review-global-initiatives-and-evolving-models
71. Carbon Credits (February 2023) California Carbon Credits (How Does It Work?) https://carboncredits.com/california-carbon-credits-how-does-it-work/
72. Arb.CA.Gov Cap-and-Trade Program: Allowance Distribution Factsheet https://ww2.arb.ca.gov/resources/documents/cap-and-trade-program-allowance-distribution-factsheet
73. Arb.CA.Gov ARB Emissions Trading Program https://ww2.arb.ca.gov/sites/default/files/cap-and-trade/guidance/cap_trade_overview.pdf
74. G0v.UK Committee on Climate Change https://www.gov.uk/government/organisations/committee-on-climate-change
75. CCC.Org Progress Reports https://www.theccc.org.uk/publicationtype/report/progress-reports/
76. Climate Action Tracker (September 2023) UK: Country Summary https://climateactiontracker.org/countries/uk/
77. Gov.UK (May 2024) Press Release https://www.gov.uk/government/news/uk-overachieves-another-carbon-emissions-target-and-rejects-rollover
78. Climate Change Committee (2022) Progress in reducing emissions 2022 Report to Parliament file:///C:/Users/urejo/Downloads/Progress-in-reducing-emissions-2022-Report-to-Parliament.pdf
79. Gov.UK (February 2024) Policy paper: Understanding climate adaptation and the third National Adaptation Programme (NAP3) https://www.gov.uk/government/publications/third-national-adaptation-programme-nap3/understanding-climate-adaptation-and-the-third-national-adaptation-programme-nap3#:~:text=The%20National%20Adaptation%20Programme%20(%20NAP%20)&text=NAP3%20explains%20the%20government%27s%20plans,our%20electricity%20networks%20and%20railways
80. CCC (March 2024) Independent Assessment of the Third National Adaptation Programme (NAP3) https://www.theccc.org.uk/wp-content/uploads/2024/03/Independent-Assessment-of-the-Third-National-Adaptation-Programme-NAP3.pdf
81. Labour (2024) Make Britain a clean energy superpower https://labour.org.uk/change/make-britain-a-clean-energy-superpower/
82. IEA (2019) Contract for Difference (CfD) https://www.iea.org/policies/5731-contract-for-difference-cfd
83. LCCC (2023) About Us https://www.lowcarboncontracts.uk/about-us/

84. Dieter Helm (2017) Cost of Energy Review https://assets.publishing.service.gov.uk/media/5a749c26ed915d0e8e39997a/Cost_of_Energy_Review.pdf
85. IIGCC (2024) Does the UK Autumn Budget 2024 deliver on climate? https://www.iigcc.org/insights/uk-autumn-budget-2024-labour-climate-action
86. Infrastructure Bank About Us https://www.ukib.org.uk/about-us
87. Financial Times (10th July 2024) 'Labour's flagship project aims to decarbonise heavy industry'
88. EcoWatch (9th July 2024) UK's New Labour Government Lifts 'Absurd' Ban on Onshore Wind https://www.ecowatch.com/uk-labour-onshore-wind-energy-ban.html
89. Financial Times (15 July 2024) 'Meloni backs return of nuclear power to cut carbon emissions'
90. The Conversation (2023) New single-use plastic ban takes effect in England – here's why its impact may be limited https://theconversation.com/new-single-use-plastic-ban-takes-effect-in-england-heres-why-its-impact-may-be-limited-214612
91. The Guardian (31st January 2023) 'Grossly irresponsible': UK hands out 24 new North Sea oil and gas licences https://www.theguardian.com/business/2024/jan/31/grossly-irresponsible-uk-hands-out-24-new-north-sea-oil-and-gas-licences
92. Carbon Brief (August 2024) UK could approve 13 new oil and gas projects despite North Sea pledge https://www.carbonbrief.org/analysis-uk-could-approve-13-new-oil-and-gas-projects-despite-north-sea-pledge/
93. Financial Times (21 June 2024) 'Court ruling threatens new fossil fuel projects'
94. PoliticoPro (23 July 2025) World's top court says climate inaction can breach international law https://subscriber.politicopro.com/article/2025/07/worlds-top-court-says-climate-inaction-can-breach-international-law-00473418
95. BBC (9 September 2024) New oil and gas ban threatens jobs, unions warn https://www.bbc.com/news/articles/c4gd1q9ejqdo
96. Energyvoice.com (15 January 2025) Labour to consult on North Sea oil and gas licensing, Miliband says https://www.energyvoice.com/oilandgas/north-sea/565575/labour-to-consult-on-north-sea-oil-and-gas-licensing-miliband-says/
97. The Guardian (8th February 2024) Labour cuts £28bn green investment pledge by half https://www.theguardian.com/politics/2024/feb/08/labour-cuts-28bn-green-investment-pledge-by-half
98. Grant Thornton (April 2022) Carbon taxes are coming – are you ready? https://www.grantthornton.co.uk/insights/carbon-taxes-are-coming-are-you-ready/
99. Gov.UK (April 2024) Climate Change Levy rates https://www.gov.uk/guidance/climate-change-levy-rates
100. SEFE-Energy UK Climate Change Levy (CCL) https://www.sefe-energy.co.uk/help-support/regulations-schemes/climate-change-levy-ccl/
101. Tax Foundation (September 2023) Carbon Taxes in Europe, 2023 https://taxfoundation.org/data/all/eu/carbon-taxes-in-europe-2023/
102. Legislation.Gov.UK (2020) The Greenhouse Gas Emissions Trading Scheme Order 2020 https://www.legislation.gov.uk/uksi/2020/1265/contents/made
103. Gov.UK (January 2024) Participating in the UK ETS https://www.gov.uk/government/publications/participating-in-the-uk-ets/participating-in-the-uk-ets
104. Statista (June 2024) United Kingdom Emission Trading System (UK ETS) allowance (UKA) futures pricing from May 2022 to June 2024 https://www.statista.com/statistics/1322275/carbon-prices-united-kingdom-emission-trading-scheme/
105. Gov.UK (March 2024) Introduction of a UK carbon border adjustment mechanism from January 2027 https://assets.publishing.service.gov.uk/media/65fc11fef1d3a0001132ac6f/Introduction_of_a_UK_carbon_border_adjustment_mechanism_from_January_2027.docx.pdf
106. Norton Rose Fulbright (May 2024) UK Carbon Border Adjustment Mechanism: How will it work? https://www.nortonrosefulbright.com/en/knowledge/publications/a3a42bda/uk-carbon-border-adjustment-mechanism
107. Ashurst (May 2024) The EU and UK Carbon Border Adjustment Mechanisms: A Comparison https://www.ashurst.com/en/insights/the-eu-and-uk-carbon-border-adjustment-mechanisms-a-comparison/

108. Federal Reserve History (2013) The Great Recession and Its Aftermath https://www.federalreservehistory.org/essays/great-recession-and-its-aftermath
109. Wikipedia Carillion https://en.wikipedia.org/wiki/Carillion
110. OECD The OECD and the G20 https://www.oecd.org/en/about/oecd-and-g20.html
111. TCFD (2017) Final Report: Recommendations of the Task Force on Climate-Related Financial Disclosures https://assets.bbhub.io/company/sites/60/2020/10/FINAL-2017-TCFD-Report-11052018.pdf
112. IFRS (2025) IFRS for SMEs https://www.ifrs.org/
113. Grant Thornton (2021) FRC: Preparing financial services for TCFD https://www.grantthornton.co.uk/insights/frc-preparing-financial-services-for-tcfd/
114. Institute for Government (June 2023) Financial Reporting Council https://www.instituteforgovernment.org.uk/explainer/financial-reporting-council
115. Gov.UK (August 2023) UK Sustainability Reporting Standards https://www.gov.uk/guidance/uk-sustainability-disclosure-standards
116. Sweep (June 2024) What is the ISSB? What do businesses need to know about it? https://www.sweep.net/insights/what-is-the-issb-what-do-businesses-need-to-know-about-it
117. Gov.UK (March 2024) Nature Markets Framework progress update March 2024 https://www.gov.uk/government/publications/nature-markets-framework-progress-update-march-2024/nature-markets-framework-progress-update-march-2024
118. Council on Foreign Relations (March 2022) How Does the European Union Work? https://www.cfr.org/backgrounder/how-does-european-union-work
119. European Commission (2024) The Digital Services Act package https://digital-strategy.ec.europa.eu/en/policies/digital-services-act-package
120. Morgan Lewis (May 2024) Two EU Directives Aim to Shape European Legal Framework for Green Claims and Greenwashing https://www.morganlewis.com/pubs/2024/05/two-eu-directives-aim-to-shape-european-legal-framework-for-green-claims-and-greenwashing
121. Feigen Baum (April 2024) The Green Claims Directive: Impact on companies and their marketing claims https://www.fiegenbaum.solutions/en/blog/the-green-claims-directive-impact-on-companies-and-their-marketing-claims
122. European Commission Labelling and packaging https://trade.ec.europa.eu/access-to-markets/en/content/labelling-and-packaging
123. Climate Action Network – Europe (October 2023) REACTION: The EU has agreed on a global phaseout of fossil fuels from the energy sector well ahead of 2050 https://caneurope.org/eu-environmental-council-cop28/
124. Clean Air Task Force (May 2024) Accounting for carbon removals in the EU: Highlights from the 4th meeting of the Carbon Removals Expert Group https://www.catf.us/2024/05/accounting-carbon-removals-eu-highlights-4th-meeting-carbon-removals-expert-group/
125. Clean Air Task Force (February 2024) A first look at the EU Industrial Carbon Management Strategy https://www.catf.us/2024/02/a-first-look-at-the-eu-industrial-carbon-management-strategy/
126. European Commission (November 2022) European Green Deal: Commission proposes certification of carbon removals to help reach net zero emissions https://ec.europa.eu/commission/presscorner/detail/en/ip_22_7156
127. IAPT.Org (may 2024) Unpacking the Eu Carbon Removal Certification Framework: Implications for EU Climate and Agriculture Policy https://www.iatp.org/sites/default/files/2024-05/CRCF%20trilogue%20article_final_0.pdf
128. Euro-Lex Directive 94/22/EC https://eur-lex.europa.eu/legal-content/EN/TXT/?uri=CELEX:31994L0022
129. Euronews (August 2021) The end of fossil fuels: Which countries have banned exploration and extraction? https://www.euronews.com/green/2021/08/12/the-end-of-fossil-fuels-which-countries-have-banned-exploration-and-extraction
130. Renew Economy (July 2024) Ten EU countries commit to stop burning fossil fuels by 2035 https://reneweconomy.com.au/eu-lines-up-behind-2035-exit-from-fossil-fuels/

131. Statista (2023) Number of operational, shutdown, and planned nuclear reactors in European countries as of August 2023 https://www.statista.com/statistics/792589/operational-nuclear-reactors-european-union-eu-28/#:~:text=As%20of%20August%202023%2C%20there,down%20reactors%2C%20with%2034%20units.
132. IAEA (September 2023) What are Small Modular Reactors (SMRs)? https://www.iaea.org/newscenter/news/what-are-small-modular-reactors-smrs
133. European Commission Delivering the European Green Deal https://commission.europa.eu/strategy-and-policy/priorities-2019-2024/european-green-deal/delivering-european-green-deal_en
134. European Commission Social Climate Fund https://climate.ec.europa.eu/eu-action/eu-emissions-trading-system-eu-ets/social-climate-fund_en
135. UN Sustainable Development Group Leave No One Behind https://unsdg.un.org/2030-agenda/universal-values/leave-no-one-behind
136. European Commission Use of international credits https://climate.ec.europa.eu/eu-action/eu-emissions-trading-system-eu-ets/use-international-credits_en
137. European Commission Our ambition for 2030 https://climate.ec.europa.eu/eu-action/eu-emissions-trading-system-eu-ets/our-ambition-2030_en
138. UNCTAD (2023) China's policy strategies for green low-carbon development: Perspective from South-South cooperation https://unctad.org/publication/chinas-policy-strategies-green-low-carbon-development-perspective-south-south
139. Brookings (June 2024) Unleashing "new quality productive forces": China's strategy for technology-led growth https://www.brookings.edu/articles/unleashing-new-quality-productive-forces-chinas-strategy-for-technology-led-growth/
140. Cambridge Industrial Innovation Policy (April 2024) China's emerging industrial vision: the significance and impact of 'New Quality Productive Forces' https://www.ciip.group.cam.ac.uk/reports-and-articles/chinas-emerging-industrial-vision/
141. IEA (2023) World Energy Outlook 2023 https://www.iea.org/reports/world-energy-outlook-2023
142. Anni Dai. Et al. (September 2024) Looking Back to Look Ahead: An Analysis of 14th Five-Year Plans (2021–2025) Berkeley Law California-China Climate Institute https://ccci.berkeley.edu/sites/default/files/Full%20Report%20%28English%29.pdf
143. The Diplomat (November 2023) The Power – and Limits – of Xi Jinping https://thediplomat.com/2023/11/the-power-and-limits-of-xi-jinping/
144. Denise Sienli van der Kamp (28 January 2023) 'Chapter 4—Blunt Force Regulation and Bureaucratic Control' Cambridge University Press (online) https://www.cambridge.org/core/books/abs/clean-air-at-what-cost/blunt-force-regulation-and-bureaucratic-control/4F745049C41F2F7C58C254C1EB440712
145. China Daily (16 March 2024) Explainer: What to know about China's high-quality development https://www.chinadaily.com.cn/a/202403/16/WS65f53a2aa31082fc043bd023.html
146. Ministry of Justice of the People's Republic of China (10 February 2023) China's roadmap for high-quality development http://en.moj.gov.cn/2023-02/10/c_857781.htm#:~:text=China%20has%20mapped%20out%20an%20outline%20to%20increase,State%20Council%2C%20was%20made%20public%20on%20Feb%206
147. Carbon Change Laws 14th Five-Year Plan and 2035 Long-Term Objectives https://climate-laws.org/document/14th-five-year-plan_0496
148. Carbon Brief (February 2024) China's energy sector carbon dioxide (CO_2) emissions increased 5.2% in 2023, meaning a record fall of 4–6% is needed by 2025 to meet the government's "carbon intensity" target. https://www.carbonbrief.org/analysis-record-drop-in-chinas-co2-emissions-needed-to-meet-2025-target/
149. Aljazeera.com (25 March 2025) China's missed emissions target poses challenge to global climate efforts https://www.aljazeera.com/news/2025/3/25/chinas-missed-emissions-target-poses-challenge-to-global-climate-efforts
150. Climateactiontracker.org (September 2024) China https://climateactiontracker.org/countries/china/

151. Phys.Org (8th November 2024) China passes energy law to 'promote carbon neutrality' https://phys.org/news/2024-11-china-energy-law-carbon-neutrality.html
152. ENR.Com (20 March 2025) China Doubles Down on Coal-Fired Power, With Record Plant Capacity Planned https://www.enr.com/articles/60469-china-doubles-down-on-coal-fired-power-with-record-plant-capacity-planned
153. Meng Jiang, et al. (2024) Tracing fossil-based plastics, chemicals and fertilizers production in China' Nature Communications v.15 (May 2023) https://www.nature.com/articles/s41467-024-47930-0#:~:text=Estimates%20show%20China%27s%20chemical%20industry,%2C39%2C41%2C42.
154. IEA 50 China: Natural Gas Supply https://www.iea.org/countries/china/natural-gas
155. IEA 50 China: Oil Supply https://www.iea.org/countries/china/oil
156. Center on Global Energy Policy (September 2021) Green Giants? China's National Oil Companies Prepare for the Energy Transition https://www.energypolicy.columbia.edu/publications/green-giants-china-s-national-oil-companies-prepare-energy-transition/
157. IEA 50 Hydrogen Industry Development Plan (2021–2035) https://www.iea.org/policies/16977-hydrogen-industry-development-plan-2021-2035
158. Weforum.org (June 2023) Green Hydrogen in China: A Roadmap for Progress https://www3.weforum.org/docs/WEF_Green_Hydrogen_in_China_A_Roadmap_for_Progress_2023.pdf
159. US State Department (November 2023) Sunnylands Statement on Enhancing Cooperation to Address the Climate Crisis https://www.state.gov/sunnylands-statement-on-enhancing-cooperation-to-address-the-climate-crisis/#:~:text=The%20United%20States%20and%20China%20decide%20to%20operationalize%20the%20Working,climate%20actions%20in%20the%202020s.
160. Wood Mackenzie (May 2024) China leads global wind turbine manufacturers' market share in 2023 https://www.woodmac.com/press-releases/2024-press-releases/global-wind-oem-marketshare/
161. CarbonBrief (November 2023) The Carbon Brief Profile: China https://interactive.carbonbrief.org/the-carbon-brief-profile-china/#renewables
162. Reuters (6 June 2024) World's biggest solar farm comes online in China's Xinjiang https://www.reuters.com/world/china/worlds-biggest-solar-farm-comes-online-chinas-xinjiang-2024-06-03/
163. DNV (April 2024) China's energy transition outlook https://www.dnv.com/news/china-energy-transition-outlook/
164. Climate Scoreboard (January 2023) China's Power Grid Systems https://www.climatescorecard.org/2023/01/chinas-power-grid-systems/
165. Haiyun Song, et al. (2020) 'The Development and Problems of China's Carbon Market' IOP Conf. Series: Earth and Environmental Science 440 (2020) (https://iopscience.iop.org/article/https://doi.org/10.1088/1755-1315/440/4/042100/pdf
166. ICAP (2022) China National ETS https://icapcarbonaction.com/system/files/ets_pdfs/icap-etsmap-factsheet-55.pdf
167. Dialogue Earth (August 2023) As China's carbon market turns two, how has it performed? https://dialogue.earth/en/climate/china-carbon-market-turns-two-how-has-it-performed/
168. Reuters (12 September 2024) What's China's carbon market and how does it work? https://www.reuters.com/sustainability/whats-chinas-carbon-market-how-does-it-work-2024-09-12/
169. ChinaCarbon.Info (2022) China Carbon Pricing Survey http://www.chinacarbon.info/wp-content/uploads/2023/07/1_2022-CCPS-Report-EN.pdf
170. Dialogue Earth (June 2023) China's carbon credit system to restart this year https://dialogue.earth/en/digest/chinas-carbon-credit-system-to-restart-this-year/
171. Dialogue Earth (September 2022) Analysis: Forest carbon credits are not a 'free pass' for forest destroyers https://dialogue.earth/en/forests/analysis-forest-carbon-credits-are-not-a-free-pass-for-forest-destroyers/
172. Dialogue Earth (July 2024) Are China's blue carbon credits a free pass to harm its oceans? https://dialogue.earth/en/ocean/are-chinas-blue-carbon-credits-a-free-pass-for-ocean-polluters/

173. Statista (2024) GHG emissions in India—Statistics & Facts https://www.statista.com/topics/8881/emissions-in-india/#:~:text=India%27s%20GHG%20emissions%20have%20roughly,of%202.7%20GtCO%E2%82%82%20in%202022.%20%20%20%20%20World%20Resources%20Institute%20(February%202020)%204%20Charts%20Explain%20Greenhouse%20Gas%20Emissions%20by%20Countries%20and%20Sectors
174. JRC (2023) GHG emissions of all world countries—JRC/IEA 2023 Report https://edgar.jrc.ec.europa.eu/report_2023#:~:text=Even%20in%202022%2C%20global%20GHG,F%2Dgases%20(2.6%25).
175. World Resources Institute (May 2023) 9 Charts Explain Per Capita Greenhouse Gas Emissions by Country https://www.wri.org/insights/charts-explain-per-capita-greenhouse-gas-emissions#:~:text=But%20among%20the%20top%2010,just%202.5%20tCO2e%20per%20person.
176. Statista (2024) Twenty countries with the largest population in 2024 https://www.statista.com/statistics/262879/countries-with-the-largest-population/#:~:text=In%202022%2C%20India%20overtook%20China,more%20than%201.44%20billion%20people.
177. Frontiers (2022) Greenhouse gas emissions in irrigated paddy rice https://www.frontiersin.org/journals/sustainable-food-systems/articles/ https://doi.org/10.3389/fsufs.2022.868479/full
178. CarbonBrief (March 2019) The Carbon Brief Profile: India https://www.carbonbrief.org/the-carbon-brief-profile-india/#:~:text=India%20is%20the%20world%27s%20third,after%20China%20and%20the%20US.&text=Coal%20power%20plants%2C%20rice%20paddies,well%20below%20the%20global%20average.
179. The Economic Times (14 October 2022) 'India has enough coal reserves to last for 50 years: Pralhad Joshi' https://economictimes.indiatimes.com/industry/indl-goods/svs/metals-mining/india-has-enough-coal-reserves-to-last-for-50-years-pralhad-joshi/articleshow/94845610.cms?from=mdr
180. The Economic Times (10 December 2023) 'Can't phase out coal power until storage, tech viable: Power Ministry https://economictimes.indiatimes.com/industry/renewables/cant-phase-out-coal-power-until-storage-tech-viable-power-ministry/articleshow/105883558.cms?from=mdr
181. ADB (October 2023) How Can We Incentivize Reducing Methane Emission in Rice Farming in Asia? https://www.adb.org/news/events/how-can-we-incentivize-reducing-methane-emission-rice-farming-asia#:~:text=Instead%20of%20continuously%20flooding%20rice,rice%20yield%20in%20a%20season.
182. Clear Center (April 2024) Why Do Cattle Produce Methane, and What Can We Do About it? https://clear.ucdavis.edu/explainers/why-do-cattle-produce-methane-and-what-can-we-do-about-it#:~:text=Feeding%20cows%20grains%20and%20fats,molecules%20in%20a%20cow%27s%20rumen.
183. NIH Chapter 7: Household Air Pollution from Solid Cookfuels and Its Effects on Health https://www.ncbi.nlm.nih.gov/books/NBK525225/#:~:text=In%20India%2C%20for%20example%2C%20an,%3B%20Lelieveld%20and%20others%202015).
184. MNRE (2018) Government making efforts to reduce dependence on traditional biomass cooking https://pib.gov.in/Pressreleaseshare.aspx?PRID=1525934#:~:text=The%20use%20of%20traditional%20biomass,a%20range%20of%20toxic%20products.
185. IEA July 2023) Transitioning India's Road Transport Sector https://iea.blob.core.windows.net/assets/06ad8de6-52c6-4be3-96fc-2bdc3510617d/TransitioningIndiasRoadTransportSector.pdf
186. Government of India (2021) India's Updated First Nationally Determined Contribution Under Paris Agreement (2021–2030) https://unfccc.int/sites/default/files/NDC/2022-08/India%20Updated%20First%20Nationally%20Determined%20Contrib.pdf
187. MEFC (2023) India achieves two targets of Nationally Determined Contribution well ahead of the time https://pib.gov.in/PressReleaseIframePage.aspx?PRID=1987752#:~:text=In%20August%202022%2C%20India%20updated,enhanced%20to%2050%25%20by%202030.
188. Green Climate Fund About https://www.greenclimate.fund/about

189. Global CCS Institute (2023) India's Carbon Credit Trading Scheme & the Indian Government's CCUS Report https://www.globalccsinstitute.com/wp-content/uploads/2023/07/India%E2%80%99s-Carbon-Credit-Trading-Scheme-the-Indian-Government%E2%80%99s-CCUS-Report.pdf
190. Ministry of Power (June 2024) Detailed Procedure for Compliance Mechanism under CCTS https://beeindia.gov.in/sites/default/files/Detailed%20Procedure%20for%20Compliance%20Mechanism.pdf
191. CarbonCredits (February 2024) India Revises Its Carbon Credit Trading Scheme for Voluntary Players https://carboncredits.com/india-revises-its-carbon-credit-trading-scheme-for-voluntary-players/
192. Center for Global Energy Policy (2024) Lessons for Structuring India's Carbon Market to Support a Cost-Efficient Energy Transition https://www.energypolicy.columbia.edu/publications/lessons-for-structuring-indias-carbon-market-to-support-a-cost-efficient-energy-transition/
193. S&P Global (December 2023) India announces domestic voluntary carbon market scheme https://www.spglobal.com/commodityinsights/en/market-insights/latest-news/energy-transition/122023-india-announces-domestic-voluntary-carbon-market-scheme
194. Wikipedia (2024) Ricardian Socialism https://en.wikipedia.org/wiki/Ricardian_socialism#:~:text=Ricardian%20socialism%20is%20considered%20to,profit%2C%20interest%20and%20rent%20were
195. Wikipedia (2024) Henry George https://en.wikipedia.org/wiki/Henry_George#:~:text=George%20famously%20argued%20that%20a,more%20productive%20and%20just%20society.&text=Philadelphia%2C%20Pennsylvania%2C%20U.S.&text=New%20York%20City%2C%20U.S.
196. Mse.gov.sg Frequently asked Questions https://www.mse.gov.sg/faq#:~:text=The%20carbon%20tax%20covers%2080,in%202026%20and%202027%2C%20and
197. Carboncredits.com (September 2022) Social Cost of Carbon in the US is $185 per ton https://carboncredits.com/social-cost-of-carbon-185/
198. Wikipedia (March 2025) Air pollution in India https://en.wikipedia.org/wiki/Air_pollution_in_India#:~:text=Of%20the%20four%20major%20Indian,close%20second%2C%20followed%20by%20Mumbai.
199. UCS.org (17 February 2025) How the Electricity Grid Works https://www.ucs.org/resources/how-electricity-grid-works
200. Wikipedia (March 2025) Electric power transmission https://en.wikipedia.org/wiki/Electric_power_transmission#:~:text=Long%2Ddistance%20transmission%20is%20typically,losses%20in%20the%20line%20conductors.
201. Tony Blair Institute for Global Change (29 April 2025) The Climate Paradox: Why We Need to Reset Action on Climate Change https://institute.global/insights/climate-and-energy/the-climate-paradox-why-we-need-to-reset-action-on-climate-change

Chapter 5
Voluntary Carbon Markets

Allowances Versus Credits

From the outset it is important to understand the distinction between emission allowances and carbon credits [3]. Emission allowances arise in the compliance markets, which account for an estimated 98% of carbon trades, and typically come under cap-and-trade Emission Trading Schemes (ETS)—see Chap. 4. These can be traded as certificates to offset *current* emissions which exceed the caps. Carbon credits arise from private-sector projects which can be traded as offsets against *future* emissions within the voluntary markets. They account for the remaining 2% of carbon trades [4]. Among many other sources, this chapter draws upon the insightful guide to the complexities of VCMs, provided by Simon Puleston Jones's *An Introduction to Carbon Markets* [5]. See Table 5.1 which lists the principal features that distinguish the MCM and VCM markets.

It is also worth remembering what is the point of carbon credits. It is to encourage private investments into projects that avoid, reduce or remove CO_2 or CO_2 greenhouse gas equivalents. This could include 'blended' finance which includes partial public funding to reduce the risk for private funding, but this should only happen when the integrity of the projects has been fully guaranteed, possibly including insurance in cases where projects could fail. The purpose is to fulfil the targets identified by the IPCC of Net Zero CO_2 emissions by 2050, Net Negative CO_2 emissions by 2070, and Net Zero emissions of all other greenhouse gases by 2100. The challenges for the private sector are finding projects of high integrity, that fulfil their claims, that offer 'additionality' meaning they add to whatever would have happened without them, and their durability is fully measured, monitored, recorded and verified (MMRV) over the lifetime of the project. Not many VCM projects are able to fully meet these criteria.

J. Ure, *Climate Change and Carbon Markets: Late or Too Late?*,
https://doi.org/10.1007/978-3-032-07619-9_5

Table 5.1 Mandatory (Compliance) Carbon Markets versus Voluntary Carbon Markets

Emission allowances versus carbon credits/offsets	Carbon offsets
A **permit** that allows the registered holder to emit one tonne of carbon dioxide or other GHG equivalent	An **instrument** representing the avoidance, reduction or removal of one tonne of carbon dioxide or other GHG equivalent
Issued by **governmental** or **regulatory** organisations	Typically issued by a **standard** in the voluntary carbon market or a **government** in the sovereign carbon market
Buyers and sellers typically have a **legal obligation** to buy or sell emission allowances	Can be sold on certain compliance carbon markets, but most likely sold on **voluntary carbon markets** or **sovereign carbon markets**—such VCMs or SCMs can operate bilaterally in OTC (Over-the-Counter) markets, or regulated exchanges and/or unregulated marketplaces
Trades are typically **reported publicly** to/by the exchange on which the sale happens, enabling a transparent current price for such emission allowances	Generated by either avoidance/reduction projects or by removal/sequestration projects (in the case of sovereign carbon offsets only reductions and removals are currently permitted). Trades may be **reported publicly** or go **unreported**, with the prices paid typically not being made public. The price for an 'avoidance' offset is typically much less than the price of a 'removal' offset
Highly regulated	Broadly speaking, **not specifically regulated**

Source 'An introduction to carbon markets' by Simon Puleston Jones[6]

Buyers and Sellers and Projects

Carbon credits arise from projects designed to avoid, reduce (mitigation) or remove CO_2 (abatement). As definitions sometimes vary, those found in *Carbon Direct* (2024) [7] are used here.

Carbon avoidance is an action that prevents a carbon-emitting activity from happening, such as the use of solar energy, hydropower or wind power in power generation, and electric vehicles in power consumption. However, the manufacture of e-vehicles does involve emissions and the electricity stored in the batteries may originate from fossil fuels. Up to the year 2023, the e-vehicle manufacturer Tesla had raised over USD9 billion by selling avoidance credits [8].

Carbon reduction is an action that decreases the amount of greenhouse gas emissions compared to prior practices, such as a *switch* from the generation of electricity using fossil fuels to solar power or producing 'green' hydrogen by electrolysis, or in industry by using more energy-efficient technologies and recycled materials [9], or the use of Sustainable Aviation Fuel (SAF) in aircraft, although exactly how clean and green SAF really is has come under question [10].

Carbon removal is the process of removing carbon dioxide from the atmosphere and locking it away for decades, centuries, or millennia, known as sequestration. Examples are nature-based projects that divide into two categories: afforestation (planting trees) and reforestation (replacing fallen trees) known as ARR and which create REDD + projects (see below), and other nature-based projects such as restoring wetlands and mangroves, or the alkalinization of seawater (also see below), all of which act as carbon sinks. Nature-based projects in 2023 were responsible for the removal of over 120 million tonnes (Mt) of CO_2, that is 40% of all carbon abatement or mitigation projects compared with around 24% by renewable energy projects [11]. It should be noted here that, while the terms 'abatement' and 'mitigation' are often used interchangeably, the former tends to refer to the compete removal of carbon, such as in CCUS projects, and the latter to when carbon is reduced. The more direct approach is carbon dioxide removal (CDR) [12] as a means of CCUS (carbon capture, utilization and storage) [13] using somewhat expensive technologies that literally suck carbon out of the air prior to sequestration. In Europe, an EU Commission report [14] (2024) by Mario Draghi, former Prime Minister of Italy and President of the European Central Bank, identifies CCUS as an area where the EU has strength and to be further developed, as does the advocacy group CO_2 Value Europe [15]. Until the election of President Trump, the US Government had also supported CDR projects, although in recent years it has come under sustained criticism as 'a dangerous distraction' by environmental groups [16], largely because fossil fuel companies can see it as a 'Get Out of Jail' card, although hardly a free one [17]. But based upon an 'anything is better than nothing' approach, it has its advocates beyond the fossil fuel interests, yet it accounts for less than 10% of all projects [18], and in the UK in 2025, the government's Public Account Committee noted previous failed attempts at CCUS and considered government investment in the technologies as high risk [19].

Carbon Credits are placed in the market by registries and are certified using methodologies from one of the standards bodies. End-user buyers are typically among the world's largest corporate entities, who can either hold them for future use or retire them as offsets to their own carbon emissions. Retired credits must not be re-used or re-issued to avoid double-counting and double-use. Prior to 2002 and up to 2022, over 800 million credits had been retired, each claiming to represent one tonne of carbon dioxide or its equivalent in other greenhouse gases [20]. See Market Metrics Box 5.1. The top providers of carbon credits are land-based projects such as reforestations, and energy-based renewable energy projects such as windfarms. The top buyers by volume, frequency and value, are fossil fuel companies, car manufacturers and technology giants. Keeping track of these transactions in a transparent way is important for the efficient functioning of markets and for the avoidance of fraud, but problematically the registries are not obliged to make public the sellers or the buyers. *Carbon Brief* (2023) [21], provides a systematic overview, and separately provides a glossary for carbon markets [22]. *Allied Offsets* is another source of data [23].

Box 5.1 Market Metrics

The most relevant metric towards climate change mitigation is how many tonnes of carbon (or carbon-equivalents as Green House Gases) have been reduced from previous levels or removed entirely from the atmosphere. Despite reductions in annual GHG emissions in some major polluting countries, such as the USA [24], globally GHGs have risen by over 50% since the 1997 Kyoto Protocols, from above 24 billion to beyond 37 billion carbon-equivalent metric tonnes by 2022 [25].

The Data

The available data covers the volume in credits and value of carbon credit transactions as given by the price per tonne of carbon. Using annual data, Bloomberg estimated the total value of the world's 30 compliance markets in 2021 at USD850 billion [26], but they only cover 20% of global emissions, meaning 80% of emissions are not addressed by emission allowances in the MCMs. VCMs pale in comparison, in 2022 accounting for little more than USD2 billion of credits by value, according to the *State of the Voluntary Carbon Market* [4], or just 0.2% of total market value. Over the entire period from pre-2005–2023, the traded value on VCMs was just under USD11 billion, and the traded volume was 2.4 billion credits [4], representing 2% of total market transactions according to the *State of the Carbon Dioxide Removal* (2nd Edition 2024) [27]. In the estimation of the *International Energy Authority* [28], carbon removal itself in the form of CCUS captures only around 0.1% of the global emissions released into the atmosphere in 2022 (490Gt) and 2023 (410Gt). Using the same metrics, at current levels all carbon reduction and removal projects combined (including CCUS) would remove less than 1% of carbon and carbon-equivalent emissions annually. Only a switch *en mass* to carbon *avoidance* and *reduction* would make significant dents in the rate of emissions.

It is challenging to determine how much worse it would be if it was not for climate action plans, carbon caps, taxes, and some trading. In the case of the USA, greater energy efficiency and lower energy intensity in areas such as electricity generation, industrial power and heating, transport and household consumption, have been responsible for keeping GHGs 18% lower than they would have been [29]. Only a few states in the USA operate an emissions capping scheme (see Chap. 4) but tax payers across the US can benefit from incentives to capture carbon emissions, such as 45Q tax credits for carbon sequestration, first introduced in 2008 [30]. In the world's greatest polluter, China, emission levels are largely determined by GDP growth rates, but the 14th 5-Year Plan (2021–2025) aims to reduce energy consumption per unit of GDP ('energy intensity') by 13.5% and CO_2 emissions per unit of GDP ('emissions intensity') by 18% [31]—but see Chap. 4.

Voluntary Market Woes

Creating standards to verify projects that claim to achieve carbon avoidance, reduction or removal are crucial to the credibility, and therefore to the sustainability of the VCMs. The standards used are set by one of several bodies, such as Verra [32], the Gold Standard [33], the American Carbon Standard [34], and California-based Carbon Action Reserve (CAR) [35]. For businesses aiming to reduce their emissions irrespective of whether or not they issue carbon credits, there are advisory bodies such as Science Based Targets Initiative (SBTi) [36] that issue guidelines. Chap. 4 has highlighted a trend for regulations in the compliance markets to influence standards setting in the voluntary markets. This is motivated, on the one hand, by the desire of governments to encourage private investment into carbon abatement and mitigation projects, which within country can contribute to the Nationally Determined Contributions (NDCs) under Article 6 of the Paris Agreement, and on the other, by the private sector wanting to lower the risk of investment by strengthening the standards in use.

The problems arise because, as distinct from the compliance or mandatory carbon markets (MCMs), there are major loopholes in the voluntary markets. First, considerable latitude has been given to project managers, their agents and the sellers of carbon credits in the implementation of the standards to authenticate the 'additionality,' of the projects, that is to say, to show they really will reduce emissions of carbon in addition to what would have happened had the project not been approved. Second, and the more insidious problem, normally neither the project managers nor their agents and sellers of the credits, nor the buyers, are under obligations to measure, monitor, report and verify (MMRV) the ongoing functioning of carbon mitigation projects over time, unless there are specific contracts or arrangements put in place to do so—for example, under the UN's *Clean Development Mechanism* (CDM), credits have to be periodically repurchased or replaced to ensure the projects are fulfilling their targets [37].

Failure of MMRV means that 'leakage' where emissions take place elsewhere to escape the restrictions of the project, and the 'durability' of the project, meaning how long-lasting is it, go undetected. This is what economists would call a *principal-agent problem* [38]. At best, the agents try to be independent in their assessments and judgements, but as paid consultants they also have an obligation to be helpful to the principal, while the principals can pick and choose the agents they use, and in some cases filter, or worse doctor, the information provided to the agents. Making money from project credits can be too easily prioritised over taking responsibility for the performance of the projects, and the incentives to commit to MMRV may only arise when serious reputational damage is at stake, such as to major international corporations, or where credits can be used to offset taxes, such as in highly regulated markets as in Singapore [39].

The negative impact that scandals in the voluntary markets have had upon buyers' willingness to buy has been overwhelming [40]. As a result of a fresh wave of allegations, the 2024 *State of the VCM* [4] report estimated that post-COVID-19 trade volumes from 2021–2023 fell year-on-year by 56% pending hopes for an improvement, while the European Union (EU) took the decision to exclude altogether carbon credits generated in the VCM from eligibility in the compliance sector [41]. Numerous scandals involving much-needed nature-based projects puts this into context.

Around 25% of all voluntary carbon market credits are nature-based REDD + projects [42], where REDDI + stands for 'Reducing Emissions from Deforestation and Forest Degradation' and + stands for 'sustainable management of forests and the conservation and enhancement of forest carbon stocks' [43]. The + was added to REDD at COP19 in Warsaw 2013 (known as the *Warsaw Framework*) to widen the scope of REDD projects to attract investment [44], but was also a response to growing concerns over the quality of REDD projects, as highlighted in a highly critical report *The great REDD gamble* by Friends of the Earth [45]. An illustration of the range of problems comes from a study in India around 2014 that reviewed 8 project-types across 10 states and concluded there was no MRV, co-benefits to the local communities were minimal, added infrastructure was limited to the projects themselves, there was no improved access to energy supply for the villagers, and the stakeholders were 'a sham' [46]. The criticisms continue to the present day, for example, *REDD-Monitor* uncompromisingly regards 'carbon credits are a scam' [47]. Box 5.2 provides the history, the context, and a perspective.

Box 5.2 Failing to see the wood for the trees—toxic disputes over the use of standards.

The Accusations

Prior to COP26 in Glasgow in 2022, the 'certificates of removal' (CERs) issued under the *Clean Development Mechanism* (CDM) of the United Nations *Kyoto Protocol* (1997) had become worthless following a price collapse in 2012, and subsequent buyers were mostly limited to China, Mexico and South Africa where they could be used as offsets against local carbon taxes [48]. Demand for CERs fell far short of supply, in part because the quality and claims of the projects were in many cases highly questionable. In 2021, the reputation of voluntary carbon credits was further shattered when a report by *Unearthed* [49] (operated by Greenpeace UK) [50] was published by *The Guardian* newspaper challenging the veracity of Verra's carbon offsetting standard, claiming 11 out of 12 projects studied showed no difference in emissions from the control groups which set the benchmarks. This was followed in 2023 by two peer-reviewed academic studies, including West et al. (2023) [51], again reported in *The Guardian* and in the German newspaper *Die Ziet,* and by *SourceMaterial*, claiming that upwards of 94% of rainforest carbon offsets approved by Verra were worthless [52]. Continuing with accusations, in 2024 *SourceMaterial* also claimed 'BP venture sells carbon offsets by "saving" trees that aren't in danger' [53].

The Response

Verra and others hit back. Verra argued that the claims were based upon the use of a 'synthetic' methodology which does 'not account for project-specific factors that cause deforestation. As a result, these studies massively miscalculate the impact of REDD projects'. Verra added 'Synthetic controls compare a project to a control scenario based on a set of variables that impact deforestation, known as covariates, whereas *Verra's approach for REDD projects compares them to real areas'* [54]. [Emphasis added by author] The DGB Group, a commercial developer of projects, accused the journalists of bias, adding that other studies 'have shown that, *when done correctly*, carbon offsetting is a valuable tool in the fight against deforestation, biodiversity loss, and ecosystem restoration' [55]. [Emphasis added by author]. Another Verra ally Everland, holding the VCM's 'largest portfolio of high-impact, community-centered, forest conservation REDD + projects' [56] called the story 'fatally flawed' and criticised the authors for not communicating the uncertainty of their findings, 'nor did they assess or account for these uncertainties, *nor the significant discrepancies between the studies'* [57]. [Emphasis added by author]. Researchers Hugh Graham, et al. (2023) argued for 'a simulation framework' that could test the findings of the synthetic approach alongside alternative approaches [58]. Among other responses were scientists from NASA and from a commercial satellite company who question the data accuracy of West et al. (2023), for example, citing improvements in satellite technology that give different aerial results over time, and the relevance of the baseline sites chosen [59]. The problem with this line of attack is that it could apply to both sides of the argument.

Methodology and Measurement Maladies

Although the debate emphasises methodological differences, which are unavoidable, what is largely missing from them is how these projects are implemented in practice. Doubts about measurement are a *real* methodological problem, not a synthetic one. Identifying suitable control sites is necessary for setting benchmarks, and for making judgments ('assessments') as to the likelihood of leakage and additionality; but measuring, monitoring, reporting, and verification (MMRV) is an ongoing process and is clearly not the *default*. Increasingly the view is that measurements are never precise because the precise is not measurable, in which case the second-best option is transparency about how judgements are arrived at. Part of the problem is that standards bodies are not regulators, they are exam boards. They are not enforcers to see projects through to their conclusion because they have no ownership of them, nor of the credits to which their stamp of approval gives rise. Indeed, credits recorded on most registries do not imply ownership, although in the UK the practice is otherwise.

It may take an investigative journalist to uncover shortcomings, such as poor project management, failure to use the capital raised from selling credits for proper purposes, and failure to deliver co-benefits if they were promised. Exactly that happened in 2023 when the media exposed the Kariba REDD + project in Zimbabwe leading to a full-length article in *The New Yorker* magazine [60]. Among the findings were that the agent, South Pole [61], acting for the project developer, had 'off-loaded twenty-three million credits from the project—eight million more than it could justify'. Both South Pole and Verra, whose standard had been used, quickly disassociated themselves for the project [62]. Soon after the CEOs of both South Pole [63] and of Verra [64] resigned, and since then both organisations have reviewed their standards and their use.

La coup de grace

A team of 14-researchers from the *UC Berkeley Carbon Trading Project* (Haya et al. 2023) [65] published a 200-page report in 2023 evaluating four REDD + projects that had used one of four Verra standard methodologies (VM0006, VM0007, VM0009, and VM0015) [66] to verify and authenticate carbon credits. Here are their findings.

1. **Baseline (Or Benchmark) Estimates**: If baselines are exaggerated by assuming a high rate of forest degradation *without the project*, then reductions in forest loss leading to higher rates of carbon absorption in the REDD + projects will be overstated. They found that in a sample study of 11 such projects, overstatement by the project developers 'resulted in credit generation 23–30% higher than our independent estimates'.
2. **Leakage:** They found that 'project developers were able to apply no leakage deduction through a number of paths, even for projects with substantial leakage risk'.
3. **Forest Carbon Accounting**: 'Best practice' by standards bodies, including Verra, advise project developers and assessment agents to use conservative estimates of the credits that a project can reasonably support. The research found that estimates for the capacity of trees above ground and below ground varies across projects by 80% and 193% respectively, and 'most developers chose equations that led to a high rather than conservative estimate of the carbon per hectare of forest'.
4. **Durability**: A period of 100 years is normally used to judge the durability of a REDD + project, and during that time the risk of losing a proportion of the trees to 'stand-clearing natural disturbance', (e.g. fires, disease, typhoons, etc.) is around 28% according to a review of the literature (i.e. 28% of the forest will not survive the full 100 years). By contrast on average REDD + projects assumed 'just 2% of credited carbon reductions'. Further, to insure

against tree loss, and therefore less carbon reduction, it is usual practice to create a buffer stock of credits, but 'more than half of all projects contributed the minimum allowed deduction, 10%, into the buffer pool to cover both natural and human risks'. This implies that carbon credit issuance is widely overplayed and risk is systematically downplayed.

5. **Social Safeguards**: Protecting or rejuvenating forests inevitably impacts upon local communities for better or for worse. Creating safeguards and opportunities for employment and livelihoods, for example, creating ranger jobs or financing the building schools from the credits sold, are referred to as co-benefits. But the researchers 'saw many instances where safeguard policies were overlooked, or only weakly carried out, yet projects were still positively verified'. Rating community engagement risk at zero from the first monitoring period was the case in 94% of the projects reviewed, 'yet our review found the project-level grievance mechanisms to be non-transparent and rarely utilised'. Although Verra and other standards bodies have recognised the rights of humans (and animals), auditors 'rarely have the time, nor the expertise, to determine whether social safeguards meet Verra's standard and relevant international rights standards'.
6. **Overall Conclusion**: The researchers 'found evidence of widespread over-crediting across all four quantification factors covered in this report'.

In his blog, green entrepreneur Paul Boeffard regards the report as *'la coup de grace'* [67] and given that 'the value of carbon credits relies heavily on trust and reputation… maintaining credibility is paramount for their effectiveness'. This implies 'meeting established standards is essential'. However, the issue becomes how to use those standards to good effect. While the ratings agency Calyx Global, cited in the blog, also found that nearly two-thirds of the 70 REDD + projects it assessed had a 'high risk of an overestimated baseline' in a later blog the agency argues that the main issue is *whether the project is additional or not* and not the measurement error as such. Estimates of emissions can never be exact, for example they can vary according to tree type [37], but should be conservative [68], a position taken by the Integrity Council of the VCM (ICVCM) [69]. There is just too much latitude given to project management in the use of standards, and no regulatory body empowered to enforce them.

Postscript 2024

In July 2024, a paper appeared in the *Global Environmental Change* journal by West, et al. [70], that again reviewed the four, now upgraded, methodologies of Verra, still finding them wanting, but mostly in terms of their failure to incorporate factors that can cause changes in the baseline assessments upon which subsequent estimates of carbon savings and of carbon credits rely. This is 'the most fundamental challenge of rigorous performance assessment by unwarrantedly disregarding the influence of confounding factors on project outcomes'. Their conclusion reiterates the need for (i) conservative offset estimates, (ii) the elimination of conflicts of interest, and (iii) 'most importantly … rigorous ex-post methods to credibly measure performance and additionality'.

If there is a problem with the conclusions of West, et al. in Box 5.2 it is they are largely beyond the current scope of the standards bodies such as Verra. Removing 'conflicts of interest' would require agents to act independently of the principals who pay their fees. Further, there are no obligations upon independent parties involving MMRV, which means however much Verra or other standards bodies encourage MMRV arrangements, for example, recurrent checking of baseline locations to eliminate or record 'confounding factors' that would change the carbon emission assessments, project managers can ignore the advice and simply not pay for ongoing MMRV, just so long as there are buyers of their credits.

REDD+ and Nature-Based Projects

Around 25% of all voluntary carbon market credits arise from REDD + projects [42], which makes them central to the future scale, and therefore liquidity, efficacy and reputation of VCMs, not only in the removal of carbon, but also in the promotion of co-benefits to local communities that depend upon forests for their livelihoods and well-being.

The frameworks within which RED + projects work securely are set out in the seven principles of the *Cancun Safeguards* (2010) [71] and the *Warsaw Framework for REDD* (2013) [72], but to undertake MMRV on a REDD + project is particularly challenging. First, a baseline is required from which to estimate the tonnage of carbon removed from the atmosphere over an extended period. This implies the baseline location will not disappear under a highway, be turned into an oil palm plantation or land for grazing cattle. It also implies the 'parameters' such as temperatures and the danger of fires, humidity levels, soil types and tree varieties are comparable to the REDD + location, which is only likely true if the baseline location is not too far away. Even local politics is important, as shown in the case of Brazil where a change of government changed policies towards deforestation [73]. Second, the complexities of measurement are endless. Frequently, satellite images are used to measure the prevailing condition of vast areas of forest, and not only can climatic conditions change dramatically, such as devastation by typhoons, but even the precision of satellite imagery changes as the technology improves, from images of 15 square metre pixels to 30 square centimetre pixels of the Earth's surface [74], revealing new data—see the reference to NASA in Box 5.2. Precise measurements in REDD + projects are therefore more often theoretical than achievable. Beyond measurement comes monitoring, and over decades data can be lost or not updated, and maintaining vigilant attention to the details as assessment personnel are replaced is an exercise that will certainly need the help of AI as well as active community engagement on the ground.

Nature-Based Removals

Ninety-nine percent of actual carbon removal by projects has ‘occurred via nature-based removal pathways, sometimes called conventional removal pathways. Conventional carbon removal is overwhelmingly driven by forest restoration and improved forest management; other nature-based removal pathways, referred to as “novel” (see Box 5.3) include wetland restoration and soil carbon sequestration’ [75].

Box 5.3 Conventional and Novel Methods of Carbon Dioxide Removal (CDR)

By 2024, around 2GtCO_2 per year of carbon dioxide removal was taking place, overwhelmingly using conventional methods widely reported by countries as part of *land use, land-use change and forestry* (LULUCF). Novel methods, very much in their early stages of development, including ocean fertilisation, ocean alkalinity enhancement, direct air carbon capture and storage (DACCS), bioenergy with carbon capture and storage (BECCS), biochar and enhanced rock weathering, were responsible for no more than 1.3 million tonnes (0.0013 GtCO_2) per year. Novel methods of CDR (Carbon Dioxide Removal) tend to be costly and difficult to scale, but supported by science. See Pura.earth below.

Novel Methods

Oceans: Alkaline seas absorb acidic carbon concentrations and store carbon [76]. Feeding seawater macroalgae (such as seaweed) and microalgae with nutrients (such as nitrogen phosphate) adds alkalinity, as does adding limestone (bicarbonate) and silicates. Protecting mangroves and accelerating sand rock weathering (see enhanced rock weathering below) add to carbon removals [77].

DACCS: Sorbents either *absorb* carbon from the atmosphere, or act as *adsorbents* to which carbon molecules attach themselves. They are then either sequestered in geological storage sites or recycled for use as carbon-based fuels or in other chemical compounds [78]. While DACCS equipment requires much greater surface areas than conventional CCUS methods, and is 3 times more energy intensive, unlike CCUS, it does not need to be location-specific [79].

BECCS: A biomass, such as a mix of wood and compost, needs converting into bioenergy as a renewable source as electricity, liquid or gas fuels, and heat stored in bricks or stone, or stored as organic matter in soil, landfills, etc. [80].

Biochar and SOM: Biochar uses pyrolysis (extreme temperatures to degrade carboniferous substances) to turn carbon ashes containing heavy metals into solid charcoal-like cakes that help soil fertility through soil nutrients, aeration and water filtration [81]. Preserving or adding to soil organic materials (SOM) [82] helps soil act as a sequester.

Enhanced Rock Weathering: Weathered rock releases elements such as calcium and magnesium, while carbon in the air reacts with rain (carbonic acid) and sea water (bicarbonates), but they can be combined to form new carbon-capture rocks like calcium carbonate (limestone) [83].

Putting CDR into Context:

1. While CDR is a feature of all the Paris Agreement 1.5 °C scenarios, most do not explicitly consider social and environmental sustainability issues such as the co-benefits to local communities.
2. Further, in all scenarios below 2 °C, over the 30 years from 2020 to 2050 a total CDR of 260Gt CO_2 is required, that is 5.7Gt CO_2 per year. Given the low rate of CDR to date, that implies a need to remove around 10Gt CO_2 annually over the remaining 26 years to 2050.
3. To qualify as CDR, the following three principles need to be observed

   (i) The CO_2 captured must come from the atmosphere, not from fossil sources.
   (ii) The storage must be durable, such that CO_2 is not soon reintroduced to the atmosphere.
   (iii) The removal must be a result of human intervention, additional to the Earth's natural processes.

Source The State of Carbon Dioxide Removal: 2nd Edition 2024 [27]

Following the adoption by the EU of its Certification Framework on permanent carbon removals and storage, the EUI (European University Institute) issued *a Policy Brief* which helpfully reviews the policies towards carbon reductions vs. carbon removals across the EU, the UK, China, India, Indonesia, and Japan [84]. This follows the *Paris Agreement Crediting Mechanism* agreed at COP-29 discussed in Chap. 4 under Article 6 of the Paris Agreement.

Nature-Based Solutions (NbS) and Natural Capital

Preserving the biodiversity of nature is seen as critical to the slowing down of climate change through the advocacy of Nature-based Solutions (NbS) supported by bodies such as the World Economic Forum (WEF) [85] which estimates 'these resources contribute more than USD125 trillion to the global economy every year'. More broadly than REDD + solutions, Nature-based Solutions (NbS) are those associated with what the UNFCCC names as 'land use, land-use-change and forestry' (LULUCF) [86] or the IPCC names as 'Agriculture, Forestry and Other Land Use' (AFOLU) [87]. Examples were given in Chap. 4 of the UK *Woodland Carbon Code* and the UK *Peatland Code* as aspects of the UK's *Nature Markets Framework* [88].

The guiding principles for the conservation of biodiversity and the protection of local communities dependent upon it were laid out in the UN's *Kunming-Montreal Biodiversity Framework* (GBF) at COP15 in Montreal, Canada, 2022 [89]. The GBF represents the completion of the work of the Task Force on Climate-related Financial Disclosures (TCFD) whose work in monitoring progress in companies financially supporting biodiversity projects has passed to the International Sustainability Standards Board (ISSB) [90]—see Chap. 6.

In this context, natural resources are referred to as natural 'capital', where nature is defined by The Nature Conservancy [91] as 'water, soil, air, and living organisms from which humans derive a wide range of goods and services'. Natural capital is a logical conceptualisation in two senses. Just as 'human capital' can be a generic reference to the inherited or learned capacity of individuals *to* produce commodities for sale in a market economy, so land and seas are regarded as a natural capital as a source *of* produce. In another sense, following the work of the eighteenth-century economist, Adam Smith [92], land can become capital whenever a fence is put around it enabling it to become either private or state property. At this point, land generates value for its owner relative to its fertility and/or its location close to a market, the closer the better. In a global marketplace, nature is regarded as potential capital primarily in this sense, which disadvantages indigenous people if they cannot establish property rights. Investment in natural capital gives rise to its financial value. However, economists, taking Adam Smith's hint into account, distinguish between economic value and financial value, where only the former includes estimates of externalities, namely the costs and benefits of investment in nature-based projects to local communities and to biodiversity. These are encapsulated in the *Cancun Safeguards* (see above).

The principal objective of NbS is to encourage private investment in nature-based projects designed to remove CO_2. The urgency was highlighted in 2023 when the UN Environmental Programme (UNEP) reported that 'private finance flows that have a direct negative impact on nature are USD5 trillion, which is 140 times larger than private investments into NbS' [93]. The UNFCCC (2023) believes that NbS have the potential to supply over one-third of cost-effective climate change mitigation needs by 2030, which on the face of it seems overly optimistic, not least because it also notes that investment opportunities in natural capital 'are highly country-specific -

investment in NbS requires a certain level of risk tolerance and adaptability from investors to avoid the risk of stranded assets' [94].

Natural Capital Accounting and the Agro-Business!

In 2022, the US government for the first time added natural capital accounting to its estimates of national economic value [95]. It may be noted that accounting for carbon reduction in agriculture consists of current or operating account and capital account entries: a carbon intensity score (CSI) refers to carbon emissions from farming measured as an annual current account issue, while soil carbon stock sequestration (SCSS) of soil organic matter (SOM) measures carbon locked away for years in the soil or in biomass as an entry on the long-term capital account [96]. At the level of investment, the WEF provides an illustration of natural capital accounting, adapted in Table 5.2. Imagine a commercial agricultural venture with 100 hectares of land, beginning with 80 units of 'healthy' land and 20 units of 'degraded' land. After one year of production which is assumed to degrade land and/or reduce biodiversity and its value, and counter-efforts to restore land, the results may be shown in an annual account.

As WEF points out, these accounts will not usually appear in company accounts. While accounting for nature-based projects is likely to bring them more into public focus (see Box 5.2), accounting approaches are often inconsistent in their handling of measurements and monitoring. However, as with everything concerned with natural resources, there is as much aggravation as there is agriculture. A story in the *New York Times* by journalist Lydia DePhillis in 2024 is a good illustration [98]. In 2017, green entrepreneur Douglas Eger, Founder and CEO of the Intrinsic Exchange Group [99], proposed creating natural asset companies (NACs) by which investors would licence land opportunities from farmers or landowners, including an 'equitable sharing benefits' clause to be built into the licences, and if registered on the stock exchange, the proceeds from the initial public offering (IPO) would go towards the investment. The

Table 5.2 Land accounts

Land account (100 hectares)	'Healthy' land (asset)	'Degraded' land (liability)
1st January (year 1)	80	20
Impact of production	−20	+20
Impact of restoration	+10	−10
31st December (year 1)	=70	=30
Land valuation = healthy land $15,000 per hectare; degraded land $5,000 per hectare		
Annual net change (Ha)	−10	+10
Annual net change ($)	−$15,000	+$5,000

Source adapted from WEF (2024)[97]

pilot project to be listed in the New York Stock Exchange involved land in Costa Rica supported by the Inter-American Development Bank. The proposal was withdrawn after a flurry of opposition, from both the Right and the Left. The Right opposition was headed up by Texas oil and mining interests who saw in the proposal a dangerous trojan horse from which environmentalists might challenge their fossil fuel ventures. The Left saw it as a trojan horse for corporate interests to override local community interests. It is reported that Intrinsic is going the private market route. One possible lesson is that schemes involving such projects are best positioned if they can first build proof-of-concept models from the ground up to demonstrate both their additionality and durability, and their co-benefits. However, such PoCs are more likely to appeal to environmentalists on the Left than to the Right.

Registries and Credits

In the voluntary sector, standards are set by private usually-not-for-profit organisations or registered charities. They each have their own approach to a set of widely-shared principles by which to judge the veracity of projects where the owners wish to issue carbon credits as a means of financing the project. They do differ in how they apply their standards, for example, in what they exclude. Excluded in some cases are non-domestic REDD and REDD + projects, which often lack rigorous MMRV, or renewables which are no longer regarded as *additional* in terms of the financial sources available to them.

The projects may be certified, often by a specialist third party as for example in cases where the project claims to offer co-benefits to local communities, and added to a registry once a validation/verification body (VVB) has recommended approval to the standards body. A distinction needs to be made between certified and verified carbon offsets. Certified refers to the carbon certificates issued under the UN *Clean Development Mechanism* (CDM) and called Certified Emissions Reductions (CERs). Those produced under standards set in the voluntary carbon markets are called Verified Emissions Reductions (VERs) [100].

A registry offers potential buyers of credits a view of the projects available and possibly a list of prices. Their purchase of original credits or existing credits most likely involves a trading agency, is done through a broker and is over-the-counter; however, there is a growing list of exchanges available for these transactions. It is important to note that credits need to be retired/cancelled from the registry once the project or the lifespan of the credit has been completed to avoid double-counting and double-usage. While exchanges have an opportunity to impose requirements, such as transparency, they too are private organisations in global competition with each other. Given the many criticisms of a failure of standards implementation in the voluntary sector (see Box 5.2) in recent years they have been at pains to revise and improve the safeguards.

An excellent coverage of standards bodies and their differences, such as whether their focus is towards nature-based or industry-based projects, or are emissions-focused or environmentally-focused, or are geographically focused upon markets in the Global North or Global South, is a report by Simon Puleston Jones published 2023 *An introduction to carbon markets* [101], which provides a guide to the structure of carbon markets. The report also draws attention to the adoption by carbon markets of digital technologies as some venture into tokenisation using a blockchain distributed technology model. While digital tokens can be traded [102], in some cases they are issued to represent an underlying non-fungible (non-tradeable) asset (the project itself or a project outcome) as an NFT (Non-Fungible Token) [103]. Examples include Moss's MCO_2 NFTs [104] linked to land in the Amazon rainforest, Toucan [105] which offers tokens in biochar markets, and Japan's KlimaDAO [106] offering a digital carbon dashboard. A parallel development is the opening of stock markets to carbon credit trading and of registry regulators such as the US *Commodities and Futures Trading Commission* (CFTC) [107] to combat fraud. However, the credits or tokens are not tradeable with each other. Credits are maintained and ultimately cancelled (or 'burned' if tokens) in separate proprietary registries. 'Fragmented' is how the report describes the VCM.

Standards Upgrades

If the lack of rigorously implemented standards held back the voluntary markets up to 2023, things began to turn when four leading standards bodies, the GHG Protocol, the Science-Based Targets Initiative (SBTi), the Integrity Council for the Voluntary Carbon Market (ICVCM) and Verra, 'joined forces to operationalize a high-integrity market to accelerate global climate action following an agreement at COP28' [108]. ICVCM and Verra act at opposite ends of the market, ICVCM provides integrity guidelines and standards for projects, and Verra does the same for the buyers of the credits created by those projects. By mid-2024 it was announced the standards of Verra and ART TREES had met ICVCM benchmarks [109]. ART TREES (*Architecture for REDD + Environmental Excellence Standard*) provides a standard for the monitoring, reporting and verification of emission reductions and removals from the forest sector.

GHG Protocol

The GHG Protocol, established jointly by the World Resources Institute (WRI) supported by corporates such as BP and General Motors, and the UN-inspired World Business Council for Sustainable Development (WBCSD) [110], produces a *Corporate Accounting and Reporting Standard* [111], for enterprises aiming at carbon abatement in their establishments [112].

Multinational companies such as the international retailer H&M use the standard, notably for supply chain Scope 3 emissions [113], but there are problems. Clean energy producers are allowed energy credits as 'renewable energy certificates' (RECs) which can be purchased as offsets under the GHG standard. The major buyers are High Tech companies, Amazon most of all. While this means that a company can pollute and offset, the argument is that as they reduce their GHG emissions over time these companies will no longer need to offset. But that's the problem. The Earth does not have time before it overheats, especially as the pace of data centre building requiring ever growing amounts of electricity driven by the AI revolution, outstrips the market for offsets. Examples of this are cited by a *Financial Times* (2024) [114] investigation that notes, by using offsets, Amazon claims it hit its 100% renewable energy targets in 2023, seven years ahead of schedule, yet its annual emissions of GHG, although down from the previous year, in 2023 were almost 69 million Mt.CO_2 [115]. Meta claimed having hit Net Zero in 2020, and in 2023 claimed emissions of just 273 tonnes CO_2, but according to the *Financial Times* it was closer to 4Mt.CO_2.

Examples could continue, but an important point is that under GHG standards, RECs do not have to relate to the grid which supplies these companies with their power, meaning MMRV is more remote and likely less rigorous, and given that prices of RECs are so low, under USD5 per tonne CO_2 since at least 2022, the financial gains (costs per tonne of CO_2 offset) far outstrip the social costs (environmental and/or community impact) of pollution. Because social costs are external to markets, without a carbon tax this is almost unavoidable. On a global scale, the funds raised through RECs may conceivably, but unlikely, cover all the costs of converting energy generation to renewable energy, so theoretically there could be a Net Zero effect worldwide, but researchers at Princeton, Harvard and the Green House Gas Management Institute, also reported by the *Financial Times* (2024), found 'buying certificates typically did not drive either a new supply of renewables or a fall in emissions'. The GHG standard has since been under review, including the global vs. local issue.

SBTi

About—Registered as a charity in the UK, the Science Based Targets initiative (SBTi) is supported by the World Wildlife Fund (WWF), the Carbon Disclosure Project (CDP), UN Global Compact, and the We Mean Business Coalition. SBTi advises companies how to reach Net Zero using *Science-Based Targets* (SBTs) [116], but due to inconsistences in the 'definition of net-zero itself, as well as the path to get there' in 2021 SBTi introduced the *Net-Zero Standard* [117], with reference to the findings of the IPCC.

Focus—The corporate sector is designed to 'halve emissions before 2030 and achieve net-zero before 2050' [117]. By 2024 SBTi claimed more than 4,000 members, with over 50% having set climate targets of either 1.5 °C or below 2 °C using the SBTi's

Science-Based Targets (SBTs) [118]. However, to put into perspective, a report by the *Environmental Resources Trust* (see ACR below) using data from Allied Offsets found over 80% of the world's largest companies lacked net zero targets [119].

Standards—Using the SBTi's *Net-Zero Standard* [116], companies design their own pathways to Net-Zero [120]. SBTi offers a Scope 1 validation process [121], in 2023 stating it followed a 'mitigation hierarchy in which companies' priority is to reduce their own GHG emissions and cut down emissions from their own organisations and value chains'. This left as an open question how and when would SBTs include Scope 2 and 3 emissions, and equally whether SBTi would give recognition to a range of third-party *Environmental Attribution Certificates* (EACs) [122]. It should also be noted that SBTi does not allow VCM credits to count against its validated net-zero targets [4].

Discussion—In 2024, SBTi pulled back from a proposal to endorse EACs for carbon credits and also from including for the first time Scope 3 emissions [117], for which keeping detailed year-to-year updated carbon accounts is obviously challenging. This is especially true for smaller businesses, and where supply chains are global, complex and often disrupted. It should be noted that in its online revision, SBTi had found that out of an initial list of over 10,000 peer-reviewed papers, only five compared the efficacy of carbon credits to direct emissions abatement. In SBTi's opinion this was insufficient to draw robust conclusions [123]. However, as reported in the *Financial Times* (2024), the withdrawal of the Scope 3 and EAC proposals were basically in response to 80 civil organisations having spoken out against the endorsement of carbon offsets, alongside a staff revolt against what they saw as opening the door to 'fossil fuel financiers'. The CEO's resignation followed shortly afterwards. Once again, a chasm opened between direct emission abatement strategies and carbon credit finance strategies. An outburst of criticism followed, notably from market makers on *Linkedin* and including a paper by academics from the Australian National University (ANU) [124], suggesting SBTi should adopt a less technical and more flexible, case-by-case approach. But principles aside, how 'flexible' is 'flexible' is the real issue.

A contribution to the SBTi's consultation paper on its Net Zero Standard 2.0 [125] from Robert Höglund of Marginal Carbon [126] is worth mentioning. Anaysing a database of 250 leading companies, he notes that around 90% of emissions arise from companies that generate only 25% of the profits, mostly industrial and transport companies. He concludes that the more profitable sectors, such as banks and commercial companies, could, by using an SBTi-approved 'internal carbon' price help pay for carbon reductions across Scope 3 supply chains, with lower carbon prices for the less profitable companies. Whereas emissions are most intense among Scopes 1 and 2, so are the steps being taken to reduce carbon, leaving Scope 3 out in the cold. He sees this as a scaling problem that could trigger CDR technologies across supply chains, and this could also include Beyond Value Chain Mitigation (BVCM), meaning companies could extend their carbon-reduction activities to third party supply chains.

ICVCM

About—The Integrity Council was established in response to the recommendations of the *Final Report of the Taskforce on Scaling the Voluntary Carbon Markets* (TSVCM) [127], initiated by the ex-Governor of the Bank of England, Mark Carney, UN Special Envoy for Climate Action and Finance, for COP26 in Glasgow, Scotland (2021) and most recently Prime Minister of Canada.

Focus—The TSVCM stressed the role of carbon credits (mandatory or voluntary) as ways to finance reduced GHG emissions, and to signal the validation of credit integrity according to the ICVCM's endorsement using its *Core Carbon Principles* (CCPs) consisting of 10 core principles. These are listed under three headings: governance, emissions impact and sustainable development [128].

Standards—In 2024 the ICVCM published its CCP eligibility decisions, including the exclusion of renewable energy projects on the grounds that the criteria of the UNFCCC's *Clean Development Mechanism* (CDM) [129] were insufficiently rigorous, but with an invitation to standards bodies such as Verra to update their own criteria [130].

Discussion—As its name implies, the ICVCM is the apex standards-setter for the VCMs [131]. During 2024 it validated various standards of ART TREES (Architecture for REDD + Environmental Excellence Standard), Gold Standard, ACR and CAR (Climate Action Reserve) and Verra [132]—see below for each.

Verra

About—Founded in the USA in 2007 by environmental and business groups, Verra's standards are the most widely used in the voluntary carbon market with an estimated 70% market share by 2024, the year it received validation from the ICVCM. Verra's founding was supported by the UK-based Climate Group [133], which acts as the secretariate of the Under2Coalition [134], the largest global alliance of state and regional governments committed to carbon Net Zero by 2050. Other supporting organisations were the International Emissions Trading Association (IETA) [135] founded in 1999 by business groups to pursue the objectives of the Kyoto Protocols, and the World Economic Forum (WEF), who were responsible for the original *Voluntary Carbon Standard* in 2005, later renamed the *Verified Carbon Standard* [136].

Focus—Verra builds 'standards for activities as diverse as reducing deforestation, to improving agricultural practices, to addressing plastic waste, and to achieving gender equality' [137].

Standards—Verra's *Verified Carbon Standard* (VCS) programme issues carbon credits named *Verified Carbon Units* (VCUs) after a project has been validated and verified, but the credits are not normally available to individuals due to the levels of investment risk in the underlying projects. Following the controversies outlined in Box 5.2, Verra issued a *VCS revision* in mid-2024 [138]. Verra's methodologies, modules and tools applied to project evaluation are denoted by the prefixes VM, VMD, or VT, such as **VM0007** for the *REDD + Methodology Framework*, **VMD0009** for an *Estimation of emissions from activity shifting for avoided planned deforestation*, **VT0002** as a *Tool for the Demonstration and Assessment of Additionality in IFM Project Activities* [139]. Besides climate change, other Verra standards address community, smallholder, biodiversity and wider environmental concerns. These include its *Climate, Community & Biodiversity Standards* (CCBS) [140], a *Sustainable Development Verified Impact Standard* (SD VISta) [141], a *Plastic Waste Reduction Standard* [142], and a sovereign carbon *Jurisdictional and Nested REDD +* (JNR) framework for harmonising with national and local government-led policies [143]. All programmes incorporate the International Standards Organisation's *ISO-14064–2/3* for 'conducting a verification of a GHG inventory or a reduction process' [144].

Discussion—Verra, unlike the Gold Standard [145], does not verify non-carbon-benefits of projects, but it can add such benefits to its registry if they are certified by other standards bodies, such as the protocols of the UNFCCC's Clean Development Mechanism (CDM) and the Climate Action Reserve (CAR) [146]. Equally, it accepts certifications by independent third parties such as Green-e Climate [147], a marketplace that endorses credits verified by standards bodies such as the American Carbon Registry, the Climate Action Reserve, and the Gold Standard. Following certification, Verra turns to the non-profit solutions provider, the Center for Resource Solutions (CRS) [148], to monitor the marketing of credits to ensure process probity.

Gold Standard and the WWF

About—The Swiss-based Gold Standard for the Global Goals (GS4GG) was founded in 2003 by the World Wildlife Fund (WWF) alongside the SBTi, Carbon Market Watch, and other leading NGOs [149]. The pivotal role of the WWF has been to link climate change and the Paris Agreement to wider environmental concerns [150], including support for the Global Environment Facility (GEF) that helps finance sustainable green development [151]. The WWF's guidance to climate change is *Beyond Net-Zero* [152].

Focus—The Gold Standard is mostly focused on developing markets in the Global South, and sustainable environmental development in support of the UN's Clean Development Mechanism, but it later added the abatement of GHGs to its portfolio with its *Beyond Value Chain Mitigation* (BVCM) pathway where companies can support projects outside of their own value chain.

Standards—The Gold Standard validates and verifies certificates for various sustainable development projects, and supports the issuance of *Verified Emission Reductions* (VERS) as carbon credits arising out of projects that reduce or remove carbon [153]. Certificates are issued for community service projects such as mini-girds, energy-efficient heating, waste management, for renewable energy projects where 'additionality' requirements are required [154], and for land use and forestry projects [155]. The Gold Standard is a member of the International Social & Environmental Accreditation & Labelling Alliance (ISEAL) and the (2024) revised ISEAL *Code of Good Practice for Sustainable Systems* [156].

Discussion—Gold Standard-certified activities must show impact for at least three of the UN Sustainable Development Goals: co-benefits, protection of the environment, and 'gender-sensitive design principles' [157] which have been recognised by the ASEAN Low Carbon Energy Programme [158], and are part of the ISEAL *Code of Good Practice* [159]. The Gold Standard was also a pioneer in providing access to pricing data in the voluntary markets, thereby helping the flow of funds [160]. ISEAL, the WWF, the Rainforest Alliance and others are partners in Evidensia [161], a platform to provide trusted evidence of sustainability.

ACR Standard

About—ACR was founded in 1996 as the American Carbon Registry, the first private voluntary GHG registry, by the environmental non-profit organisation Environmental Resources Trust (ERT). In 2007, ERT and its registry became a subsidiary of Winrock International, a philanthropic development agency of the Rockefeller family [162].

Focus—ACR promotes globally the GHG Protocol, but its registry is exclusively American [163]. It is used by, among others, the California cap-and-trade programme (see Chap. 4) and the International Civil Aviation Organization (ICAO) to supply units to the *Carbon Offsetting and Reduction Scheme for International Aviation* (CORSIA) widely adopted by airlines [164]. ARC, along with others such as Verra, the Gold Standard, CAR, and Plan Vivo, adheres to the *Code of Practice* of the *International Carbon Reduction and Offset Alliance* (ICROA) to promote the integrity of the voluntary market [165].

Standards—ACR issues Emission Reduction Ton (ERTs), as credits named after the Environmental Resources Trust (ERT). Following its publication in 2023, the *ACR Standard (Version 8.0)* is to be updated every three years [166]. The standard emphasises that conservatism of emission estimates 'is embedded into all ACR methodologies'. Projects with co-benefits or 'non-carbon' benefits are not directly certified by ACR but are evaluated before a project is registered; for example, projects must show that co-benefits are not a regulatory requirement, one of ACR's three tests of additionality. The other two are additional to common practice, and going beyond existing financial, technological, or institutional barriers. Buffer stocks, insurance or covenants are required to cover the risk of sequestration project reversals, but one difference with Verra is that ACR allows project managers to buy credits from another project to add to the buffer stock [167].

Discussion—ACR requires adherence to conflict-of-interest policies among all ACR-approved third-party registries and the staff of Validation/Verification Bodies (VVBs). The Standard does not require VVBs to visit every site, but to use 'their own discretion' and at minimum subject all sites to desk-based validation and verification. It further requires project proponents to justify any plans to exclude regular MMRV. As a safeguard, usually ERTs can be issued for a maximum of 10 years, after which new ERTs may be issued. ACR excludes hard-to-check international REDD and REDD + projects due to issues of additionality, permanence, leakage, and the risk of double counting. But other nature-based *Agriculture, Forestry, and Other Land Use* (AFOLU) projects are admitted, those with a project-reversal risk being assigned a minimum project life of 40 years as a safeguard of permanence.

Carbon Action Reserve (CAR)

About—The Climate Action Reserve (CAR) was founded as a non-profit organisation in California, USA in 2001, as the California Climate Action Registry, helping agencies and municipalities to voluntarily calculate and publicly report their GHG emissions, which later translated into expertise in emissions reductions and carbon accounting methods for the North American market [168].

Focus—The American market for standards, issuing carbon credits, third-party verifications and registrations for sustainable environmental and community co-benefits.

Standards—Operates a registry for offsetting Climate Reserve Tons (CRTs).

Discussion—One of ten standards bodies that adhere to the Code of Practice of the International Carbon Reduction and Offset Alliance (ICROA) to promote carbon market integrity—see ICROA. The Annual Reports of the CAR stress its protocols pay special attention to the costs and benefits of fully listed carbon credits [169].

Plan Vivo

About—The Scotland-based Plan Vivo arose in 1994 in collaboration with the University of Edinburg, El Colegio de la Frontera Sur, other local partners, when it helped communities plant trees in Chiapas, Mexico. The project was named Scolel'te and became the world's first voluntary market carbon credits when they were sold to the World Bank [170]. In 2014, Mikoko Pamoja (Kenya) 'became the world's first Blue Carbon project by receiving Plan Vivo Certificates for the conservation of its mangrove forests' and won the UN Equator Prize award [171].

Focus—Standards and guidelines for smallholders and rural communities, mostly in the Global South.

Standards—The *Plan Vivo Standard* (v1.0) was released in 2001, issuing Plan Vivo Certificates. v3.0 was released in 2008 to permit REDD/conservation projects to obtain certification. v5.0 was released in 2022. From 2023, the standard was renamed *Plan Vivo Carbon Standard* (officially known as PV Climate) and a *Plan Vivo Biodiversity Standard* (PV Nature) was launched [172].

Discussion—Plan Vivo has combined academic support with local NGOs and action groups, and as such has become a pioneer in local community engagement and environmental standards, with a strong focus upon the sustainable development goals, such as Goal No.1 to relieve poverty by supporting local community livelihoods and their environment and helping to build their capacity by transferring skills and knowledge.

Puro.Earth

About—Puro.earth is a Finnish start-up founded by two women and a male colleague in 2019 to provide the world's first B2B marketplace for carbon removal projects, a standard and a registry [173]. The stock exchange Nasdaq acquired a majority stake in Pura Earth in 2021 [174].

Focus—The certification of nature-based carbon removal projects, notably using 'novel' methods such as the creation of biochar and rock weathering (see Box 5.3) as a listing of its clients shows [175].

Standards—Puro.earth verifies carbon removals by issuing CO_2 Removal Certificates (CORCs), and showing registered retired CORCS online [176]. CORCs can be purchased through a Pre-Purchase Agreement where the buyer agrees to purchase CORCs from a future issuance at a pre-agreed price.

Discussion—Carbon removal projects are controversial for two reasons. One is the criticism that they might let corporate polluters off the hook by selling them as offsets, and the other is that the technologies and processes involved are not cost-effective. A third could be added that 'novel' methods are especially difficult to operationalize—see Box 5.3. Despite this, the State of Carbon Removal 2nd Edition (2024) [177] sees a pathway to removing 7 to 9Gt CO_2 per year by 2050. Puro.earth is out to prove this is possible.

The Global Carbon Council

About—The Global Carbon Council (GCC) is the Middle East & North Africa (MENA) region's first voluntary carbon-offsetting programme and an initiative of Gulf Organization for Research and Development (GORD) in 2016. GCC aims to contribute to a vision of sustainable and low carbon economy of the region [178].

Focus—To direct climate finance to projects which mitigate climate change and contribute towards the UN Sustainable Development Goals (SDGs). This can help governments in their efforts to meet Nationally Determined Contributions (NDCs) and trade Internationally Transferred Mitigation Outcomes (ITMOs) [179].

Standards—The GCC issues Approved Carbon Credits (ACCs) under its GCC *Project Verification Standard* once projects have been validated and verified, and additionally they may qualify under the GCC's *Environmental and Social Safeguards Standard* for the certification label E + where there is 'Environmental No-net-harm', and S + where there is 'Social No-net-harm.' GCC rules are also aligned with ISO *standards 14,064–2 and 14,064–3*.

Discussion—Unlike Verra, and closer to the Gold Standard which also has a Global South focus, GCC does regard renewable energy projects as 'additional' when it uses the baseline methodology of the UNFCCC's Clean Development Mechanism (CDM) guideline, and has a 'Tool for Demonstration and Assessment of Additionality'. This applies to projects covering solar photovoltaics, on-shore or off-shore wind, tidal, and wave power. But it also applies restrictions, such that when a project involves not a utility power plant (power generation to the grid) but a distributed power plant (DPP) that serves other customers in addition to the grid, the project must show that the grid connection pre-exists the project [180].

ICAO/CORSIA

About—The International Civil Aviation Organization (ICAO) is a United Nations agency which helps 193 countries to cooperate and share their skies to their mutual benefit [181]. In 2016 the UN agreed the *Carbon Offsetting and Reduction Scheme for International Aviation* (CORSIA) [182].

Focus—CORSIA is the first global market-based measure for any sector to move away from a 'patchwork' of national or regional regulatory initiatives to worldwide cooperation [183].

Standards—the CORSIA standards apply to all aircraft, except humanitarian, medical, and firefighting, with a take-off mass of 5,700 kg or above, conducting international flights, producing carbon emissions of at least 10,000 tonnes. Details of the updated 2023 *Standard Volume IV, 2nd Edition* are available online [184].

Discussion—CORSIA offers offsetting from the carbon market of CO_2 emissions that cannot be reduced through the use of technological improvements, operational improvements, and sustainable aviation fuels.

ICROA

About—The International Carbon Reduction & Offset Alliance (ICROA) is UK-based membership association launched in 2008 by eight companies to promote standards, high-quality credits and lobby for the recognition of VCMs. By 2024, ICROA members were the British Columbia Offset Program, Climate Action Reserve (CAR), Emissions Reduction Fund (ERF) of the Australian Government, Global Carbon Council (GCC), Gold Standard, Plan Vivo, UNFCCC Mechanisms, the UK Woodland Carbon Code (UK WCC), and Verra's Verified Carbon Standard (VCS).

Focus—A self-regulating body hosted within the International Emissions Trading Association (IETA), providing credit auditing services as it lobbies for proactive corporate investors to 'be recognised and rewarded' [185].

Standards—Members must be IETA members and are required to undergo an annual independent audit to assure compliance with the ICROA *Code of Best Practice.* Only its members as 'Accredited Organisations' can offer its endorsements of carbon credits as offsets on behalf of a client.

Discussion—ICROA operates strict rules to trading schemes before a member can meet the *Accreditation Eligibility Criteria*, including annual independent audits. Before using the ICROA Accreditation Label members must have been trading for at least a year, have reached a minimum trading volume of 10,000 tonnes of CO_2 equivalent, and shown a science-aligned pathway to the Paris Agreement goals. The strong promotion of codes of conduct by its members may be regarded as an alternative to inviting compliance regulations.

IETA

Agenda—The Switzerland-based International Emissions Trading Association (IETA) founded in 1999 was the first multi-sectoral international 'purely business group' building on the Kyoto Protocols and devoted to pricing and trading greenhouse gas reductions [186]. It has steadily expanded its work to cover the EU's Emissions Trading Scheme (ETS—see Chap. 4) and initiatives in North America, East Asia, New Zealand, and emerging markets in Latin America.

Focus—To establish effective market-based trading systems for GHG emissions and removals that are 'environmentally robust, fair, open, efficient, accountable, and consistent across national boundaries'.

Standards—Hosts ICROA, and founded, in partnership with the World Bank and Singapore, the Climate Action Data (CAD) Trust [187], a decentralised metadata platform 'that links, aggregates and harmonises all major carbon credit registry data to enhance transparent accounting in line with Article 6 of the Paris Agreement' [188]. The platform is open to the public, listing credits by registry, by sector and by country.

Discussion—IETA is highly active in organising events, such as carbon forums, and procedures to promote the VCM, including digital projects such as the Climate Action Data (CAD) Trust (see above), carbon removal projects, and the International Air Transport Association (IATA) Aviation Carbon Exchange (ACE) for trading CORSIA eligible emission units [189].

Voluntary Carbon Markets Integrity Initiative (VCMI)

About—VCMI is a not-for-profit philanthropic organisation announced at COP26 in March 2021, with funding from the Children's Investment Fund Foundation, the Ballmer Group, the Bezos Earth Fund, Google LLC, the Packard Foundation and the UK Department for Business, Energy and Industrial Strategy (BEIS) [190]. VCMI is hosted by the Rockefeller Philanthropy Advisors in the USA whose programmes include climate and environmental projects [191].

Focus—On the supply-side, a VCM *Access Strategy Toolkit* provides guidance for countries to engage in high-integrity VCMs in support of national climate and economic prosperity. On the demand-side, a VCMI *Claims Code of Practice* guides companies towards making use of voluntary carbon credits as part of 'science-aligned' net-zero decarbonization pathways [192].

Standards—VCMI (with ICVCM endorsement) validates and verifies company Carbon Integrity Claims, in line with its *Claims Code of Practice* issuing silver, gold, and platinum certificates, subject to compliance with its *Monitoring, Reporting and Assurance (MRA) Framework* [193]. In 2024 VCMI put the challenge of incorporating Scope 3 supply chain claims into a public consultation document [194].

Discussion—VCMI adds to a concerted effort to ensure the integrity of carbon credits, including an insistence that under the *Claims Code* carbon credits may not be used for offsetting emissions, or for making carbon neutral claims, as an alternative to reducing emissions within corporate value chains. 'Instead, companies can use carbon credits to go "above and beyond" the decarbonisation of activities in their own value chain to 'further contribute' to cutting emissions' [195].

Climate, Community and Biodiversity Alliance (CCBA)

About—The US-based Climate, Community and Biodiversity Alliance (CCBA) was founded in 2003 to 'bring together diverse stakeholders through a transparent and inclusive participatory process to develop standards and tools that stimulate, identify, and promote high quality multiple-benefit land management activities'. Its members consist of CARE, Conservation International, The Nature Conservancy, Rainforest Alliance, and the Wildlife Conservation Society [196].

Focus—Nature-based conservation and sustainable community development as co-benefits by aligning NGOs and local engagement.

Standards—Since 2014, CCB Standards, which evaluate land management projects from the early stages of development through implementation, such as its *Sustainable Landscapes Rating Tool* [197], have been managed by Verra. Besides providing offsets for buyers and project managers, CCB Standards 'identify projects that simultaneously address climate change, support local communities and smallholders, and conserve biodiversity'.

Discussion—CCBA could be seen as bridging a gap between NGO/community work and business alliances such as IETA/ICROA as its list of projects on Verra's registry shows [198]. CCB has, for example, at its disposal the *Carbon Footprint Calculator* available from Conservation International [199].

The Changing Carbon Credit Ecosystem

The Ecosystem Marketplace [200] in 2022, identified 170 different types of carbon credits under the categories: renewable energy, forestry & land Use, household & community, chemical/industrial, energy efficiency, waste disposal, agriculture, transport—the relevant diagram is reproduced in *An introduction to carbon markets* [6]. By 2024, the MSCI platform listed over 15,000 projects spanning more than 15 registries, offering quality assessments for over 5,000 of them [201], yet these are but a fraction of the projects and registries offering credits in the carbon markets globally. According to data analysed by MSCI, the USD2 billion voluntary market offset around 300Mt CO_2 annually, of which only 27Mt CO_2 was verified according to ICVCM standards, that is less than 10% [202]. Despite the small scale of VCMs, there is a proliferation of standards and certification bodies and registries, such as • Bureau Veritas [203] (USA with a focus on certifying industrial carbon reductions) • South Pole [204] (Singapore with a focus upon new business opportunities) • Nori [205] (USA with a focus on farms) • Carbonibus [206] (Liberia-based auction house for carbon offset projects focused on households and communities) • Climate Impact X (CIX—a Singapore-based 'public–private initiative to establish the country as a climate services and carbon trading hub' [207].

More recently, multinational companies which run high-energy-consuming data centres and face problems finding a sufficient volume of high-quality offsets are themselves entering the market for standards and registries. Salesforce (2022) and Amazon (2023) have been the first. Salesforce opened a Net Zero Marketplace [208], with ten partners to connect buyers with 'ecopreneurs' in an open carbon credit platform [209]. Amazon, previously the largest client of Verra, has created the Abacus marketplace [210], and while continuing to rely upon Verra for issuing VCUs, will independently seek projects that more strictly meet its criteria through its *Amazon Sustainability Data Initiative* (ASDI) [211], using an open data registry [212]. To assist it has developed a *Customer Carbon Footprint Tool* [213].

Beyond standards bodies and registers, the ecosystem includes *inter alia*: project managers and their agents, most of whom are small scale and REDD + focused [214]; Validation and Verification Bodies (VVB)—see indicative lists at Verra [215], and the Gold Standard, which leaves the responsibility for MMRV up to the buyers [216]; legal, insurance and financial service agents; brokers, dealers and ratings agencies such as BeZero [217], Calyx Global [218], Syvera [219]; specialised media and databases such as a 'Jobs in Carbon' dashboard [220], Allied Offsets [221], the World Bank's carbon-pricing dashboard [222], Platt S&P Global carbon markets [223], and sector specialists such as Carbon Minds [224] which follows the chemical industry; those with an R&D focus such as Mission 2.0 [225], a global initiative of 23 countries; and additional to the potential role of regulators and local authorities, are stakeholders such as local communities and business associations. Between them they promote the trading of credits, traditionally private non-regulated 'over-the-counter' (OTC) transactions, but stock markets, such as New York-based Xpansiv CBL [226], Singapore's AirCarbon Exchange [227], and the Chicago Mercantile Exchange [228], are increasingly playing a role in three ways. First, for publicly-traded stocks they require ESG (Environmental, Social, and Governance) statements attached to annual reports, partly in support of the Paris Agreement and NDCs, and partly as investor information. Second, they are starting to open their own doors to trade in carbon credits. Third, alongside several private marketplaces (as discussed above) they are turning to distributed digital technology (blockchain) to register trades and to create tradable tokens which are derivatives of the underlying asset, namely a carbon credit [229]. And just as credits must at some stage be retired to avoid double-counting, tokens too need to be 'burned'. A thorough account of the growing role of tokens, their uncertain legal status when serving as Non-Fungible Tokens (NFTs), and a list of key participants in the ecosystem, including exchanges, can be found in the previously mentioned *An introduction to carbon markets* [6].

Conclusions

Compared with the mandatory carbon markets (MCM) which cover most of a nation's heavy polluting industries, the voluntary carbon markets (VCM) account for a tiny fraction of global emissions. The best estimates for the VCMs for 2023 are 41Mt CO_2

covered by the removal of credits [230], that is 0.4% of the total annual requirement by 2030 estimated by the IPCC [231]. The carbon credits price per tonne for avoidance and reduction flatlined below USD4 for a long time before rising marginally to below USD6 by December 2024. Carbon removal credits favoured by fossil fuel companies rose to USD11 per tonne by December 2024 [232]. The bad reputation of voluntary carbon credits (see Box 5.2) has suppressed demand even as supply has increased, and in turn this has pushed the standards bodies for explore ways (through consultations) to raise the integrity of their standards. One standards-body, Compensate in 2020 found 90% of the projects it looked at did not meet its integrity criteria [233]. Integrity implies closing the more obvious loopholes in the standards, such as a greater insistence upon monitoring and measuring, reporting and validating (MMRV) project approvals. One measure that might ensure greater integrity is the listing of projects on stock exchanges, a step that was approved along with guidelines by the US Commodities Futures Trading Commission (CFTC) in 2024 [234], but that seems like a lot of work given only 3 such listings up to the 1Q 2025, and the CFTC can only give advice. And undermining the process, in March 2025 the Securities Exchange Commission (SEC) revoked its requirement for companies to report their GHGs [235].

But there remains an insuperable problem, economists call the principal-agent dilemma [38]. This arises because the agents of the verification and validation bodies (VVBs) are mostly chosen and employed directly by the principals who own and manage the projects. In other words, there is always a potential conflict of interest between sellers and buyers potentially giving rise to ‘moral hazard’ which falsely presents the project as conforming to the standards. Only around 10% of all carbon credit projects are validated. Around 40% of all VCM projects involve forestry [236], and at stake are the assurances that the baseline benchmarks (similar forested areas) against which projects are compared are genuine and sustainable, and the projects themselves cover additionality, leakage, permanence, and co-benefits to the local communities such as workers and residents. So much is dependent upon whether the driving force is a genuine commitment to the preservation of nature, or simply making money as a seller, or buyers simply offsetting their own emissions at a lower cost than avoiding them altogether.

Fragmentation

Despite the efforts of standards bodies to better ensure the integrity of the projects they validate, the results have created an even more fragmented market as each takes a slightly different approach-see the ‘discussion’ sections above. One of the unresolved and controversial issues is whether to support or not carbon credits offsetting. This split opinions within the SBTi, for example. While most VCM participants support offsetting, the VCMI excludes offsets, as does UN advice [237], and the EU ETS, but not the UK ETS—see Chap. 4. The ACR takes a stronger approach to avoid conflicts of interest when registering VVBs, as this has been a problem especially for hard-to-monitor non-local REDD + projects. The Global Carbon Fund along with the Gold Standard takes a different approach to Verra when establishing project baselines. Verra outsources assessments of co-benefits whereas both Plan Vivo and the Gold Standard emphasise alignment with the UN SDGs, covering co-benefits, environmental protection, and gender-sensitivity. Puro-Earth promotes CDR projects. All of the standard bodies are grappling with whether and how to extend credits to Scope 3 emissions abatement along extended supply chains, and even to third party Beyond Value Chain Mitigation (BVCM) as these are difficult to measure and go largely untouched; but some of the larger business chains such as H&M apply the GHG Protocol directly to their supply chains. Finally, each of the standards bodies issues there own separate certificates which mostly cannot be directly compared and therefore traded with each other.

These examples illustrate the diversity of approaches and standards criteria. Some in the industry look towards a creeping overlap between the regulations that govern the compliance markets and the unregulated voluntary sector, including a growing interest in government supporting projects in developing counties using Article 6 ITOS (Internationally Transferred Mitigation Outcomes) of the Paris Agreement to count towards their own NDCs [238]. ICROA offers accreditation to its members in the absence of a regulatory approach. By-passing VCMs altogether are NGOs like Conservation International [239] and the Rainforest Alliance [240] who appeal to donors to support their work directly with local communities alongside supply-side and demand-side supply-chains to marry conservation and sustainability with market solutions.

Besides reputational problems associated with the VCMs, the politically-motivated backlash against 'woke' and ESG has driven many corporate companies to retreat from ESG commitments, at least publicly, even though ESG statements are now a requirement of most stock markets. It is what one NGO has termed 'green-hushing [241].' Among larger corporate buyers there has also been a move away from third party carbon credits, for example, by major IT corporates such as Google [242], Meta, Microsoft and Salesforce as they focus more upon direct investments in nature conservation and direct carbon capture projects which are more easily measured and monitored. In May 2024 they jointly formed a consortium named the *Symbiosis Consortium* [243] to pursue an 'advanced market commitment' (AMC) for higher-quality projects [244]. This comes against a background of growing emissions from their data centres, and the formation of an *iMasons Climate Accord* among data centre operators and equipment suppliers [245], as investment surges in power-hungry AI development [246]. With growing public attention to their emissions, these corporates need better assurance that their investments in carbon avoidance, reduction, and removal are genuine..

The Future for the VCM

The twin core problems for the VCMs are credibility and relevance. MMRV is vital for their credibility and lies at the heart of authentic project development, and therefore voluntary market investment. In practical terms this means the search for generally-accepted principles and practices for MMRV is important, but equally so is an understanding that for many projects their circumstances are unique, their nature, their scale, their location, their management, etc. Ideally, this calls for a ground-up approach whenever projects do not fit neatly into an established common category with well-defined parameters.

But in most cases the ground-up approach is likely to be impractical, because it is labour-intensive, travel intensive, and dependent upon the right skill sets being available at a reasonable cost. That leaves VCMs with few alternatives but to accept higher investment risks and enduring uncertainties. But there may be alternatives, such as the proposal from 100% Together Cooperative [247] to create cooperation between investors, or what might be referred to as coalitions, to fund mutual initiatives. The proposal as it stands refers specifically to Scope 3 emissions, but the concept of such coalitions of stakeholders could be extended to a variety of projects, achieving a scale of investment and funding for MMRV that might not otherwise be available. Training local communities as stakeholders to manage MMRV could also be a way to internalise an important externality.

With regard to relevance, it is worth repeating that at current rates, by 2030 the voluntary market will account for only 0.4% of the total annual carbon removals specified as needed by the IPCC to be on target for the Paris goals. Using the same metrics, at current levels all carbon reduction and removal projects combined (including CCUS) remove less than 1% of carbon and carbon-equivalent emissions annually. Only a switch to carbon avoidance technologies and processes where it is possible, and otherwise to carbon reduction among the major polluters, will make significant dents in the rate of emissions, and where these are not voluntary they ought to become mandatory.

References

1. Carbon-direct.com (2024) Types of carbon credits: Reduction, Removal, and Avoidance https://www.carbon-direct.com/insights/how-do-carbon-credits-actually-work-removal-reduction-and-avoidance-credits-explained
2. Climateactiontracker.org Net zero targets https://climateactiontracker.org/methodology/net-zero-targets/
3. Carbon Credits (August 2024) Investor's Education Center https://carboncredits.com/what-is-the-voluntary-carbon-market/
4. Ecosystem Marketplace (2024) State of the Voluntary Carbon Market https://3298623.fs1.hubspotusercontent-na1.net/hubfs/3298623/SOVCM%202024/State_of_the_Voluntary_Carbon_Markets_20240529%201.pdf

5. Simon Puleston Jones(2023) An introduction to carbon markets Climate Solution and Simons & Simons LLP https://www.stroock.com/uploads/An-introduction-to-carbon-markets.pdf
6. Simon Puleston Jones (2023) An introduction to carbon markets Climate Solution and Simons & Simons LLP https://www.stroock.com/uploads/An-introduction-to-carbon-markets.pdf
7. Carbon Direct (2024) Types of carbon credits: Reduction, Removal, and Avoidance https://www.carbon-direct.com/insights/how-do-carbon-credits-actually-work-removal-reduction-and-avoidance-credits-explained
8. CarbonCredits.com (February 2024) Tesla Hits Record High Sales from Carbon Credits at $1.79B https://carboncredits.com/tesla-hits-record-high-sales-from-carbon-credits-at-1-79b/
9. Normative (2024) How manufacturers can reduce carbon emissions https://normative.io/insight/reduce-manufacturer-carbon-emissions/
10. EASA (2024) How sustainable are SAF? https://www.easa.europa.eu/eco/eaer/topics/sustainable-aviation-fuels/how-sustainable-are-saf
11. Climate Focus (2024) Voluntary Carbon Market January 2024: 2023 Review https://climatefocus.com/wp-content/uploads/2024/01/VCM-2023-Review-Report.pdf
12. IPCC AR6 Factsheet (2023) Carbon Dioxide Removal https://www.ipcc.ch/report/ar6/wg3/downloads/outreach/IPCC_AR6_WGIII_Factsheet_CDR.pdf
13. IEA 50 (2024) Carbon Capture, Utilisation and Storage https://www.iea.org/energy-system/carbon-capture-utilisation-and-storage
14. EU Commission (September 2024) The future of European competitiveness https://commission.europa.eu/document/download/97e481fd-2dc3-412d-be4c-f152a8232961_en
15. CO_2 Value Europe (2023) The non-profit association representing the Carbon Capture and Utilisation (CCU) community in Europe https://co2value.eu/
16. CIEL.org (July 2021) Letter to President Biden: Re: Carbon capture is not a climate solution https://www.ciel.org/wp-content/uploads/2021/07/CCS-Letter_FINAL_US-1.pdf
17. Carbon-Based Commentary (June 2023) 7 reasons carbon removal startups shouldn't partner with fossil fuel companies https://carbonbasedcommentary.substack.com/p/7-reasons-carbon-removal-startups
18. The State of Carbon Removal (2024) The State of Carbon Removal Report 2024 Second Edition https://static1.squarespace.com/static/633458017a1ae214f3772c76/t/665ed1e2b9d34b2bf8e17c63/1717490167773/The-State-of-Carbon-Dioxide-Removal-2Edition.pdf
19. UK Parliament Committee (February 2025) Carbon capture: High degree of uncertainty whether risky investment by Govt will pay off https://committees.parliament.uk/committee/127/public-accounts-committee/news/205139/carbon-capture-high-degree-of-uncertainty-whether-risky-investment-by-govt-will-pay-off/
20. International Federation of Accountants (November 2023) Understanding the Voluntary Carbon Markets https://ifacweb.blob.core.windows.net/publicfiles/2023-11/23-3284%20-%20Carbon%20Credit%20Teaser_EN_final.pdf
21. Carbon Brief (September 2023) Analysis: How some of the world's largest companies rely on carbon offsets to 'reach net-zero' https://interactive.carbonbrief.org/carbon-offsets-2023/companies.html
22. Carbon Brief (September 2023) Glossary: Carbon Brief's guide to the terminology of carbon offsets https://interactive.carbonbrief.org/carbon-offsets-2023/glossary.html
23. Allied Offsets (2023) Data Transparency for Carbon Offsetting https://alliedoffsets.com/
24. EPA Climate Change Indicators: Greenhouse Gases https://www.epa.gov/climate-indicators/greenhouse-gases
25. Statista (2024) Annual global emissions of carbon dioxide 1940–2023 https://www.statista.com/statistics/276629/global-co2-emissions/

26. BloombergNEF (September 2022) The Untapped Power of Carbon Markets in Five Charts https://about.bnef.com/blog/the-untapped-power-of-carbon-markets-in-five-charts/
27. Smith School of Enterprise and the Environment (2024) The State of Carbon Dioxide Removal: 2nd Edition 2024 https://static1.squarespace.com/static/633458017a1ae214f3772c76/t/665ed1e2b9d34b2bf8e17c63/1717490167773/The-State-of-Carbon-Dioxide-Removal-2Edition.pdf
28. IEA (2023) Executive Summary https://www.iea.org/reports/co2-emissions-in-2023/executive-summary
29. CarbonBrief (2017) Analysis: Why US carbon emissions have fallen 14% since 2005 https://www.carbonbrief.org/analysis-why-us-carbon-emissions-have-fallen-14-since-2005/
30. Congressional Research Service (August 2023) The Section 45Q Tax Credit for Carbon Sequestration https://crsreports.congress.gov/product/pdf/IF/IF11455
31. CarbonBrief (March 2021) What does China's 14th 'five-year plan' mean for climate change? https://www.carbonbrief.org/qa-what-does-chinas-14th-five-year-plan-mean-for-climate-change/
32. Verra (2024) Who We Are https://verra.org/about/overview/
33. Gold Standard (2024) Carbon Market Regulations Tracker https://www.goldstandard.org
34. ACR (2024) About Us https://acrcarbon.org/about-us/
35. Carbon Action Reserve (2022) Climate Forward https://www.climatcactionreserve.org/
36. SBTi (2023) SBTi Monitoring Report 2023 https://sciencebasedtargets.org/reports/sbti-monitoring-report-2023
37. odi.org.uk Carbon offset methodologies, standards and monitoring: infosheet 5 https://media.odi.org/documents/6082.pdf
38. Investopedia (2024) Principal-Agent Problem Causes, Solutions, and Examples Explained https://www.investopedia.com/terms/p/principal-agent-problem.asp
39. NCCS (2023) Carbon Tax https://www.nccs.gov.sg/singapores-climate-action/mitigation-efforts/carbontax/
40. CarbonCredits (March 2023) Inside Carbon Markets: Problems, Causes, and Potential Solutions https://carboncredits.com/inside-carbon-markets-problems-causes-and-potential-solutions/
41. The Guardian (17th January 2024) EU bans 'misleading' environmental claims that rely on offsetting https://www.theguardian.com/environment/2024/jan/17/eu-bans-misleading-environmental-claims-that-rely-on-offsetting#:~:text=Terms%20such%20as%20climate%20neutral,crackdown%20on%20misleading%20environmental%20claims.
42. Berkeley Public Policy (2023) Quality Assessment of REDD+ Carbon Credit Projects https://gspp.berkeley.edu/assets/uploads/page/Quality-Assessment-of-REDD+-Carbon-Crediting.pdf
43. UN Climate Change What is REDD+? https://unfccc.int/topics/land-use/workstreams/redd/what-is-redd
44. UNFCCC (2013) Warsaw Framework for REDD-plus https://unfccc.int/topics/land-use/workstreams/redd/redd-resources#Warsaw-Framework-for-REDD
45. FOE (2014) The great REDD gamble https://foe.org/wp-content/uploads/2018/10/The-great-REDD-gamble.pdf
46. CarbonMarketWatch.Org (2014) Mapping & Analysis of CDM Projects in India from a Sustainable Development Perspective https://carbonmarketwatch.org/wp-content/uploads/2014/09/A.-Tiwari-Mapping-Analysis-of-CDM-Projects-in-India-from-a-Sustainable-Development-Perspective.pdf
47. REDD-Monitor (2024) About https://reddmonitor.substack.com/about
48. CEPR (March 2022) Collapse of the Clean Development Mechanism scheme under the Kyoto Protocol and its spillover: Consequences of 'carbon panic https://cepr.org/voxeu/columns/collapse-clean-development-mechanism-scheme-under-kyoto-protocol-and-its-spillover
49. Wikipedia (2021) Unearthed https://en.wikipedia.org/wiki/Unearthed_(publication)
50. Wikipedia (July 2024) Greenpeace https://en.wikipedia.org/wiki/Greenpeace

51. West at al. (2023) 'Action needed to make carbon offsets from tropical forest conservation work for climate change mitigation' pre-publication paper https://arxiv.org/pdf/2301.03354
52. Wikipedia (July 2024) Verified Carbon Standard https://en.wikipedia.org/wiki/Verified_Carbon_Standard
53. SourceMaterial (July 2024) BP venture sells carbon offsets by 'saving' trees that aren't in danger https://www.source-material.org/
54. Verra (January 2023) Verra Response to Guardian Article on Carbon Offsets https://verra.org/verra-response-guardian-rainforest-carbon-offsets/
55. DGB Group (January 2023) Why is the Guardian so negative about conserving rainforests? https://www.green.earth/blog/carbon-offsetting-guardian-verra-redd-response
56. Everland Who We Are https://everland.earth/who-we-are/
57. Everland (January 2023) The science behind the Guardian piece is fatally flawed https://everland.earth/news/the-science-behind-the-the-guardian-piece-is-fatally-flawed/
58. Hugh A. Graham et al. (2023) Response to West at al. (2023) https://permianglobal.com/wp-content/uploads/2024/01/response-to-west.pdf
59. Ecosystem Marketplace (2023) Press Release: Top scientists issue a rebuttal to West at al paper https://www.ecosystemmarketplace.com/articles/press-release-top-scientist-sissue-rebuttal-to-west-et-al-paper/
60. The New Yorker (23 October 2023) The Great Cash-for-Carbon Hustle https://www.newyorker.com/magazine/2023/10/23/the-great-cash-for-carbon-hustle
61. South Pole (2024) What We Do https://www.southpole.com/what-we-do
62. Swissinfo.ch (October 2023) Swiss carbon offset giant pulls plug on Zimbabwe project amid allegations https://www.swissinfo.ch/eng/sci-tech/swiss-carbon-offset-giant-pulls-plug-zimbabwe-project-amid-forest-protection-allegations/48927908
63. South Pole (November 2023) South Pole announces senior leadership change https://www.southpole.com/news/south-pole-announces-senior-leadership-change
64. Verra (May 2023) Verra boss steps down after criticism of its carbon credits https://www.climatechangenews.com/2023/05/23/verra-boss-steps-down-after-criticism-of-its-carbon-credits/
65. Berkeley Public Policy: The Goldman School (2023) Quality Assessment of REDD+ Carbon Credit Projects https://gspp.berkeley.edu/assets/uploads/page/Quality-Assessment-of-REDD+-Carbon-Crediting.pdf
66. Verra (2024) Revisions to Avoiding Unplanned Deforestation and/or Degradation Methodologies https://verra.org/methodologies/revisions-to-avoiding-unplanned-deforestation-and-degradation-methodologies/
67. Paul Boefford (April 2024) SBTi policy and lessons from the 2023 redd+ controversy https://illuminem.com/illuminemvoices/sbti-policy-and-lessons-from-the-2023-redd-controversy
68. Calyx Global (July 2024) The biggest problem with carbon credits is not measurement error https://calyxglobal.com/blog-post?q=166
69. ICVCM (2024) Leading the way to high integrity in the Voluntary Carbon Market https://icvcm.org/
70. Thales A.P. West et al. (2024) 'Methodological issues with deforestation baselines compromise the integrity of carbon offsets from REDD+' Global Environmental Change v.87 July 2024 https://www.sciencedirect.com/science/article/pii/S0959378024000670
71. UN-REDD.Org (2024) Cancun safeguards https://www.un-redd.org/glossary/cancun-safeguards
72. UNFCC Warsaw Framework for REDD+ https://redd.unfccc.int/fact-sheets/warsaw-framework-for-redd.html
73. Nature (December 2023) Politics and the environment collide in Brazil: Lula's first year back in office https://www.nature.com/articles/d41586-023-04042-x
74. Maxar (2024) Spatial resolution: distance and detail https://explore.maxar.com/Imagery-Leadership-Spatial-Resolution
75. GreenBiz (2024) The state of carbon removal in 3 charts https://www.greenbiz.com/article/state-carbon-removal-3-charts?mkt_tok=MjExLU5KWS0xNjUAAAGUfnEQmRnxL4qVjxz2CpNf6dx_kmNbyaWcwCKDN93d3nntzr_I1D2DPAOT7KiM_MqHSDc6_Pc-S_ulSvE8OiwidIkWoSooBZevPYaVhppprsRIMNc

76. Ocean Visions Ocean Alkalinity Enhancement https://www2.oceanvisions.org/roadmaps/ocean-alkalinity-enhancement/
77. Ocean Visions (2023) Accelerating Ocean and Climate Restoration: Ocean-Based Climate Solutions https://oceanvisions.org/wp-content/uploads/2023/09/Ocean-Based-CDR-factsheet_Sept-2023.pdf
78. Ajay Gambhir and Massimo Tavoni (2019) 'Direct Air Carbon Capture and Sequestration: How It Works and How It Could Contribute to Climate-Change Mitigation' Science Direct, Volume 1.4, 20 December 2019, Pages 405–409 https://www.sciencedirect.com/science/article/pii/S2590332219302167
79. Tom Terlouw, et al. (2021) 'Life Cycle Assessment of Direct Air Carbon Capture and Storage with Low-Carbon Energy Sources' Environmental Science & Technology Vol 55.16 August 2021 https://pubs.acs.org/doi/ https://doi.org/10.1021/acs.est.1c03263
80. PSCI (2020) Preventing Climate Change with BECCS: Bioenergy with Carbon Capture and Storage https://psci.princeton.edu/tips/2020/11/15/preventing-climate-change-with-beccs-bioenergy-with-carbon-capture-and-storage
81. Thanh-Binh Nguyen, et al. (2023) 'Biochar for soil remediation: A comprehensive review of current research on pollutant removal' Environmental Pollution v.377 November 2023 https://www.sciencedirect.com/science/article/abs/pii/S0269749123015737
82. Ontl, T. A. & Schulte, L. A. (2012) 'Soil Carbon Storage' Nature Education Knowledge 3(10):35 https://www.nature.com/scitable/knowledge/library/soil-carbon-storage-84223790/
83. Climate Portal (MIT) Enhanced Rock Weathering https://climate.mit.edu/explainers/enhanced-rock-weathering
84. EUI (January 2025) Policy Developments in Carbon Markets Rewarding High-Quality Carbon Removals STG Policy Brief https://cadmus.eui.eu/entities/publication/b26e21a7-cdf6-5029-bdf5-faf5cac29a85
85. WEF (2021) Put a price' on nature to increase biodiversity incentives, experts say https://www.weforum.org/agenda/2021/05/nature-capital-climate-change-biodiversity-loss-environment/#:~:text=Natural%20capital%20can%20be%20defined%20as%20the%20world%E2%80%99s,US%24125%20trillion%20to%20the%20global%20economy%20every%20year.
86. UNFCCC Land Use, Land-Use Change and Forestry (LULUCF): Key documents related to LULUCF Past considerations of issues relating to LULUCF https://unfccc.int/topics/land-use/workstreams/land-use--land-use-change-and-forestry-lulucf
87. IPCC (2024) Agriculture, Forestry and Other Land Use (AFOLU) https://www.ipcc.ch/report/ar5/wg3/agriculture-forestry-and-other-land-use-afolu/
88. UK.Gov (2024) Nature Markets Framework progress update March 2024 https://www.gov.uk/government/publications/nature-markets-framework-progress-update-march-2024/nature-markets-framework-progress-update-march-2024
89. UNEP (2022) Convention on Biodiversity https://www.cbd.int/doc/decisions/cop-15/cop-15-dec-04-en.pdf
90. King&Wood Mallesons (2023) TNFD publishes final framework as TCFD disbands https://www.kwm.com/global/en/insights/latest-thinking/tnfd-publishes-final-framework-as-tcfd-disbands.html
91. The Nature Conservancy (2019) Investing in Nature: Private finance for nature-based resilience https://www.nature.org/content/dam/tnc/nature/en/documents/TNC-INVESTING-IN-NATURE_Report_01.pdf
92. Library of Congress Adam Smith:The wealth of nations https://www.loc.gov/item/2002564559/
93. UNEP (2023) State of Finance for Nature https://www.unep.org/resources/state-finance-nature-2023
94. UNFCCC (2023) Mobilizing Private Capital for Nature to meet Climate and Nature Goal S https://climatechampions.unfccc.int/wp-content/uploads/2023/09/Nature-finance-paper.pdf
95. US Department of the Interior Natural Capital Accounting https://www.doi.gov/ppa/integrative/natural-capital-accounting

96. Neil Havermale & Philip Mulvey (March 2024) Operational and Capital Ecoservice Accounting Standards for Farm Carbon https://cdn.prod.website-files.com/6445d9835d64c65f3a182901/65e1320f6aa347b9184191c4_Operational%20and%20Capital%20Ecoservice%20Accounting%20Standards%20for%20Farm%20Carbon%20%E2%80%93%20Ryzo%20Whitepaper%202024.pdf
97. WEF(2024) How to spot a nature crisis – and why we should use natural capital to combat it https://www.weforum.org/agenda/2024/06/3-traits-nature-crisis-natural-capital/
98. New York Times (18 February 2024) Nature has value. Could we literally invest in it? https://www.nytimes.com/2024/02/18/business/economy/natural-assets.html
99. Green Finance Institute (2024) Douglas Eger, Founder and CEO, Intrinsic Exchange Group https://www.greenfinanceinstitute.com/podcasts/episode-22-douglas-eger-founder-and-ceo-intrinsic-exchange-group/
100. Native.Eco (2017) Certified or Verified Carbon Offsets: Which Should You Choose? https://native.eco/2017/12/certified-or-verified-carbon-offsets-which-should-you-choose/#:~:text=Carbon%20offsets%20that%20are%20produced%20under%20a%20voluntary,schemes%2C%20and%20are%20called%20Verified%20Emissions%20Reductions%20%28VERs%29.
101. Simon Puleston Jones (2023) An introduction to carbon markets Carbon Solution + Simmons & Simmons https://img1.wsimg.com/blobby/go/76665bf9-5b22-49a8-8e5c-709d9eecc8d1/downloads/An%20introduction%20to%20carbon%20markets.pdf?ver=1710417376815
102. Carboncredits.com (2024) Carbon Crypto Guide 2024: KlimaDAO, Carbon NFTs, and Carbon Tokens https://carboncredits.com/carbon-crypto-guide-2023-klimadao-carbon-nfts-and-carbon-tokens/
103. Kaspersky (2024) What are NFTs and how do they work? https://usa.kaspersky.com/resource-center/definitions/what-is-an-nft#:~:text=NFT%20stands%20for%20%27non%2Dfungible,which%20makes%20each%20one%20unique.
104. Agfundernews.com (February 2022) Moss.Earth is using NFTs & crypto carbon credits to fight climate change in the Amazon https://agfundernews.com/nfts-crypto-carbon-credits-moss-earth-is-using-to-fight-climate-change
105. Toucan (2024) The first liquid market for biochar credit https://toucan.earth/
106. Klimadao Finance State of the Digital Carbon Market https://data.klimadao.finance/
107. CFTC About the CFTC https://www.cftc.gov/
108. ICVCM (June 2023) ICVCM and VCMI join forces to operationalize a high-integrity market to accelerate global climate action https://icvcm.org/icvcm-and-vcmi-join-forces-to-operationalize-a-high-integrity-market-to-accelerate-global-climate-action/
109. ICVCM (May 2024) Integrity Council confirms Verra and ART meet its high-integrity benchmark https://icvcm.org/integrity-council-confirms-verra-and-art-meet-its-high-integrity-benchmark/
110. WBCSD (2024) Our History https://www.wbcsd.org/who-we-are/our-history/
111. GHGProtocol Corporate Standard https://ghgprotocol.org/corporate-standard
112. GHG Protocol (2024) Greenhouse Gas Protocol provides standards, guidance, tools and training for business and government to measure and manage climate-warming emissions. https://ghgprotocol.org/about-us
113. H&M Group (2024) Climate reporting https://hmgroup.com/sustainability/circularity-and-climate/climate/climate-reporting/
114. Financial Times (14 August 2024) 'The battle over disclosing pollution' https://ft.pressreader.com/v99f/20240814/281741274726840
115. Bloomberg (2024) Amazon Says Its Carbon Emissions Fell in 2023 Amid Post-Pandemic Pullback https://www.bloomberg.com/news/articles/2024-07-10/amazon-carbon-emissions-fell-in-2023-amid-post-pandemic-pullback
116. SBT.Org The Corporate Net-Zero Standard https://sciencebasedtargets.org/net-zero
117. CDP (2024) Setting science-based emission reduction targets through the Science Based Targets initiative (SBTi) https://www.cdp.net/en/campaigns/science-based-targets

118. Firstgreen Consulting Pvt Ltd (June 2024) Understanding SBTi and Net Zero Targets: How Companies are Committing to Climate Action https://www.linkedin.com/pulse/understanding-sbti-net-zero-targets-how-companies-owwdc/?trackingId=YzBo8GX5RwyG9kn9mri0LA%3D%3D
119. Carbon Herald (17 April 2024) IETA Issues New Guidelines For Corporates On Carbon Credits https://carbonherald.com/ieta-issues-new-guidelines-for-corporates-on-carbon-credits/
120. Plana Earth (2023) Science-based targets and SBTi, explained https://plana.earth/academy/science-based-targets-sbti-explained
121. SBT.Org About target services https://sciencebasedtargets.org/set-a-target
122. SBT.Org (November 2023) Statement on use of environmental attribute certificates, including but not limited to voluntary carbon markets https://sciencebasedtargets.org/news/sbti-statement-on-eacs
123. SBTi (July 2024) Statement on Evidence-led systematic literature review to understand the effectiveness of corporate carbon offsetting as an alternative to direct emissions abatement https://sciencebasedtargets.us1.list-manage.com/track/click?u=02d7a943deeb0be5c375f4552&id=e8d224e00e&e=db5103f67a
124. Reisinger et al. (2024) 'Science-based targets miss the mark' Communications Earth Environment 5.383 (2024) Nature Briefing https://www.nature.com/articles/s43247-024-01535-z
125. SBTi (March 2025) SBTi Corporate Net-Zero Standard https://sciencebasedtargets.org/resources/files/Net-Zero-Standard-v2-Consultation-Draft.pdf
126. Marginal Carbon (18 March 2028) What needs to change in the SBTi Net Zero Standard 2.0 to scale carbon removal https://marginalcarbon.substack.com/p/what-needs-to-change-in-the-sbti
127. TFVCM (January 2021) Taskforce on Scaling Voluntary Carbon Markets https://www.iif.com/Portals/1/Files/TSVCM_Report.pdf
128. ICVCM (2024) The Core Carbon Principles https://icvcm.org/core-carbon-principles/
129. UNFCCC Clean Development Mechanism (CDM) https://cdm.unfccc.int/index.html?_gl=1*3jji0p*_ga*NTI1MDc0MzczLjE3MjE2ODkxNDI.*_ga_7ZZWT14N79*MTcyMzkxNjgwNC4yLjEuMTcyMzkxNjk
130. Verra (August 2024) New ICVCM Methodology Decisions Announced https://verra.org/new-icvcm-methodology-decisions-announced/
131. ICVCM Leading the way to high integrity in the Voluntary Carbon Market https://icvcm.org/
132. Eco-Business (May 2024) Verra validated by carbon market watchdog after 'significant' procedural changes https://www.eco-business.com/news/verra-validated-by-carbon-market-watchdog-after-significant-procedural-changes/
133. Wikipedia (July 2024) Climate Group https://en.wikipedia.org/wiki/Climate_Group
134. Wikipedia (July 2024) Under2 Coalition https://en.wikipedia.org/wiki/Under2_Coalition
135. IETA (2024) About https://www.ieta.org
136. WEF (2024) Verra https://www.weforum.org/organizations/verra/
137. Verra (2023) About https://verra.org/about/overview/
138. Verra (2024) VCS Standard https://verra.org/wp-content/uploads/2024/04/VCS-Standard-v4.7-FINAL-4.15.24.pdf
139. Verra (2024) Methodologies https://verra.org/methodologies-main/
140. CCBA (2024) CCB Standards https://www.climate standards.org/ccb standards/
141. Verra (2024) Sustainable Development Verified Impact Standard https://verra.org/programs/sd-verified-impact-standard/
142. Verra (2024) Plastic Waste Reduction Standard https://verra.org/programs/plastic-waste-reduction-standard/
143. Verra (2024) Jurisdictional and Nested Redd+ Framework https://verra.org/programs/jurisdictional-nested-redd-framework/
144. Jay Wintergreen and Tod Delaney ISO 14064, International Standard for GHG Emissions Inventories and Verification https://www3.epa.gov/ttnchie1/conference/ei16/session13/wintergreen.pdf

145. IBBN (2024) Comparing Gold Standard and Verra Certification for Biochar Carbon Credits: Key Features and Differences https://ibbn.org.in/news-related-posts/comparing-gold-standard-and-verra-certification-for-biochar-carbon-credits/
146. CAR (2024) About Us https://www.climateactionreserve.org/about-us/
147. Green-e.org (2024) A global third-party certification program for carbon offsets https://www.green-e.org/programs/climate
148. CRS (20240 Policy and Market Development https://resource-solutions.org/programs/policy-initiatives/
149. Gold Standard (2024) Funding Beyond Value Chain Mitigation https://www.goldstandard.org/publications/funding-beyond-value-chain-mitigation
150. WWF (2024) What the WWF is Doing https://www.worldwildlife.org/initiatives/climate
151. WWF (2024) WWF-GEF https://www.worldwildlife.org/topics/wwf-gef
152. WWF Beyond Net-Zero: A business pathway to spur urgent climate action towards 2030 https://wwf.panda.org/wwf_news/?5482441/beyond-net-zero
153. Gold Standard (2024) Gold Standard for the Global Goals https://globalgoals.goldstandard.org
154. Gold Standard (2024) Renewable Energy Activity Requirements https://globalgoals.goldstandard.org/202-ar-renewable-energy-activity-requirements/
155. SustainCert (2024) Carbon credits 101: what is the Gold Standard? https://sustain-cert.com/news/carbon-credits-101-what-is-the-gold-standard
156. ISEAL (2024) ISEAL Community Member Documents https://www.isealalliance.org/get-involved/resources/iseal-community-member-documents
157. Gold Standard Our Standard https://www.goldstandard.org/gold-standard-for-the-global-goals/our-standard
158. UK Gov et al. (2022) Integrating a Gender Lens in Voluntary Carbon Markets https://www.sddirect.org.uk/sites/default/files/2022-11/ASEAN%20Integrating%20gender%20into%20VCMs%20-%20Volume%20I%20-%20FINAL.pdf
159. ISEAL (2024) ISEAL Community Members https://www.isealalliance.org/iseal-community-members?field_code_compliant=1
160. Simon Puleston Jones(2023) An introduction to carbon markets Climate Solution and Simons & Simons LLP https://img1.wsimg.com/blobby/go/76665bf9-5b22-49a8-8e5c-709d9eecc8d1/downloads/An%20introduction%20to%20carbon%20markets.pdf?ver=1710417376815
161. Evidensia Using evidence to inform sustainability action https://www.evidensia.eco/
162. Carbon Offset Guide American Carbon Registry https://offsetguide.org/understanding-carbon-offsets/carbon-offset-programs/voluntary-offset-programs/american-carbon-registry/#:~:text=The%20American%20Carbon%20Registry%20(ACR,profit%20based%20in%20the%20USA.
163. ACR (2024) A Global Leader in Regulatory and Voluntary Carbon Markets https://acrcarbon.org/
164. ARC (2024) Home https://acrcarbon.org/
165. ICROA ICROA Code of Best Practice https://icroa.org
166. ACR (July 2023) The ACR Standard https://acrcarbon.org/wp-content/uploads/2023/10/ACR-Standard-v8.0.pdf
167. Sylvera (2022) Guide to Carbon Credit Buffer Pools https://www.sylvera.com/blog/carbon-credit-buffer-pools
168. CAR (2024) Climate Forward https://www.climateactionreserve.org/about-us/
169. CAR (2024) 2023 Annual Report https://www.climateactionreserve.org/wp-content/uploads/2024/06/Reserve-2023-annual-report-final-post.pdf
170. Plan Vivo Our History https://www.planvivo.org/history#:~:text=The%20concept%20of%20Plan%20Vivo%20is%20born%20in,Colegio%20de%20la%20Frontera%20Sur%2C%20other%20local%20partners.
171. Mikokopamoja About https://www.mikokopamoja.org/

172. Plan Vivo (December 2023) Plan Vivo launch pioneering Biodiversity Standard https://www.planvivo.org/news/plan-vivo-launch-biodiversity-standard
173. Forbes (September 2020) Women Launch First Voluntary Scalable Marketplace For 'Carbon Removal' To Help Corporations Reach Net Zero https://www.forbes.com/sites/rebeccaponton/2020/09/22/woman-owned-finnish-carbon-capture-and-storage-company-puroearth-created-to-help-corporations-reach-net-zero/
174. Pura Earth (June 2021) Nasdaq Acquires Emerging Carbon Removal Market Puro.earth https://www.nasdaq.com/press-release/nasdaq-acquires-emerging-carbon-removal-market-puro.earth-2021-06-14
175. Puro.earth Supplier Listings https://puro.earth/CORC-co2-removal-certificate/supplier-listings
176. Puro.earth Puro Registry https://registry.puro.earth/carbon-sequestration/retirements
177. Static.squarespace.com (2024) The State of Carbon Removal 2nd Edition https://static1.squarespace.com/static/633458017a1ae214f3772c76/t/665ed1e2b9d34b2bf8e17c63/1717490167773/The-State-of-Carbon-Dioxide-Removal-2Edition.pdf
178. GCC (2023) GCC Program Framework v.3 2023 https://www.globalcarboncouncil.com/wp-content/uploads/2024/01/GCC-program-framework-v3.pdf
179. GCC (2024) Who We Are https://www.globalcarboncouncil.com/about/who-we-are/
180. GCC (2020) Methodology for Grid Connected Renewable Energy Generation Projects v2–2020 https://globalcarboncouncil.com/wp-content/uploads/2021/10/GCCM001_Methodology_for_Grid_Connected_Renewable_Energy_Generation_Approved-v2.pdf
181. ICAO About ICAO https://www.icao.int/about-icao/Pages/default.aspx
182. IATA (2024) Offsetting CO_2 Emissions with CORSIA https://www.iata.org/en/programs/sustainability/corsia/#:~:text=In%202016%2C%20Governments%20adopted%20CORSIA,from%20international%20aviation%20from%202021.
183. ICAO Carbon Offsetting and Reduction Scheme for International Aviation (CORSIA) https://www.icao.int/environmental-protection/CORSIA/Pages/default.aspx
184. ICAO (2024) International Standards and Recommended Practices https://elibrary.icao.int/reader/229739/&returnUrl%3DaHR0cHM6Ly9lbGlicmFyeS5pY2FvLmludC9wcm9kdWN0LzIyOTczOSUzRl9nbCUzRDEqMnljanM5Kl9nYSpNVFkwTkRrMU16RXlOeTR4TnpJMk1UZzFOelV6Kl9nYV85OTJOM1lETEJRKk1UY3lOakU0TlRjMU1pNHhMakV1TVRjeU5qRTROelF3TWk0d0xqQXVNQS4u?productType=ebook
185. ICROA (2024) About https://icroa.org/about/
186. IETA (2024) History https://www.ieta.org/about/
187. Climate Action Data Trust (2024) Connecting Carbon markets Through Open Data https://climateactiondata.org/
188. IETA (2024) Decentralised Digital Infrastructure for Carbon Markets https://www.ieta.org/initiatives/cad-trust/
189. IATA.Org (2024) Aviation Carbon Exchange https://www.iata.org/en/programs/sustainability//ace
190. VCMI (2024) About https://vcmintegrity.org/about/
191. RPA (2024) Featured Projects: Climate & Environment https://www.rockpa.org/projects/issue-areas/?member=environment-global-commons-alliance
192. VCMI (2024) VCMI Claims Code of Practice https://vcmintegrity.org/vcmi-claims-code-of-practice/
193. Linklaters (July 2023) VCMI launches Claims Code of Practice to enhance integrity in the use of carbon credits and associated claims https://sustainablefutures.linklaters.com/post/102iihg/vcmi-launches-claims-code-of-practice-to-enhance-integrity-in-the-use-of-carbon-c
194. VCMI (2024) Beta Scope 3 Claim https://vcmintegrity.org/scope-3-claim-upcoming-public-consultation/
195. VCMI (2023) Carbon Market Opportunities in the Agriculture Sector in Latin America and the Caribbean https://vcmintegrity.org/wp-content/uploads/2023/12/BVE23119401i-1.pdf
196. CCBA (2024) About the CCBA https://www.climate-standards.org/ccb-standards/

197. CCBA (2024) Sustainable Landscapes Rating Tool https://www.climate-standards.org/sustainable-landscapes-rating-tool/
198. Verra (2024) Welcome to the Verra Registry https://registry.verra.org/#/ccb
199. Conservation International Carbon Footprint Calculator https://footprint.conservation.org/en-us/
200. Ecosystem Marketplace (August 2022) The Art of Integrity https://app.hubspot.com/documents/3298623/view/433338095?accessId=3abc8b
201. MSCI (2024) MSCI Carbon Markets https://www.msci.com/our-solutions/climate-investing/carbon-markets
202. Devdiscource.com (January 2024) Amazon Launches Abacus to Revolutionize Carbon Offset Standards https://www.devdiscourse.com/article/technology/3002035-amazon-launches-abacus-to-revolutionize-carbon-offset-standards
203. Group.beauveritas.com (2024) Expertise for sustainability https://group.bureauveritas.com/expertise-sustainability-0
204. Southpole.com (2024) The South Pole 2024 Net Zero Report https://www.southpole.com/
205. Nori.com (2024) Nori in 5 minutes https://nori.com/whitepaper
206. Carbonibus About us https://www.carbonibus.org/about
207. Climateimpactx Who We Are https://www.climateimpactx.com/
208. Netzero.salesforce.com (2024) Emissions Factors Data https://netzero.salesforce.com/s/
209. Salesforce.com (2022) Salesforce Announces First-of-Its-Kind Carbon Credit Marketplace, Empowering Any Organization to Take Climate Action on Their Journey to Net Zero https://www.salesforce.com/news/press-releases/2022/09/20/salesforce-announces-carbon-credit-marketplace/
210. Fastcompany.com (January 2024) Amazon sidesteps a Bezos-funded carbon offsets standard for a more ambitious one https://www.fastcompany.com/91149605/amazon-bezos-carbon-offsets-standard-abacus
211. aws.amazon.com (2024) Now find open data from the Amazon Sustainability Data Initiative catalog in AWS Marketplace https://aws.amazon.com/marketplace/solutions/sustainability/#asdi
212. Aws.amazon.com (2024) All products https://aws.amazon.com/marketplace/search/results?ref_=awsmp_sol_esg_aws_opndt&FULFILLMENT_OPTION_TYPE=DATA_EXCHANGE&CONTRACT_TYPE=OPEN_DATA_LICENSES&filters=FULFILLMENT_OPTION_TYPE%2CCONTRACT_TYPE
213. Aws.amazon.com (2024) Customer Carbon Footprint Tool https://aws.amazon.com/aws-cost-management/aws-customer-carbon-footprint-tool/
214. Carboncredits.com (2024) The 5 Top Carbon Offset Project Developers https://carboncredits.com/top-carbon-offset-project-developers/
215. Verra (2024) Active VVBs and Pending https://verra.org/validation-verification/
216. Globalgaols.globalstandard.org (2024) Validation & Verification Bodies https://globalgoals.goldstandard.org/verification-validation-bodies/
217. BeZero (2024) The carbon ratings agency https://bezerocarbon.com/
218. Calyxglobal.com Navigate carbon markets with confidence https://www.calyxglobal.com/
219. Sylvera.com (2024) The end-to-end carbon data platform for the 'net' in your net zero goals https://www.sylvera.com/
220. Jobsincarbon.niceboard.com (2024) Jobs in Carbon https://jobsincarbon.niceboard.co
221. Alliedoffsets.com (2023) Data Transparency for Carbon Offsetting https://alliedoffsets.com/
222. World Bank Group (2024) State and Trends of Carbon Pricing Dashboard https://carbonpricingdashboard.worldbank.org/
223. Platt (2024) Specifications Guide Carbon Markets https://www.spglobal.com/commodityinsights/PlattsContent/_assets/_files/en/our-methodology/methodology-specifications/method_carbon_credits.pdf
224. Carbon-mids.com Good decisions come from good data https://www.carbon-minds.com/
225. Mission 2.0 (2024) Overview https://mission-innovation.net/about-mi/overview/
226. Epansiv Joining the Exchange https://xpansiv.com/cbl/

227. ACX Our Vision Is a World Where Sustainability and the Global Economy Work in Harmony https://acx.net/
228. Cmecompany.com Where Risk Meets Opportunity https://www.cmegroup.com/
229. Lexology.com (2022) What exactly are carbon tokens? https://www.lexology.com/library/detail.aspx?g=731712ed-155b-4fd3-8824-562e5110965c#:~:text=As%20the%20voluntary%20carbon%20market%20continues%20to%20scale,by%20a%20smart%20contract%20on%20a%20blockchain%20network.
230. Oliverwyman.com (2024) Unlocking The Potential Of Carbon Dioxide Removal https://www.oliverwyman.com/our-expertise/insights/2024/jun/carbon-dioxide-removal-solutions.html
231. Carbon-direct.com (2024) A look at the 2023 voluntary carbon market https://www.carbon-direct.com/insights/state-of-the-voluntary-carbon-market-2023
232. Carboncreits.com (10 February 2025) Carbon Markets in 2025: A New Era of Accountability, Quality, and Transparency https://carboncredits.com/carbon-markets-in-2025-a-new-era-of-accountability-quality-and-transparency/
233. Compensate (2021) Reforming the voluntary carbon market https://downloads.ctfassets.net/f6kng81cu8b8/5vgGIHhsrTAbMnqaDYNGYJ/25a7d0e148a6d15cd10e2409107d7f3d/Reforming_the_voluntary_carbon_market_-_Compensate.pdf
234. CFTC (20 September 2024) CFTC Approves Final Guidance Regarding the Listing of Voluntary Carbon Credit Derivative Contracts https://www.cftc.gov/PressRoom/PressReleases/8969-24
235. ESGDrive.com (28 March 2025) SEC withdraws defense of climate-risk disclosure rule https://www.esgdive.com/news/sec-withdraws-climate-risk-disclosure-rule-defense-eighth-circuit-reactions/743860/?utm_source=Sailthru&utm_medium=email&utm_campaign=Issue:%202025-03-28%20ESG%20Dive%20%5Bissue:71729%5D&utm_term=ESG%20Dive
236. ImpactOne (28 August 2023) The Role of Forestry in the Voluntary Carbon Market https://impact.one/insights/the-role-of-for estry-in-the-voluntary-carbon-market/#:~:text = In%202021%2C%20forestry%20constituted%2040%%20of%20all,been%20purchased%20(referred%20to%20as%20%E2%80%9Cretired%E2%80%9D%20projects).&text=Many%20existing%20forestry%20projects%20primarily%20involve%20single%2Dspecies,do%20not%20contribute%20to%20overall%20ecosystem%20health.
237. UN.Org (2 December 2023) Secretary-General's remarks at Global Climate Action High Level Event: Towards a Turning Point for Climate Action [as delivered] https://www.un.org/sg/en/content/sg/statement/2023-12-02/secretary-generals-remarks-global-climate-action-high-level-event-towards-turning-point-for-climate-action-delivered
238. Morgan Lewis (2024) Article 6 of the Paris Agreement and Its Impact on Investments in Clean Cookstoves Empowered https://www.morganlewis.com/pubs/2024/06/article-6-of-the-paris-agreement-and-its-impact-on-investments-in-clean-cookstoves
239. Conservation International (2024) About Us https://www.conservation.org/about
240. Rainforest Alliance (2023) Annual Report 2023 https://www.rainforest-alliance.org/annual_report/2023/
241. Future-insight.green (2024) The Voluntary Carbon Market: A Tipping Point in the Fight Against Climate Change https://www.future-insight.green/2024/07/31/the-voluntary-carbon-market-a-tipping-point-in-the-fight-against-climate-change/#comment-5
242. Carbon Credits (2024) Google Ditches Carbon Offsets, Here's Its New Net Zero Focus https://carboncredits.com/google-ditches-carbon-offsets-heres-its-new-net-zero-focus/
243. Symbiosis (2024) We're spurring the next-generation of high-impact nature restoration projects https://www.symbiosiscoalition.org/
244. Carbon Herald (23 May 2024) Google, Meta, Microsoft, And Salesforce Establish World's Largest Advance Market Commitment For Nature-Based Projects https://carbonherald.com/google-meta-microsoft-salesforce-largest-advance-market-commitment-nature-based-projects/
245. iMasons Climate Accord (2024) Achieving carbon neutrality in digital infrastructure https://climateaccord.org/#

246. ESGDrive (17 July 2024) AWS, Google, Meta, Microsoft, more push for digital infrastructure emissions transparency https://www.esgdive.com/news/imasons-climate-suppliers-issue-epds-scope-3-ghg-data-centers-aws-google-meta/721629/
247. Future-insight.green (2024) We solve the indirect emissions – 100% Together https://www.future-insight.green/2024/07/10/discover-ecoadvis/

Chapter 6
Green Finance

Mind the Gap

International green finance from public and private sources *combined* reached USD2.1 trillion by 2024 according to BloombergNEF [3], who forecast USD5.6 billion being required *annually* by 2030 to achieve Net Zero targets. See also Fig. 6.1. BloombergNEF sees investment mostly occurring in established sectors such as electrified transport, renewable energy and power grids, and less in emerging technologies such as carbon capture and clean industrial processes. The last is especially worrying as cleaning up existing production processes is a vital step towards C02e avoidance and reduction, where C02e refers to carbon and carbon-equivalent GHGs.

Forecasts of what is needed differ according to sources, but they all far exceed the USD1.3 trillion annually by 2035 decided at COP-29 in Baku, Azerbaijan, November 2024-as discussed below. The independent, not-for-profit Climate Policy Initiative's (CPI) 2023 report on *Global Landscape of Climate Finance* [4] concluded that by 2030 global green funding of between USD8 and 9 trillion *annually* was required. From 2030 to 2050 this rises to USD10 trillion annually. These figures include direct investments from both public and private sectors in climate-specific physical assets such as direct carbon avoidance, reduction or removal projects, although the effectiveness of the latter two remains to be determined, but *exclude* 'transition-related unabated fossil-fuel finance' such as the purchase of carbon credit offsets. The CPI cites the estimate of the *Climate Scenarios for Central Banks and Supervisors* [5], from the 2022 report of the Network of Central Banks and Supervisors for Greening the Financial System (NGFS-also see below) of a global green financial gap of over USD260 trillion 2025–2050 if the pathway to 1.5 °C is to be achieved. As the CPI points out, this gap is an under-estimate because it excludes estimates of losses from stranded assets, biodiversity threats, human conflicts, and migration. In particular, the importance of biodiversity cannot be overstated. These figures speak for themselves.

J. Ure, *Climate Change and Carbon Markets: Late or Too Late?*,
https://doi.org/10.1007/978-3-032-07619-9_6

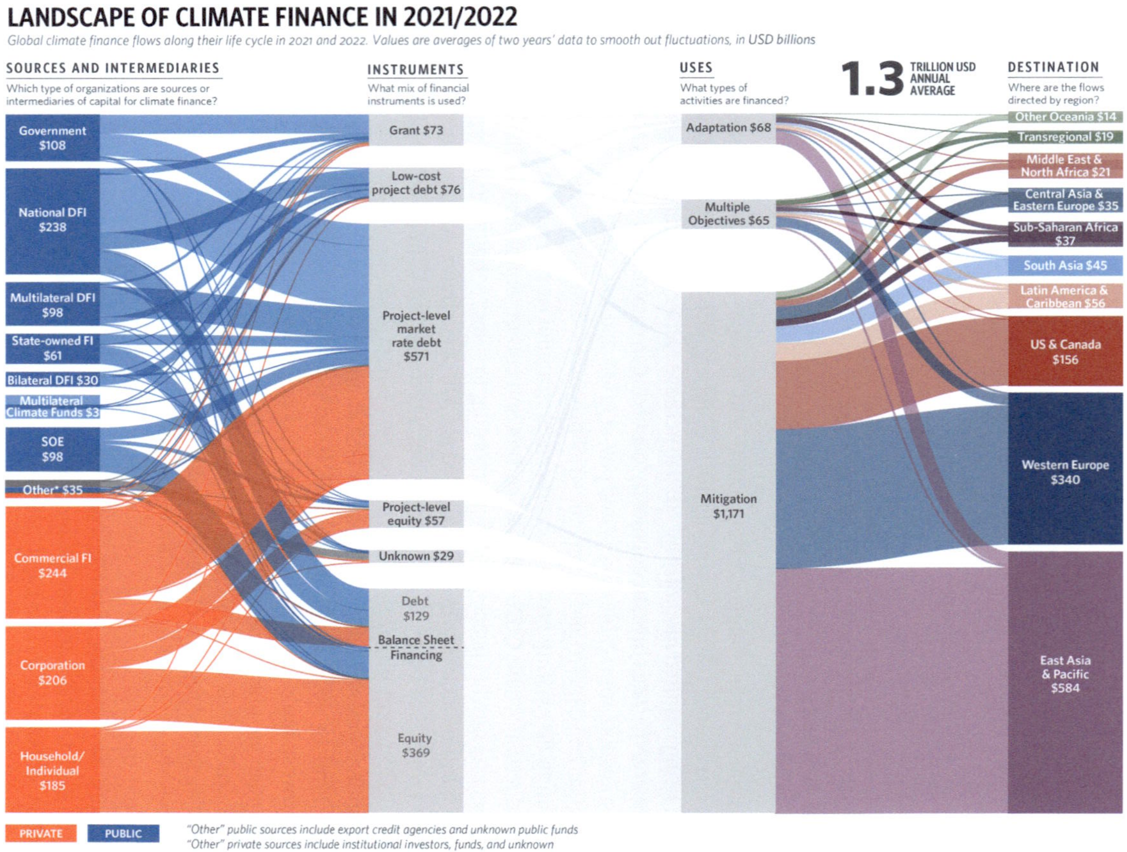

Fig. 6.1 The Climate Finance Landscape 2021/22. Global landscape of climate finance 2024: insights for COP29. *Source* Climate Policy Initiative (2024) [24]

A Just Transition

Green finance is, or should be, first and foremost a way to provide a Just Transition to a low-carbon economy, for 'all communities, all workers, all social groups' in the words of the United Nations Development Programme (UNDP) [6]. This could be broadened: 'to sustain projects protecting and conserving the natural environment and/or mitigate, or help communities adapt to, the emission of Green House Gases (GHGs) and climate change' as an extended definition of the one offered by the United Nations Environment Programme (UNEP) [7]. It recognises that biodiversity, as a carbon sink, and climate change, as a threat to biodiversity, are complementary and should be treated as such.

The concept of a Just Transition originates from the campaigns of the trade unions in the USA in the 1970s and 1980s to improve water supplies and combat air pollution [8], but it has broadened out to refer to an international Just Transition in which the Global South is fully included, as outlined in a paper by the Union Institute of Work, Environment and Health (ISTAS) [9]. The principle is included in the UN's Independent High-Level Expert Group on Climate Finance (IHLEG), appointed at COP-26 (Glasgow, Scotland) in 2021 [10]; and in the Climateworks Foundation of philanthropic organisations which launched a *Global South Just Transition Fund* [11]. However, the outcome of COP-29 (Baku, Azerbaijan) in 2024 was widely regarded as falling far short of a just transition. Amnesty International called it 'an affront to human rights' [12], while others saw some very small signs of progress [13].

Green Finance and the Conference of Parties (COPs)

Green finance covers financial flows from a variety of banking, micro-credit, insurance, and investment sources, and from public, private, and not-for-profit sectors, to tackle the rise of Green House Gases (GHG) of which carbon dioxide (C02) makes up 74% and 92% of which comes from fossil fuels [14]. In 2023, as reported by the UNEP annual global GHG emissions were measured at a record high of 57.1 GtCO2e (gigatons of carbon dioxide equivalents) [15]. Most of that, upwards of 60%, came from the industrialising countries of the Global South. The USA was responsible for adding 11% to global GHGs, but much less than China's 29%, with 9% being added by the 27 EU countries, and 8% by India [16]—see also Fig. 7.1 in Chap. 7. The good news is that most of the industrial countries have already passed their peak year for emissions.

The Global South of less developed countries, as classified by UNCTAD (UN Commission on Trade and Development) [17], is still in the process of industrial development, a process associated with GHG emissions, although it need not always be so; see Chap. 3 which cites the IEA showing that GDP growth is freeing itself from GHG growth as renewable energy replaces fossil fuels. But given the power of international markets, the model of development of the Global South is a replica of the Global North, namely, capitalism. The choices offered to the peoples of the Global South are therefore also replicas: either sacrificing economic growth and the materials that are taken for granted by the Global North, or making a transition to renewable sources of energy. But a transition requires massive investments which most of them lack.

China

The outlier is China, which accounts for 29% of global GHGs, followed a long way behind by India at 8%. China does have the financial clout to invest in renewables, is a world leader in doing so, and has aggressive plans to cap its emissions [18]. At COP-29 one of the signalled messages was to China to do more to help finance less developed countries [19]. However, although China is the world's largest overseas investor, overtaking the USA, the EU, and the UK combined through its Belt and Road Initiative [20], the sad fact is that, despite pledges to change direction, of that investment that has gone into energy and transportation infrastructure projects, over 90% has gone into fossil fuels, mostly coal [21]. Complementing the Belt and Road Initiative is the BRICS (Brazil, Russia, India, China, South Africa) New Development Bank (NDB) which has clean energy as one of its five priority sustainable development projects [22].

Conceivably China has been left some space to become the world leader in combatting global warming following the decision of President Trump in January 2025 to withdraw the US from the Paris Agreement for the second time. The *Putting America First in International Environmental Agreements* Executive Order, 'to limit the United States' financial contributions to other countries in furtherance of mitigating climate change and adapting to climate change globally' [23], together with the curtailing of USAID programmes, seems to open the door to China which could bring China a windfall of diplomatic gains. The messaging from COP-29 encourages China to do so.

This text predates COP-30 held in Belem, Brazil, November 2025, but the outcomes were mostly muted, and like COP-29 was subject to the many fault lines that go back to previous COPs. Table 6.1 selects and outlines several of these. Their resolution will be the test of progress.

Climate Finance Landscape

As listed in Table 6.1, challenges for green finance on an international scale are as daunting as the environmental threats they are designed to overcome. Tracking the sources of that funding, and identifying ways to rachet it up, is a global challenge.

Table 6.1 COP fault lines

Funding issues	Fault lines
Global North versus Global South	1. How much are the industrial nations willing to finance the Global South by updating their Nationally Determined Contributions (NDCs) to a New Collective Quantified Goal (NCQG)? 2. How big a part is China willing to play in global financing? 3. What is reasonable to expect from the Global South in the way of self-funding?
Public versus private funding	1. Can state/public sector finances act as a catalyst for the private sector to invest more to help the Global South? 2. Can the Multilateral Development Banks (MDBs) take a more active role in climate change financing? 3. What conditionalities are justified by the Global North to ensure funds to the Global South are used correctly?
Article 6 (Paris) versus carbon credits	1. Can trading carbon credits between countries (Article 6.2) accelerate green finance to support projects in the Global South, and if so which country offsets these credits against their Nationally Determined Contributions (NDCs)? 2. Can an UN-supervised regime of global carbon credit trading (Article 6.4) improve the quality and volume of such credits? 3. Can voluntary carbon market (VCM) self-regulation ever succeed to close the gap in international green finance?
Mitigation versus adaptation	1. How should funds be divided between mitigation and adaptation and can the Global South develop and sustain a disaster insurance sector? 2. Should the *Loss and Damage Fund* be increased according to a given scale? 3. Can financial regimes shift away from fossil-fuel subsidies to capital funding of, and lines of credit to, renewable energy sources?

(continued)

Table 6.1 (continued)

Funding issues	Fault lines
COP on biodiversity versus COP on climate change	1. Does the separation of COP (diversity) from COP (climate change) help keep them focused, or does it divert funding away from one or the other? 2. Biodiversity COPs emphasise the importance of local community co-benefits, how can Climate Change COPs do likewise? 3. All 'natural capital' is energy in a different form, and subject to the Second Law of Thermodynamics (entropy), so how can COPs deal with the environmental issue of recycling of waste?
Small Island Developing States (SIDS)	1. Is there a climate mitigation solution to rising sea levels threatening SIDS, or is adaptation and depopulation the ultimate answer for the SIDS? 2. Are there adaptation solutions available to SIDS and who will fund them? 3. Apart from paying for their attendance at COPs, what practical steps can the international community take to support the long-term survival of the SIDS?

Source Author

Prepared for COP-29, Chart 6.1 provides a graphic overview of international climate-related finance 2021–2022, assembled by the Climate Policy Initiative (2024), an independent organisation supported by philanthropic and government development finance [25].

Figure 6.1 lists international green finance into eight categories, dominated by three types of financing: (1) project financing: low-cost debt, market-level debts, project-level equity financing; (2) non-project financing: concessionary debt, private company balance-sheet financing, and private equity which includes household ownership of energy-saving installations; and (3) grants. For information on which types of renewable energy sources received this investment, see IRENA (2023) [26].

Mitigation and Adaptation

Ninety percent of these international flows went to mitigation, and less than 6% to adaptation. On the face of it, mitigation is prioritised because that directly addresses the causes of global warming, but as pointed out by Prof. Michael Oppenheimer of Princeton University [27], while the benefits of mitigation, if successful, may take decades to show themselves, the risk is that public responsiveness may diminish, whereas the more immediate benefits of adjustment projects, such as shoreline defences, housing improvements, and reinforced infrastructure, may prove more compelling to local communities. Given the obscuration of climate change by vested interests, mostly by the fossil-fuel industry, the immediate relevance of adaptation may prove an easier political pathway to public spending on climate change. An interesting point.

Loss and Damage Fund

Global warming from mid-2023 exceeded 1.5C [28] for the first time, and as the cost of lives lost and property damage owing to violent weather events continues to rise, the urgency of committing more resources to adaptation becomes ever more pressing. Figure 6.2 maps the sharp increase in such events from 1980 to 2019.

The EMDEs (emerging markets and developing economies) of the Global South are far more vulnerable to the impacts of climate change and more violent natural events due to their inadequate infrastructure, which frequently is too little or of inferior design and construction. But even for the better protected USA, the total

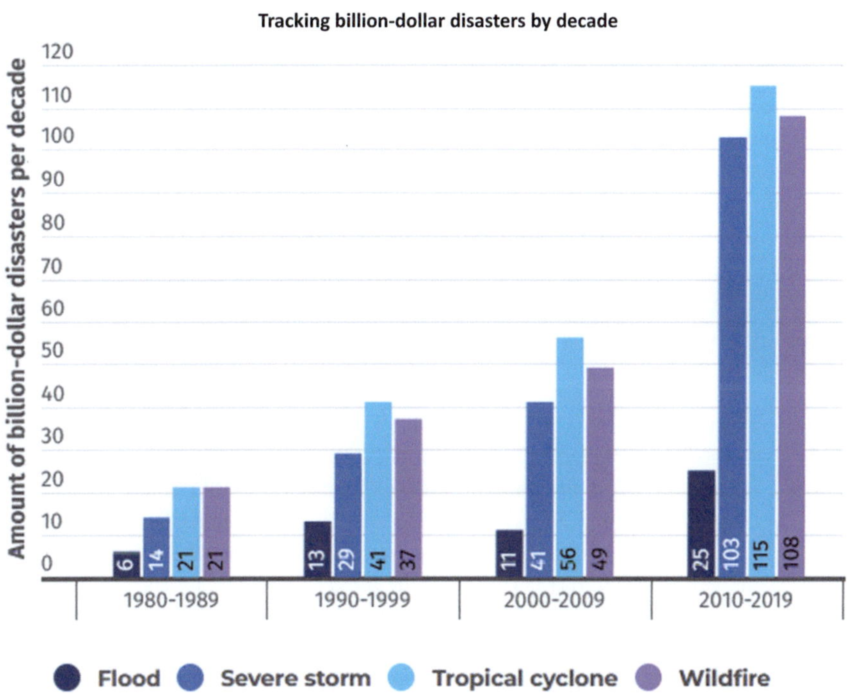

Fig. 6.2 Weather-related disasters 1980-2019. Billion-dollar weather-related disasters, 1980–2019. *Source* NQAA Centers for Environmnetal Enformation (NCEI) U.S. billion-dollar Weather and climate disasters (2024)

annual insurance costs by mid-2024 were estimated by the NCEI (National Centers for Environmental Information) [29] as USD2.2 trillion. The main global channel involved of financial assistance is the USD700 million *Loss and Damage Fund*, administered by the World Bank, first proposed in 1991 by the Pacific Island-state of Vanuatu [30] on behalf of the SIDS (Small Island Developing States) but agreed only 31 years later, and finally enacted in 2023 by COP-28 (Dubai, United Arab Emirates) [31]. Even given higher property prices in the US, the Fund is clearly woefully inadequate-see below. Some progress was made towards improved funding for adaptation projects at COP-30.

New Collective Quantified Goal (NCQG) and COP-29 Outcomes

The 'Conference of Parties' (COPs) are the apex events organised by the UNEP (UN Environment Programme) [32], responsible for commitments to, and finding ways to fund, the principal goal of the Paris COP21, held in 2015, to keep global warming under 2 °C, ideally to no more than 1.5 °C above pre-industrial levels by 2050. However, the UNEP's *Emissions Gap Report 2022* (EGR) presented to COP27 (Sharm el-Sheikh, Egypt) warned that the best estimates for 2100 were between 2.4 and 2.8 °C [33], but revised upwards in ERG 2024 [15] to 3.1°C by the end the century if existing Nationally Determined Contributions (NDCs) were not significantly increased. The financing of projects to achieve the goals of the Paris Agreement is contained in Article 2.1, but without agreement as to how this could be achieved: 'making finance flows consistent with a pathway towards low greenhouse gas (GHG) emissions and climate-resilient development' [34].

COP-29

COP-29 (Baku, Azerbaijan) in 2024, widely referred to as 'the financial COP', was 'one of the most bitter negotiations in the UN climate convention's three-decade history… to at least triple the annual outflows from UNFCCC climate funds from 2022 levels by 2030 - from USD1.74 billion [35] in 2022 to at least USD5.2 billion per year by 2030' [36]. Early on the G77 + China group,which since the 1960s has grown to represent over 130 EMDEs [37], rejected the *Baku Framework* as a negotiating text, claiming it ignored wider commitments to the UN Sustainable Development Goals [38]. Supporting developing country alliances included the Like-Minded Developing Countries (LMDCs), the Independent Alliance of Latin America and the Caribbean (AILAC), the African Group of Negotiators, the Alliance of Small Island Developing States (AOSIS), and the Group SUR (Brazil, Paraguay, Uruguay, Argentina) [39].

The reluctance of the Global North to commit more and sooner in large part reflected the uncertainty at COP-29 of the position of the USA post-January 2025 under the new presidency of Donald Trump, either a climate denier or a climate ditherer-see Chap. 1. The three main issues, the NCOG, the revised collective NDCs (including the 'voluntary agreement'), and the tripling of the UNFCCC climate funds, were known as Triple Finance [40] which are explained as follows:

1. The **New Collective Quantified Goal** (NCQG) [41] aiming to raise USD1.3 trillion by 2035 [42] was seen as an acknowledgement of the proposal in 2023 of USD1 trillion from the COP-26 appointed [43] Independent High-Level Expert Group on Climate Finance (IHLEG) [44].
2. The **Nationally Determined Contributions** (NDCs) to rachet up to a *collective* USD300 billion annually by 2035 came after intense lobbying by EMDEs. Numerous alliances of developing countries (see above) lobbied for an annual agreement of at least USD500 billion [45]. To increase the chances of reaching USD300 billion annually, a 'voluntary agreement' was made to include finances from Multilateral Development Banks (MDAs) and for 'developing countries to voluntarily count finance they provide towards the goal'. This was seen as being directed at China [46].
3. **UNFCCC Climate Funds** [40] to triple to USD5.2 billion by 2030 followed by an inventory report on developing countries (EMDE ex-China) consisting of 'a range of financial, technology development and transfer, and capacity-building needs' [47]. Twenty-eight EMDEs submitted LT-LEDS (*Long-Term Low-Emission Development Strategies*) reports cost their needs at USD2.4 trillion—see the UNFCCC *Synthesis Report* [48]. In other words, a wide-ranging development package. Estimated needs for adaptation were costed upwards of USD2.4 billion annually, far beyond the USD700 million *Loss and Damage Fund* agreed at COP-28 (Dubai, United Arab Emirates). The need for such funds is amplified by the fact that private capital considers returns on investments on adaptation projects, such as drainage systems or shoreline barriers, 'less clear' and they only become clearer when these projects are funded publicly [49].

Article 6 as 'Shred of Progress'

The comment of the Executive Director of the Cambridge Institute for Sustainability Leadership (CISL) on the outcome of COP-29 was that it 'limped over the line with some shreds of progress' [50]. One of those shreds was countries agreed standards for a centralised carbon market under Article 6.4 of the Paris Agreement Crediting Mechanism [51]. The ambition is to impose greater integrity upon the rather shoddy voluntary carbon markets-see YouTube video [52], (8–12 min in) and also Chaps. 4 and 5. Otherwise the reactions of many EMDE spokespersons and climate change activists hit decibels. Intensifying the reactions were the obtuse opening remarks by the President of Azerbaijan that oil and gas were 'gifts from God' [53]. Tempers were further angered by the choice of a 'petrostate' to host the event, but such is the reality that petrostates are both part of the problem and part of the solution. As the novelist and humourist Kurt Vonnegut would have concluded with sardonic wit, 'And so it goes' [54].

The Anomalies of Public Sector Green Finance

Successive COPs have taught the world that public finances are never enough to cover the costs of climate mitigation, not because the finances could never be sufficient but because of competing priorities. These are set by capitalist imperatives, priorities that privilege profits and the resulting accumulation of capital over other objectives. Governments of industrialised countries are fully aware that under normal circumstances, meaning non-wartime conditions, they cannot raise the resources to meet the global demands for green finance. They cling to the idea that public finances and public financial institutions can act as a catalyst to generate funds from the private sector. They can, but never enough to meet the required levels to stop global warming.

The choice governments have to make is between, on the one hand, fine-tuning market mechanisms such as improved financial and sustainability disclosures, although that runs against the ethos of most banks, improving the integrity of carbon credits to attract more private capital, providing tax incentives to companies to adopt cleaner technologies, and eliminating subsidies on fossil fuels; on the other hand, going all out to direct funding into mitigation and adjustments, such as ramping up public finances and imposing carbon taxes, domestically and cross-border, that truly reflect the social costs of GHG emissions. Carbon taxes also raise revenues. Politically, this approach will be very difficult to achieve between governments on an international scale, or within countries. However, international agreements are essential, otherwise capital and investments will simply migrate to the lowest tax countries. Without governments grasping the nettle on these issues, funding NDCs and other obligations under COP will be forever a circuitous game of 'ifs and 'maybes' until it is simply too late.

Political Economy of Financing Zero Emissions

Most governments in the Global North basically have the same financial playbook as they are prevented from 'going it alone'. Due to their need to remain competitive in world markets, and in terms of domestic resources management, countries are inhibited from switching resources on a mass scale to renewables, as if treating climate change as a national emergency to achieve Net Zero C02, or better still as Net Negative C02. (see Box 6.1) The possible exceptions, China and the USA, the two major polluters, owing to the scale of their economies, could if they chose, redirect enormous resources to renewables. Insofar as China has become the world leader in solar and wind power sectors, this is a viable commercial strategy, tariffs and sanctions notwithstanding. Politics within the USA is trending in the opposite direction, and the choice is less about resources than whether the US will renege on, or fulfil their international commitments. As far as the private sector is concerned, countries such as China have more room for manoeuvre with policy instruments such as political, administrative, and economic pressure on polluting industries, such as carbon taxes, tax incentives, and subsidies. This seems unlikely in the USA, at least until 2028. But relying upon private sector investment will never be enough.

Box 6.1 A Green Agenda

There is a reason why countries are reluctant to go all out in spending money on a green agenda, despite the real risks involved with climate change, but it is not a valid reason. It is, in a word, the global capitalist market. This needs explanation.

Any government that embarks upon a radical green agenda must incur upfront capital investments in areas such as upgrading the national electricity grid, building battery storage units for renewable energy supplies, subsidising the construction of windfarms, especially if ocean based, and seashore defences as part of an adaptation programme, and so forth. This implies governments create debts to cover these expenses prior to receiving back taxes from green growth further down the road. If governments also create green jobs at scale, with apprenticeships, digital skills training schemes in AI for monitoring and control centres, and subsidise the installation of insulation and heating and cooling systems for homes, then the net result is a redistribution of national wealth towards civil society. This on top of expenditures on social services, and possibly a universal basic income to lift everyone out of poverty-the argument is set out in John Ure (2024) [55]. A radical programme of this sort would appeal to working class families who could see the benefits, and would unify the top-down green policies with bottom-up civil support.

But! There would be push-back from conservative thinkers and capitalists who would regard state expenditure as crowding-out private capital, especially if the green agenda was in part financed by taxes on the earnings from wealth, otherwise known as unearned incomes. They would also cry foul as government debt rose to pay for the green agenda. Actually, debt (a government liability) is not a problem as long as the central bank issues bonds (a bank asset) covering the debt. The problem arises if a low rate of interest (set by the central bank), which keeps debt servicing charges down, is met by capital flight overseas where rates of interest may be higher. If this results in a rise in the rate of interest, this is unpopular and increases the debt burden leaving less finance from the budget for green investment. Only USA, where the dollar is the world's most traded currency, is somewhat protected. But even there, capital flight from

US Treasury bonds manifest itself in 2025 following President Trump's erractic imposition of tariffs on trade, sending the dollar south.

Most governments are tied into a non-radical neo-liberal agenda, which sees debt as either unsustainable or inevitably associated with high taxation. But debt is only unsustainable if it is used in unproductive ways. Used productively, it pays for itself over time. Therefore, the problem is not debt but how it is used, and who it is used for. Politically nevertheless it is difficult for any government to survive capital flight and the opposition of the fossil-fuel coalitions unless there is strong political support coming from civil society. But one possible way forward would be to create an International Green Bank (IGB) if it could gain wide support from enough leading economies. An IGB could aggregate green investible funds to overcome separate and possibly competing national interests, especially if exclusively focused on the Global South - see Chap. 7.

Public Sector as a Catalyst

Besides arm-twisting by the state to persuade the private sector to participate, the possibilities of public sources of finance being able to catalyse private finance depend partly upon whether they can act as investment partners. A much-favoured approach is 'blended finance' [56] which is a combination of public and private finance that spreads the risk of investing in projects. From Chart 6.1 it can be seen that the principal source of public sector green finance is the national DFIs (Development Finance Institutions). They were responsible for 36% of public green finance, during the financial year 2021–2022. Examples would include British International Investment (BII), which was subject to a highly critical review by the UK Parliament's International Development Committee [57], and on a much larger scale, the China Development Bank-not to be confused with the BRICS New Development Bank, see above-which provides loans to the BRICS (Brazil, Russia, India, China, South Africa) countries, for example, a loan of USD690 million announced during COP-29 to Brazil's National Bank for Economic and Social Development (BNDES) [58].

An additional source of public sector investments comes from State-Owned Enterprises (SOEs). Most SOE financing is 'balance sheet' financing, although perhaps better called 'off-balance sheet financing' [59], such as R&D funding, partnership funding, or leasing arrangements. The CPI notes that in the renewable energy sector with its high upfront capital outlays, SOE 'balance sheet' financing shows signs of replacing private corporate 'balance sheet' financing, suggesting a lower risk-appetite in the private sector. The risk factor is exactly the problem governments face when trying to catalyse private investments, so sharing it is widely seen as a way forward.

A further 26% of public green finance came from a mix of national government grants, loans, investments and debt relief, and from multilateral 'DFIs' otherwise known as MDBs (Multilateral Development Banks) such as the World Bank and its offshoots, the IFC (International Finance Corporation) and the IBRC (International Bank for Reconstruction & Development), and regional development banks such as the IDB (Inter-American Development Bank), the ADB (Asian Development Bank), and the China-led AIIB (Asian Infrastructure Investment Bank). According to a *Joint Report on Multilateral Development Banks Climate Finance*, 60% of their funds went to EMDEs [60]. As the climate emergency increases, and across the smorgasbord of acronyms, there has been a concerted push to give MDAs (Multilateral Development Agencies) a greater role—see below. International funding from these agencies embraces various forms of risk sharing and opportunities for project investments by private capital. This is especially the case where development aid is made on a concessionary basis which often involves contracts for private companies from the donor countries.

According to the CPI (see Chart 6.1) approximately 49% of international climate change financing comes from the private sector, but total funding remains woefully behind the trillion of dollars required, as identified by COP-29 (see above), and public finances are not seen as filling the gap. Hence the continued searching for mechanisms to catalyse far greater private financing. Pressure is being brought to bear upon state-owned financial institutions (SOFIs), for example, Germany's KfW Bankengruppe [61], and bilateral DFIs, such as the US International Development Finance Corporation (DFC) [62], to become more proactive. Currently they provide less than 8% of the total funding. This is highlighted in a report on 23 European DFIs by the Climate Bonds Initiative [63], which poses two issues. First, the involvement of DFIs in green finance as a means of mobilising private capital markets remains at too low a level, and second, this may in part be because their focus is more upon transactional than institutional activity. This is illustrated in Fig. 6.3.

Only 0.4% of international green financing came from climate-specific multilateral funds, and 50% of these, according to the CPI report, came from the UNFCCC's *Green Climate Fund* (GCF) [64] and a just over 25% from the *Global Environment Facility* (GEF) [65]. Of these, 47% went to adaptation, not mitigation, and 40% to AFOLU (Agriculture, Forestry and Other Land Use) projects. This pattern of more multilateral climate-related finance going into adaptation and nature-based projects appears to be the policy of the UNFCCC [66]. By 2022, 35% climate-related finance from the MDBs was devoted to adaptation projects [67], a grim fact that reflects the growing threats facing less developed countries of biodiversity loss and damage consequential of the climate crisis.

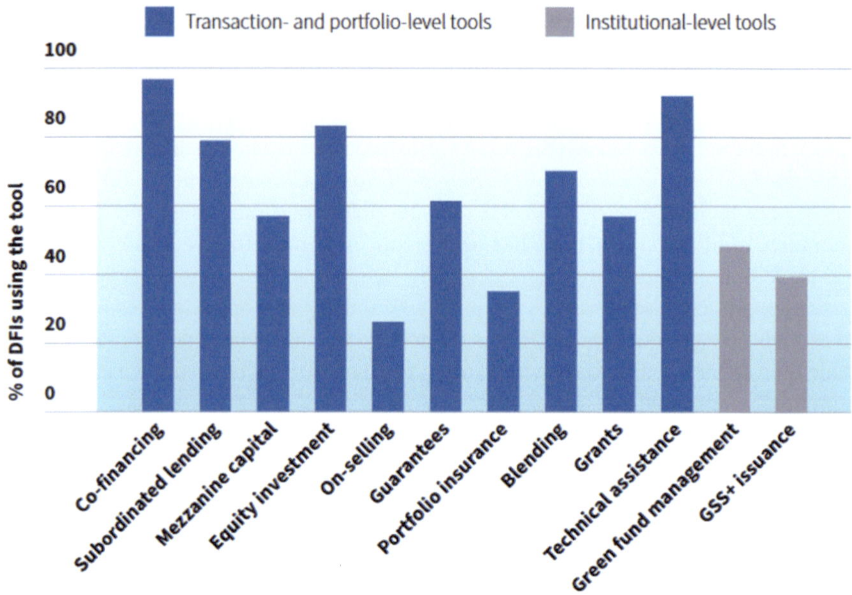

Fig. 6.3 Transactional and Institutional Tools of DFIs. The main transactions and portfolios of European Development Financial Institutions (DFIs). *Source* Climate bonds initiative

Public–Private Finance into Nature-Based Projects, Carbon Removal, Nuclear Power

Besides the three questions of how much climate-related finance for the Global South comes from the Global North, how might public financial flows catalyse private financial flows, and how should they be distributed between mitigation and adjustment, two further issues arise. First, what priority to give to, and how to finance, nature-based or 'natural capital' projects? The role of mangroves, peatlands, and forests in providing natural sinks that trap carbon dioxide and methane is well understood, as is the need to maintain a biodiversity of fauna and flora to enable the wild to sustain its natural cycles. A report from the World Wildlife Fund (WWF) in 2024 [68] warns 'wildlife populations, which are essential for the preservation of natural vegetation, have shrunk by an alarming 73% over 50 years (1970–2020)' as measured by the *Living Planet Index* (LPI) [69]. In 2019, an UN-backed Intergovernmental Science-Policy Platform on Biodiversity and Ecosystem Services (IPBES) [70] reckoned that along with co-benefits from biodiversity, nature-based solutions (NbS), such as reforestation and restoring wetlands, will be needed to provide 37% of climate change mitigation.

Second, can investment in carbon dioxide removal (CDR) projects using emerging technologies such as DACCS (Direct Air Capture with Carbon Capture and Storage)

play a significant role in reducing GHGs, assuming they are ready, or are they no more than expensive and ineffective palliative offsets to be used by fossil-fuel companies to prolong their commercial operations—see also Chap. 5. The answer to the last question will influence whether these technologies are supported or not by public authorities, for example, through tax breaks.

Nuclear Power

The voice of nuclear power on a global scale is the World Nuclear Association (WNA) with 44 member countries, operating 440 nuclear reactors, generating around 9% of the world's electricity, with many more planned [71]. At COP-29 the WNA signed up a further six countries to its *Declaration to Triple Nuclear Energy* [72], following its approval at COP-28 to triple nuclear energy by 2050 [73]. But is nuclear power green? Insofar as it emits no GHGs, it is a green way to generate electricity, notwithstanding the emissions of GHGs in the construction and management of nuclear plants. Insofar as it generates radioactive waste which has to be buried anywhere up to 5,000 m underground, depending upon its levels of toxicity [74], and needs to remain sealed for up to 10,000 years, with isotope plutonium-239 having a half-life of 24,000 years [75], it is a controlled [76], yet potentially major environmental hazard [77]. But, when is it reasonable to expect renewable energy supplies to make nuclear unnecessary? The most optimistic, but perhaps not the most realistic study would suggest it could be by 2035 [78]. Another more circumspect study suggests it could be by 2050 [79]. In reality, like Samuel Beckett's *Waiting for Godot*, it may be a lifetime away. That is, assuming no commercial breakthrough in nuclear fusion.

The cost of a nuclear plant of any size is measured as its Levelized Cost of Electricity (LCOE) or cost per megawatt per hour, which is the total cost of building and operating the plant, of which capital costs are typically 60% [80], divided by the total energy output over the lifetime of the plant, including the costs of financing (not just the 'overnight' costs). But as the US Department of Energy points out, LCOE 'does not fully account for the value of reliable, carbon-free heat for industrial steam, which is critical to many industrial processes and has few decarbonized alternatives' [81]. In other words, in additional to financial returns, the economic returns to industry should take into account the reliability and sustainability of nuclear at times when renewables may not be available.

The key problem to which nuclear is being promoted as an answer is that the world is facing a dilemma: the greater efficiency in electricity consumption is being outstripped by an increased demand for electricity. This renders measurements such as energy-intensity per capita meaningless because even if everyone consumes less on routine tasks, they are taking on a wider range of energy-consuming tasks (uses) at an ever-increasing rate. This is a direct outcome of digitalisation, whether it is social media or the use of Generative AI, Large Language Models (LLMs), and task-driven AI agents. For example, 'with just 100 million weekly active users making

Table 6.2 AI and energy usage

GWs energy usage ±	LLM pre- and post-training	AI agents and applications
Upstream energy consumption	+data centres/R&D	+data centres/R&D
Downstream energy consumption	+enterprise R&D	− commerce/consumers

Note + an increase in total energy used; − a reduced, more efficient use of energy as a consequence of AI

five queries per week, GPT-3.5's annual energy consumption reaches 44,200 MWh—equivalent to powering 4,150 U.S. households for a year' [82]. An estimate by the Electric Power Research Institute (EPRI) suggests data centres in the USA by 2030 could be consuming as much energy as the entire country of Indonesia [83].

The competitive race between Generative AI service providers is transforming stand-alone data centres providing cloud-based services into mega-clusters of hyper-scale data centres and energy consumption from Megawatts to Gigawatts and, collectively, towards Terawatts. Further, the competition to pre-train LLMs, post-train them as self-correcting models, and then spawn AI agents, which is where AI's commercial future lies, is so fierce that energy costs, which are typically less than 20% of the total expenditure on hundreds of thousands of GPUs chips, are being traded-off against rapid technological development [84]. Upstream investment in AI is therefore making a net-negative contribution to a net-negative or net-zero target. Downstream, the possibilities vary. According to an Accenture survey of 2,000 of the world's largest companies, by 2024 only 'one in five European businesses (20%) are using AI for decarbonization purposes, compared to 14% in the Asia Pacific and 10% in North America' [85]. However, in the social balance the benefits arising from AI in areas such as medicine are offsets. An assessment by Nicholas Stern et al. from the Grantham Institute in 2025 'Green and intelligent: the role of AI in the climate transition' is more optimistic - https://doi.org/10.1038/344168-025-00262-3. Table 6.2 imagines where AI could be part the problem or part the solution.

Multilateral Green Finance and EMDEs

In the view of the Independent High-Level Expert Group on Climate Finance (IHLEG) [10], MDBs 'are key to both unlocking investment opportunities and mobilising finance' and need to 'play a much stronger role in reducing, managing and sharing risk and in reducing the cost of capital. To deliver on the Paris targets, their role will need to change fundamentally and the scale of their support'. While this sounds logical, it implies that commercial banking interests are insufficient to address the green finances gap, especially for the Global South, and thus public finances must step in, a shift from the view that the task of public finance is to unlock private finance. This is evident in the following statement from the Asian Development Bank (ADB): 'Commercial financial institutions are limited in providing affordable finance especially for activities with uncertain returns and positive externalities. As such, recognition is growing that public financial institutions need to play a greater role in scaling up investment in climate action and the Sustainable Development Goals (SDGs)' [86]. To explore this logic, see Box 6.2.

Box 6.2 Green Financial Gap Versus Financial Governance

The inner logic of the ADB's statement, while arguing that financial institutions should do more, is that clean energy industrial processes need to take the lead, financed by *the economic surpluses they create*. There are no other sufficient sources of funds. This is not just a play on words. The green financial gap as discussed above is far too great to be bridged by either public or private funds. Further, even a mobilisation of more funds towards and within the economies of the Global South will, in many if not in *all* cases, require major reforms in the financial governance in those economies. There is a substantial literature on the need for these reforms, although they should not be thought of as necessary only in EMDEs. They range from cleansing corruption, accountability, and transparency to more effective tax revenue collections and the creation of a national Green Bank [87]. In other words, to take action to slow down climate change calls for a much wider range of reforms across economies and societies alike. Climate change mitigation is only an outcome of a chain of reforms that start with a change of industrial, economic, and social practices. That is actually the good news, because not all such changes require green funds, rather they require willful decision-making.

The sentiments of the IHLEG echo those expressed in 2023 by the G20 Expert Group on Capital Adequacy Framework's *G20 Independent Review* [88], which sees MDBs playing 'a significant role as policy tools, enabling shareholders to leverage scarce fiscal resources for development, tackling climate and other public goods and responding to crisis'. This can work so long as it is understood, and indeed driven by a recognition that MDB resources will never be sufficient for the job. Such a recognition will help if it adds urgency to doing whatever *is* achievable. This builds on efforts by the World Bank from 2017 onwards to coordinate climate-related policies with eight MDBs in four areas: direct investment lending operations, policy-based lending operations, intermediated financing, and general corporate purpose finance [89]. The call for MDBs to play a more active role in climate-related projects was again echoed in 2024 by the World Resources Institute (WRI) [90].

The World Bank's advisory on *Sustainable Finance and Environmental, Social, and Governance (ESG)* [91] categorises its lending as shown in Fig. 6.4.

The World Bank issues bonds to its investors who provide the finances for the emerging economies. Likewise various UN agencies provide green development loans led by the UN Environment Programme (UNEP) [92], supported by the work

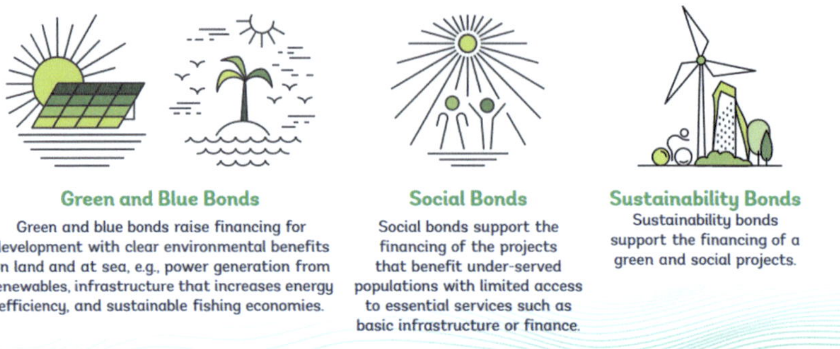

Fig. 6.4 The World Bank's Green and Blue, Social and Sustainable Bonds. World Bank's 4 ESG lending categories. *Source* The World Bank

of various other UN bodies: the High-level Political Forum on Sustainable Development, the Commission on Sustainable Development (CSD), the Intergovernmental Panel on Climate Change (IPCC), and the UN Forum on Forests [93]. Moving down the financial ladder come regional development banks and national aid agencies sponsoring green grants and offering green financial loans, often at concessionary rates, supported by sovereign wealth funds, and global and regional philanthropic agencies.

Commercial Sectors

Beyond the public and NGO sectors come the commercial sector, private global and regional banks, pension funds, asset management companies, insurance companies, and many other financial institutions in addition to green investments by multinational companies. In the *World Energy Investment* (WEI) 2022 report [94], shown in Table 6.3, the International Energy Authority (IEA) provides the following breakdown of funding sources, instruments, and capital markets.

At the ground level, as it were, is the voluntary carbon market, buying and dealing in carbon credits—see Chap. 5. For example, the 2024 report of the VCMI (Voluntary Carbon Markets Integrity Initiative) proposing offsetting claims based upon Scope 3 emissions recommends that claims need to include a demonstration 'that they are making progress on financial allocation, governance, and strategy towards meeting a near-term emissions reduction target' [98]. The involvement of the private sector in green finance and climate-related and nature-based projects has undoubtedly been struggling to bridge the gap between for-profit and not-for-profit investments, and to re-establish the integrity of carbon markets, an effort strongly supported by UN think-tanks at the COP meetings. One outcome has been the creation of new forms of both private and listed 'public benefit companies'. This is examined below.

Table 6.3 A Breakdown of Funding Sources, Instruments and Capital Markets. WEI methodology breaks down the following areas

Funding for energy	Financial actors
Funding sources	Household, retained earnings, banks, public finance, capital markets
Funding instruments	Equity, debt, corporate finance, project finance, grants, guarantees
Capital markets	1. **Institutional investors**: Asset managers, infrastructure funds, insurance companies, pension funds, private equity and venture capital, sovereign wealth funds *Note: WEI noted 90% of energy investments come from the balance sheets of companies and consumers, and a smaller amount from project finance, mostly from banks* 2. **Fixed-income investments**: Sustainable finance (debt securities) including green bonds, green loans, etc., certified under schemes such as the Green Bond Principles [95] or the Climate Bonds Standard and Certification Scheme [96] 3. **Equity markets**: A combination of institutional investors (see above) and ESG (Environmental, Social and Governance) investment such as ESG exchange-traded funds designed to focus upon corporate responsibility and sustainability [97]

Source IEA-World Energy Investment 2022

Financial Taskforces

COPs are mini-UN gatherings, where country commitments to finance the Paris Agreement and NDCs are difficult to realise, but by agreeing on the principles for new financial initiatives they open the door to a series of diplomatic and financial deal-making, both in their lead-up and in their aftermath. In 2020, in preparation for COP26 (Glasgow 2021), the Institute of International Finance (IIF) established a private-sector *Taskforce on Scaling Voluntary Carbon Markets* (TSVCM) initiated by Mark Carney, former Governor of the Bank of England, UN Special Envoy for Climate Action and Finance [99], and in 2025 Canada's Prime Minister. The report of the TSVCM noted that, if financing carbon reduction projects were to be achieved solely through the trading of carbon credits, then achieving 2Gt CO_2 sequestration and removal by 2030 would require a '15-fold scale-up of voluntary offsetting in 2030 versus 2019' [100].

Mark Carney was also the Chair of the Financial Stability Board (FSB), created by the G20 countries following the global financial crisis of 2008. In the footsteps of COP21 (Paris 2015), and as a means to leverage private sector funds, the FSB created a *Taskforce on Climate-Related Financial Disclosures* (TCFD) chaired by Michael Bloomberg, 'to develop recommendations on the types of information that companies should disclose to support investors, lenders, and insurance underwriters in appropriately assessing and pricing a specific set of risks—risks related to climate change' [101]. As with the TSVCM, the TCFD was focused upon the management

Fig. 6.5 TCFC Recommendations for Climate-Related Disclosures. Types of disclosures recommended by the TCFD. *Source* TCFD (2017)

of financial risk in private markets. Its *Final Report* delivered in 2017 [102] cited a 2015 report on climate risk to global capital assets at between USD4.2 trillion and USD43 trillion by the end of the century. But given the role of the insurance industry and the uneven regional distribution of climate damage, it is not clear this would be a sufficient incentive to move the dial. Figure 6.5 is the TCFD's own summary of its recommendations, with an emphasis upon the quality of data available.

COPs 26 and 27

The TSVCM report was received at COP26 (Glasgow 2021) which became the arena for financial ways to meet the goals of the Paris Agreement. The private route placed the emphasis upon emissions disclosures, risk management by banks, insurers and investors, and the role to be played by carbon credits in the voluntary carbon markets (VCMs). The public route was to boost funds for climate change mitigation and adaptation, with mitigation receiving the greater support; the other way around would be tantamount to admitting defeat. Some will feel that is the outcome of COP 30.

Following its final report, the TSVCM was disbanded, and the setting of financial standards for climate-related projects was absorbed into a wider standards-setting framework for sustainability projects. The International Financial Reporting Standards (IFRS) body [103] also absorbed the Climate Standards Disclosure Board (CSDB), an international consortium of business and environmental NGOs [104], and other bodies to create the International Sustainability Standards Board (ISSB) with four objectives: standards for sustainability disclosures, investor information,

information for global capital markets, and facilitation of interoperability with disclosures that are jurisdiction-specific and/or aimed at broader stakeholder groups [105]. Under the aegis of the IFRS, the ISSB in 2023 issued two IFRS standards, *IFRS-S1* as a generic set of disclosure standards to communicate sustainability risks, and *IFRS-S2* for climate-related disclosures [106].

Following the public route, COP26 gathered pledges of over USD350 million to an *Adaptation Fund*, and USD600 million to the *Least Developed Countries Fund* (LDCF) [107]. For each COP the Standing Committee on Finance (SCF) presents the findings of a *Biennial Assessment and Overview of Climate Finance Flows* (BA) [108]. In 2020, the BA estimated globally climate-related spending had reached almost USD87 billion in 2017–2018, a 16% increase over 2015–2016. This included, for example, spending on electric vehicle infrastructure, but the world lacked a common standard or even a definition of what constituted climate-related finance. The report confirmed that support for funds for mitigation were nearly 4 × those for adaptation. It is not unreasonable to assume that any shift of emphasis towards funds for adaptation would imply a growing disbelief in the ability to halt climate change. That said, the growing immediacy of climate change effects, such as rising sea levels and a greater frequency of violent storms, was recognised in an *Implementation Plan* at the COP27 (Sharm El-Sheik, Egypt, 2022) which called for the mobilisation of between USD140 and USD300 billion in additional funding annually from public and private sources, with a minimum target of 50% for adaptation. It was noted that funds for Africa were only 3% (USD11.4 billion) of the total and needed to be increased seven times by 2030 [109]. The International Finance Corporation (IFC) and its online Green Banking Academy (GBAC) concluded that the *Implementation Plan* for a transition to low-carbon economies would need between USD4 and USD6 trillion *annually*, and 'will require a transformation of the financial system and its structures and processes, engaging governments, central banks, commercial banks, institutional investors and other financial actors' [110]. A tall order by any standard—see also Box 6.2.

Included in the IFC assessment was the need for banks to increase their share of climate finance from 7% in 2016 to 30% by 2030. A key part of the proposed strategy was 'blended finance' whereby government 'concessionary capital' funding (meaning at a lower cost of capital) combines with private sector funding, reducing the investment risk for the latter. According to the IFC/GBAC report *COP 27-Implications for Financial Institutions* (see above) the 'IFC has catalyzed $11 of commercial funding from sponsors and other lenders for every $1 of blended climate financing deployed' and by reducing risk also encourages smaller investors to participate [111]. However, according to a 2019 report by the Overseas Development Institute (ODI) [112], each dollar invested by MDBs and DFIs mobilised on average only USD0.75 of private finance for developing countries, and less for lower income countries, and concluded 'Expectations that blended finance can bridge the SDG financing gap are unrealistic: "billions to billions" is more plausible than "billions to trillions"' [113]. Nevertheless, a call for the scaling up of blended finance came from the UN-backed Net-Zero Asset Owner Alliance in 2022 [114].

GZANZ: Banks and Climate-Related Finance

During COP26 (Glasgow, 2021) the UN-backed Glasgow Financial Alliance for Net Zero (GFANZ) was established, with assets of over USD130 trillion committed to using science-based guidelines to reach net-zero carbon emissions by mid-century. Chaired jointly by Mark Carney and Michael Bloomberg, pre-Donald Trump, GFANZ membership grew from 450 banks and other financial institutions across 6 sectors in 2021 to 700 financial institutions across 50 countries and 8 industrial sectors in 2024. The Network of Central Banks and Supervisors for Greening the Financial System (NGFS) [115] was placed under GFANZ (see above), as were regional networks of MDBs and development financial institutions (DFIs) covering Africa, Asia Pacific, and South America [116]; and renewable energy projects under the Just Energy Transition Partnerships (JETPs) which by-pass for EMDEs the use of LNG (liquid natural gas) as a transitionary stage to renewables. But from the very start the capital available was totally inadequate, by up to USD90 billion according to the UN-backed International Institute for Sustainable Development (IISD) [117]. That was before President Trump withdrew US bank support for GFANZ in 2025.

From 2025 it all started falling apart, including doubts about the future of US commitments to JETPs with Indonesia in 2022 [118], and the Philippines, Senegal and Vietnam in 2024. As widely reported by the media, right-wing Republicans, climate deniers, and sceptics in the USA had been leading the charge to stop 'woke' banks supporting ESG initiatives [119], and Republicans in the House of Representatives challenged US support for GFANZ, partly on the grounds that it involved non-US banks supporting 'models of development financially backed by U.S. adversaries' [120]. Under this pressure, and in anticipation of the presidency of Donald Trump, from early December 2024 onwards six major US banks found it expedient to withdraw from GFANZ and its private banking division the Net-Zero Banking Alliance (NZBA): Goldman Sachs, Wells Fargo, Citigroup, Bank of America, Morgan Stanley and JPMorgan Chase, along with BlackRock, the US largest asset management group [121]. Following the inauguration of the President in January 2025, the US Federal Reserve Bank (FRB) and the Federal Deposit Insurance Corporation (FDIC) also announced their withdrawal [122]. The implication of these withdrawals being these institutions would no longer make explicit commitments to, and certainly would not prioritise, net zero by 2050. At the same time, the US Treasury announced its withdrawal from the GFANZ-linked NGFS [123]. For the diminishing of GFANZ and NGFS, see Box 6.3.

Box 6.3 The Diminishing of GFANZ

In response to the US withdrawal of support, GFANZ has shrunk its ambitions, and lost membership, although it still exceeds 500 financial institutions with over USD100 trillion in balance sheet assets. Under a reconstruction plan, the CEOs within GFANZ are to take on responsibility for the following issues:

- First, barriers to capital mobilisation for the transition 'including sovereign wealth funds, financial institutions, and market participants in countries with longer transition pathways' to net zero.
- Second, to promote the use of International Sustainability Standards Board (ISSB) standards for financial data.
- Third, to support the Industrial Transition Accelerator (ITA) [124], which was launched at COP28 (Dubai, United Arab Emirates, 2023) to accelerate funding by up to USD700 billion by 2030 into 450 large-scale low-carbon projects across six sectors that account for 30% of global emissions (aluminum, cement, chemicals, steel, shipping, and aviation) [124], but if it risks low or not existent financial returns, it will be reliant upon substantial public policy intervention.
- Fourth, 'the scaling of high-integrity voluntary carbon markets by engaging with governments and regulators to help remove regulatory barriers to private sector investment and develop enabling policy environments' [125].

On paper, this is a skeletal range of commitments which pull back from proactive capital-raising initiatives by the banking and financial sector, settling upon removing 'barriers to capital', handing the baton very much to the voluntary carbon markets which are still struggling to establish a record of integrity (see Chap. 5) and ultimately clinging onto the hope that markets can succeed without unsustainable public capital.

Not Fit for Purpose

As it stands the international financial system with regard to funding climate-related projects and programmes is not fit-for-purpose. This is already implied in the NGFS early report, *Acute physical impacts from climate change and monetary policy* [126], which found that expenditure on disaster recovery far exceeds that on risk reduction and adaptation. The report looked somewhat hopelessly towards the insurance markets to carry the risks which naturally they will do by raising premiums and reducing coverage. Needless to say, the traditional role of central banks has been to focus upon monetary policy, but the NGFS notes that central bank mandates may limit the actions they can take with regard 'to the transition rather than managing risks'. In other words, it is not part of their mandate as set by governments.

Large private banks as commercial institutions also do not have a climate mandate, but many of them on paper at least made commitments before the collapse of US support. The outgoing the Chair of NZBA in 2024 noted over two-thirds of its 140 member banks had set science-based 2030 targets to limit warming to 1.5C, which leaves many not having done so [127], as confirmed more generally by the *State of Transition in the Banking Sector* (2024), a report of the Transition Pathway Initiative, a research project based at the London School of Economics (LSE) which found that 'On average, the 38 banks assessed in 2024 score on only 15% of the 72 sub-indicators that make up the NZBAF, and no bank scores on more than half of the sub-indicators … Banks' Carbon Performance is similarly weak, with low levels of alignment with the 1.5 °C and Below 2 °C benchmarks, particularly in the medium and long term' [128].

The overall picture is that international finance cannot close the gap to fund climate change mitigation, adaptation, and damage coverage. This was foreshadowed in the finding of the GFANZ *Progress Report* for 2023 [129], that although 75% of new energy capacity added in 2022 to EMDEs was low-carbon, 'the International Energy Agency (IEA) estimates that annual investment in low-carbon energy supply must grow more than *five-fold* from 2022 to 2030' [130]—emphasis added (see Box 6.4). Equally troubling for GFANZ, the Industrial Transition Accelerator (ITA) was struggling 'with demand for green products still too weak to justify investment' [131]. Clearly, the commercial sector will not be a leader without substantial public policy intervention.

Box 6.4 Not All Is Lost

To keep a balance between what GFANZ has achieved and its insufficiency, the record includes the following:

1. The Climate Finance Leadership Initiative (CFLI) was created in 2019 by Michael Bloomberg to work with financial regulators and others on NGFS-initiated climate-related projects, as in India and Columbia [132].
2. GFANZ supports a Nexus Water, Food and Energy (NWFE) country platform in Egypt.
3. The Global Energy Alliance for People and Planet (GEAPP) [133] is devoted to spreading access to green energy within emerging economies [134].
4. The Tropical Forest Alliance's Innovative Finance for the Amazon, Cerrado, and Chaco (IFACC) [135].
5. The **F**inance to **A**ccelerate the **S**ustainable **T**ransition-**Infra**structure (FAST-Infra) was conceived in 2020 by the Climate Policy Initiative (CPI), the OECD, and others [136].
6. **Mobilis**ing **I**nstitutional Capital **T**hrough **L**isted Product **S**tructures (MOBILIST) is a UK government programme supporting 'development and the climate transition' [137].

7. An activist NGO Forests & Finance [138] reviews the role of banks and others [139], and provides a database of projects to reverse deforestation in developing regions of the world [140].

GFANZ also set up in 2024 a Global Capacity Building Coalition (GCBC) [141] to bring expert advice to green financial initiatives 'by building an ecosystem of financial institutions, policymakers and regulators, multilateral organisations, professional bodies, trade associations, universities and business schools, climate NGOs, and others to accelerate learning and scale climate finance' [142]. In light of the withdrawal of US support, the GCBC could prove its worth by becoming a global voice.

Private Capital Markets

With debt levels as high as they are, government bodies inevitably look somewhat desperately towards the private capital markets, including those in the Global South, for financing climate-related projects. This has been the focus of the UN-supported think-tanks cited above. The alternative approach, but seen as complementary, is to focus upon the role of public finances to leverage or catalyse private finance, such as the issuing of green bonds or blended capital which reduces risk to private loans and investments. The WEF [143] provides an overview of blended finance options and financial instruments available from the private sector:

- Sustainability-linked green bonds/loans.
- Nature-based solutions (loans, project finance).
- Green insurance.
- Nature impact funds.
- Nature-venture capital and equity investments.
- Blended finance/public–private-philanthropic partnerships (4P)/debt-for-nature swaps/nature infrastructure/conservation bonds (e.g. blue bonds for ocean preservation).

Markets work principally according to price-setting mechanisms that reflect issues such as the cost-of-capital and risk-adjusted rates of return on investments, which in turn implies giving rights of ownership and imputing market values to natural capital. Any one of these can be influenced by changes in law, by government policies, but historically the operation of 'free' (unregulated) markets has been given policy priority over the conservation of nature.

Prior to the industrial revolution, the loss of wildlife habitat was mostly witnessed locally, but the globalisation of markets, largely driven by colonisation, was the start of a new era of international capitalism that witnessed the systematic subjugation of the natural world to the commercial world. This has been comprehensively covered in *The Economics of Biodiversity: The Dasgupta Review* (2021), a report commissioned by the British Government (HM Treasury):'In recent decades eroding natural capital has been precisely the means by which the world economy has deployed for enjoying what is routinely celebrated as "economic growth"' [144].

Natural Capital and the Pricing of Public Goods

Reducing the speed and impacts of global warming may be considered a public good as it benefits everyone to the exclusion of no-one. Two issues arise from this. First, ways to place a value on public goods, such as natural capital, where positive and negative externalities arise from the actions of private capital in open markets. Second, the emerging business financial models that try to overcome negative externalities by bridging private markets with conservation for the public good.

Historically, different ways have been used to place value on public goods, such as the cost of supporting public parks and nature reserves, but because climate change and global warming cross borders, international political economy comes into play [145]. As the degrees of financial and regulatory freedom of countries become more limited, there is a greater need to coordinate a collective response to negative externalities (GHG emissions) across markets, whether controlled by private or state capital. In theory, there could be answers agreed between states, such as a carbon or Pigou Tax, named after the 1920s British economist Aurthur Pigou who proposed a tax on negative externalities, where pollution would count, and a negative tax (subsidy) for positive externalities, where renewable energy would count [146]. The underlying question though is how to place a value on either. One answer economists often pursue is 'revealed preference' which refers to the preference of consumers or the general public as revealed, either through their market activity (what items they spend on at different prices), or through what elected governments spend on public services such as environmental conservation.

However, there are too many abstract assumptions behind these approaches, such as high levels (if not perfect levels) of information and transparency. Other answers include 'shadow pricing' [147], for example, through pricing the costs of mitigation, such as health costs and loss of productivity resulting from days off through sickness. However, assessing how far the management of a public park contributes to GHG mitigation is likely to be somewhat arbitrary. Whatever techniques may be used to assign a value (through the use of prices) or to tax an asset, including a natural asset, there always remains the 'free-loader' problem, discussed with regard to climate change by Nobel Prize-winning economist, William Nordhaus in *The Spirit of Green* (2021) [148]. He cites a US Government survey in 2020 which assigned a USD40 per tonne CO2 as the estimated social cost of carbon (SCC) which he quite rightly views as a serious underestimate if global warming is to be kept below 2C by 2050 (see also Chap. 3). But there is a cost to any state adopting a carbon tax when others do not. As a paper from the International Monetary Fund (IMF) points out, the COP21 (2015) Paris Agreement did not spell out how it could be enforced [149].

Innovation and New Business Models

In recent years, new business models have arisen to straddle private capital markets and the public goods sector, especially for climate-change and nature-based projects-see Box 6.5. These business models rely upon private capital to finance nature-based projects, but they either have to involve an income or revenue-stream, or they are essentially philanthropic. At their core lie two conceptually different ways in which to value nature. One places a price upon the value of nature as a public good, and one prices the value of natural capital as a marketable commodity. However, there is what might be called a quasi-value which arises from selling carbon credits related to a project. If the project fails to deliver a profitable business, then the credits are similar to a philanthropic contribution even if the motive is less philanthropic and the buyers really want to use the credits as an offset against their own polluting. On the other hand, if the project becomes profitable, then the credits can also be retired as recognizable capital outlays. Examples of profitability could include benefits such as ecotourism which create jobs (co-benefits) for local communities, or the products of nature that can be processed into homeopathic medicines or foods. But, as a review of the new business models and case studies presented by the WEF suggests (Box 6.5), in the short-run investment arising from the sale of carbon credits seems the most likely source of private capital.

Box 6.5 Private Capital Markets and Climate-related ESGs

Public benefit corporations (PBCs), as they are called in the US state of Delaware, are a relatively new form of corporation and distinct from non-profit organisations [150]. Their purpose is to serve the triple interests of profitability, community interests, and environmental responsibility, or 'People, Planet and Profits', as attributed to John Elkington in 1994 [151]. By 2024, PBCs were recognised in 43 US states, plus the District of Columbia and Puerto Rico, but under slightly different names and with slightly different legal requirements, such as Benefit Corporations (B-Corporations), and in California, Social Benefit Corporations [152]. When they issue shares they must make their purpose clear to buyers. In other words, they are vehicles for investors who are not singularly focused upon profitability. A key driver of the concept of a socially responsible corporation has been B-Lab [153], which provides certification to B-Corporations authenticating their triple focus.

An even more recent form of social organisation is the Natural Asset Corporation (NAC) which is specifically focused upon preserving and restoring 'natural capital' in the form of biodiversity on land, wetlands, mangroves, or in the oceans. The concept has been promoted by entrepreneur and conservatist, Douglas Eger [154], CEO of Intrinsic Exchange Group (IEG) [155]. The role of IEG forms one of nine case studies in WEF (2024) *The Future of Nature and Business* [156]. While IEG is a private corporation, the US Securities and Exchange Commission (SEC) had plans in January 2024 for the listing of NACs on the NYSE [157]; but that ran into a brick wall of opposition from a mixed coalition of Far Right objectionists who saw NACs as a challenge to the freedoms of land owners, even as a backdoor entry of hostile foreign powers, and others who questioned the role of carbon credits as a way to fund conservation [158]. Such are the ideological divisions that spice climate politics in the twenty-first century, in which everything takes on a duplicitous and toxic aspect.

The key question is, despite official attitudes and an inability to list publicly, can natural capital projects and ventures succeed in attracting sufficient private capital and nature-conscious investors to supplement public financing of nature conservation and renewal in the wider effort to combat climate change. Nature provides carbon sinks but these are currently overwhelmed by human activities that emit CO2 and its equivalent GHGs. Can a combination of industrial carbon abatement and natural carbon sinks as outlined in the WEF case studies contribute an answer? At least it's a start.

Takeaways from the Case Studies

- Innovations in technologies and methodologies can help reduce the costs of projects that sustain natural capital, and may possibly lead to natural capital products, such as the discovery of new homeopathic medicines, biochar as a local community renewable energy source, ecotourism as a way to sustain and empower local communities. Examples of innovations include the use of satellite technologies such as LIDAR (Light Detecting and Ranging) [159] which is now widely used in archeology [160], and 'convolutional neural networks' (CNNs) that provide algorithm-generated ground-level views from satellite images [161]. These technologies can reveal at low cost the potential commercial value of natural resources. Other technologies, such as blockchain apps, can offer more cost-effective management and marketing tools. But is the consequence anything more beneficial than the commodification of nature?
- In cases where natural capital products are available, the potential for financing frequently includes the voluntary market for carbon credits, but this is controversial both for the integrity reasons discussed in Chap. 5, and

because this threatens to make the sustainability of NACs dependent upon the uncertain sustainability and scale of voluntary carbon markets (VCMs). Nevertheless, this was the vision of the TSVCM and remains the vision of the GFANZ and other market makers.

- Scaling of such projects, which tend to be small and often community based, requires accelerator capital beyond the initial angel and venture capital. This adds to the uncertain long-term sustainability of such projects. Ultimately, capital 'patience' is the ubiquitous message coming from the case studies. If the expected returns on commercial capital in competitive markets is around 4–5 years, for natural capital projects it is rarely less than 8–10 years, and often runs into many decades as is the case with reforestation and wetlands. As a contributor to LinkedIn put it: 'we don't have decades'.

Price versus Value

Shareholders in PBCs and NACs have de facto accepted lower levels of return in the early years, with returns from natural capital ventures coming much later if at all, and with high average levels of risk and uncertainty, where risks likely involve higher insurance premiums against bad weather, tree disease, and fire damage. The question is whether this cohort of responsible and socially conscious 'patient' wealthy people can be grown. While the inducement is unlikely to be financial, research in 2024 by BeZero does show that credits with 'beyond carbon' sustainability goals do sell at a premium: industrial processes by 20%, household devices by 19%, and nature-based solutions by 7% [162]. Maybe that ratio of close to 4:1 is about right, if it can be fully scaled by 2030.

The issue of what most motivates investors, higher returns to finance capital or higher returns to natural capital, is somewhat ontological. Do 'animal spirits' drive people as 'economic actors' to create markets and to trade at exchange values (prices) or does the existence of commercial markets as the dominant institutional and structural form of exchange drive the 'animal spirits' in people. The answer to this slightly philosophical question is relevant because it will influence the success of efforts to engage investors in PBCs, NACs, and nature-based projects which have uncertain and lower rates of profitability. It brings to mind the concept of the 'tragedy of the commons' which suggests that without ownership there is no responsibility to conserve nature, a concept that totally overlooks the fact that on the contrary so many people *do* conserve nature.

Concluding Observations

Natural capital projects seem to want to straddle two very different spaces: sustainability versus profitability; what is profitable may or may not be sustainable financially, and what is sustainable in nature may or may not be profitable. If the two realms compete for resources, society needs to determine its preferences and ways to signal them.

- **Accessibility versus property**: Traditionally, whether a natural area is private or public property has often determined whether or not it is accessible to conservation. PBCs, NACs, and other business models suggest ways to resolve this dilemma. But, as acknowledged in the foreword to the WEF report, over 80% of total financial flows supporting the protection, restoration, and enhancing of nature come from public sector sources.
- **Commonwealth versus private wealth**: The distribution of wealth-generating income will influence the ability of society to finance the conservation of nature, while private capital has a choice to finance nature projects or not. Ultimately, it comes down to placing a social value on nature, irrespective of whether it is or is not underpinned by a pricing of natural capital.

A key element in most of these new business models is their commitment to co-benefits (benefits shared with others), especially attractive to philanthropic investors. In this regard, the GFANZ's Global Capacity Building Coalition (GCBC) as outlined in Box 6.3 may offer a template for such an approach if sufficient weight is granted to the opinions and involvement of NGOs, academic and local communities. One example that arose before the inauguration of President Trump was the US Green Bank 50 (GB50) initiative, launched during Climate Week in New York City, September 2024 [163]. GB50 is a membership of 'not-for-profit' banks across the 50 states of the US, with the mission of green banks playing 'a crucial role in creating jobs, boosting local economies, addressing climate change and making clean energy accessible to all Americans' [164]. However, a large part of the funding was expected to come from the USD7 billion *Gas Reduction Fund* [165], created under the Inflation Reduction Act (IRA) [166]. As an Act of Congress, the IRA was contested terrain within the 'separation of powers' as Donald Trump 'tests the boundaries of Presidential power' [167]. But the IRA looked doomed by 2025.

Financing Biodiversity

Biodiversity runs parallel to, but is also an intricate part of climate change abatement. Biodiversity is a public good in its own right, climate crisis or not. There is no clearer or more emphatic statement of this than in the *Foreword* by Sir David Attenborough to the *Dasgupta Review* (2021) of the *Economics of Biodiversity* [144]. Biodiversity is also the guarantee of sustainability, because monocultures (plant and animal species) are vulnerable to eradication by diseases. The climate crisis simply makes biodiversity all the more urgent, and while UN conventions on biodiversity are also Conference of Parties (COPs), they are run separately to those devoted to climate change.

Intergovernmental efforts to address the challenges of biodiversity degradation date back to the Rio Earth Summit or Convention on Biological Diversity (CBD) in 1992 [168]. The only major country not signed up to it is the USA [169]. In 2005 at COP11, the Biodiversity Finance Initiative (BIOFIN) was developed by the UNDP and the European Commission to channel finances towards global and national biodiversity goals, and now involves 40 countries with reports on their biodiversity financial needs and expenditures [170]. The Rio Earth Summit led up to the UN Climate Change Conference (Cancun, 2010) and the adoption of the *Strategic Plan for Biodiversity 2011–2020* [171]. A *Post-2020 Global Biodiversity Framework* was issued in 2022 [172], and adopted as the *Kunming-Montreal Global Biodiversity Framework* (GBF) at the UN Biodiversity Conference (Montreal, 2022) [173]. At Cancun also, the UNFCCC agreed a *Green Climate Fund* (GCF), although sovereign pledges to the GCF by 2030 were only USD13.5 billion, viewed by Friends of the Earth Europe as 'not worth cheering' [174], with an additional USD38.4 billion of co-financing [175].

Biodiversity funding began on an international scale with the Global Environment Facility (GEF), established at the Rio Convention as a multiagency fund including NGOs and community organisations. The GEF consists of a family of funds: the *Global Environment Facility Trust Fund*, created at Rio by the World Bank to provide blended finance across various nature-related projects [176]; the *Global Biodiversity Framework Fund* (GBFF) launched 2023 at the GEF Assembly in Vancouver to scale up national funding for biodiversity [177]; the *Least Developed Countries Fund* (LDCF), created in 2021 by the UNFCCC specifically to fund climate-change-related projects in LDCs [178]; the *Special Climate Change Fund* (SCCF), created 2021 in light of the UNEP's 2021 *Adaptation Gap report*, that adaptation costs in developing countries are expected to rise from USD70 billion per year to USD300 billion by 2030 and up to USD500 billion by 2050 [179]; the 2012 *Nagoya Protocol Implementation Fund* (NPIF), focused upon funding access to genetic resources [180]; and the *Capacity-building Initiative for Transparency Trust Fund* (CBIT), created in 2016 to promote greater transparency in developing countries [181].

Climate-Biodiversity-Environment

There are two additional bodies that need a mention as they approach biodiversity from opposite ends of the spectrum. The Taskforce on Nature-related Financial Disclosures (TNFD) was created in 2021 by 'financial institutions, multinationals and market service providers' to 'bring transparency to capital markets on how climate change and nature loss can impact upon organisations' financial positions' [182]. Similar to the TCFD but not affiliated with any UN agency, its scope extends beyond combatting climate change to biodiversity and to the wider environmental issue of single-use plastics. The TNFD closely replicates the financial standards framework [183] of the TCFD (see Fig. 6.5) including use of the *Low Emissions Analysis Platform* (LEAP) [184], one of the most widely used 'software tools for energy policy analysis and climate change mitigation assessment developed at the Stockholm Environment Institute' [185]. With over 400 international companies in support with 'USD15.9 trillion in Assets under Management' and listing guidelines for multiple industry sectors [186], it seems well positioned to gain traction. That said, some NGOs such as the Rainforest Action Network (RAN) [187] have criticized it for lack of engagement with local communities in its standard-setting processes and lack of transparency despite being in receipt of public funds [188].

There are many other initiatives being adopted by the private sector. One such launched in 2021 is the publicly listed *ASN Biodiversity Fund* bankrolled for €15 million seed money by the ASN Bank's parent Volksbank. By 2024 it was co-financing biodiversity projects globally [189]. Another, launched in 2023, is the *Biodiversity Pledge* [190], promoted by the Finance for Biodiversity Foundation, along with their private sector guide: *Finance for Biodiversity Guide on biodiversity measurement approaches* [191]. While the TCFD and the TNFD are designed to mobilise support from private capital markets for climate change and biodiversity action programmes, the IMF staff Climate Note (2024) *Embedded in Nature: Nature-Related Economic and Financial Risks and Policy Considerations* found that 'nearly 38% of bank loans of the 100 largest global banks are to harmful subsidies-dependent sectors and 44% are exposed to conservation areas under the Global Biodiversity Framework, and that industries most exposed to nature degradation are not well prepared to manage these risks' [192]. The problem with the narratives around capital markets is that they are always Janus-faced, a tug-of-war between returns to capital and returns to nature.

Coming from the other end of the environmental spectrum is the *Nature Positive Initiative* (NPI) [193], a group of environmental organisations, sustainable business platforms, and research institutions convened by the WWF in 2019 with the defined task ahead to 'halt and reverse nature loss measured from a baseline of 2020, through increasing the health, abundance, diversity and resilience of species, populations and ecosystems so that by 2030 nature is visibly and measurably on the path of recovery' [194]. The NPI is developing a series of science-based metrics for measuring and monitoring the restoration of natural habitats according to species, ecosystems, and natural processes at global, national, and landscape levels [195]. The creation of the NPI anticipated the UN's Biodiversity Conference (Montreal, 2022) which agreed the Kunming-Montreal *Global Biodiversity Framework* (GBF) [196], also calling for updated national framework reports. These were published as *National Biodiversity Strategies and Action Plans* (NBSAPs) in 2024 [197], which includes an *Online Reporting Tool* for countries to submit their national targets, plans, and an online dashboard. Identifying the threats to biodiversity, the losses and the strategies for recovery has inevitably been accompanied by a range of estimates of the funding that is still required to reach the 2030 goals.

Table 6.5 Investments by Big Tech companies in carbon removal and environmental restoration

Company	Investments in carbon-removal and environmental restoration
Apple *Carbon Removal Strategy* (2024) [221]	Supports two funds: 2021 created the *Restore Fund* for carbon credit purchase for carbon removal in partnership with Goldman Sachs and Conservation International; in 2023 created a USD200 million additional nature-based fund managed by Climate Asset Management [222]. Apple does invest in renewable energy, but draws its own energy from the grid which is a mix of fossil fuel, renewables, and nuclear, so it is not carbon neutral
Amazon *Amazon Sustainability Report, 2023* [223]	In 2019, co-founded the *Climate Pledge Fund* (with 500 + members by 2023) to achieve net-zero carbon emissions by 2040; Amazon's USD2 billion venture fund to help hard-to-abate sectors such as aviation, building, construction, and shipping; in 2022, Amazon introduced its own 'Abacus' standard for credits, to be administered by Verra [224], (see Chap. 5) and in 2024 committed, as part of the Leaf Coalition [225], to buy 5 million credits in a USD180 million deal to save the Amazon rainforest in the Brazilian state of Para [226]
Google *Google Environmental Report, 2024* [227]	From 2023 stopped buying cheap carbon credits to offset emissions but remains a participant through its membership of Symbiosis; committed USD10 million to support the carbon removal purchase programme of the DOE to purchase direct air capture credits from startup Holocene to be delivered by the early 2030s [228]
Meta *Sustainability Report (2024)* [229]	Uses credits as part of its carbon-zero targets; in 2024, Meta bought 6.75 million carbon removal credits [230], and committed to a minimum of USD35 million for carbon removal projects in 2025 as a response to the removal purchase programme the DOE launched in March 2024 [231]
Microsoft *Environmental Stability Report* (2023) [232]	MSFT is heavily involved in nature-based projects, such as a rainforest restoration project in the Calakmul region of southeastern Mexico [233], in Brazil in 2024 MSFT agreed to buy 8 million nature-based carbon removal credits by 2043 [234], agreed to buy 40,000 agricultural soil carbon credits (the largest ever) from a sustainable agricultural company, Indigo Ag. [235], in 2023 MSFT's first ever 'bankable' CDR *Enhanced Rock Weathering* (ERW) [236] credits to remove 5,000 tonnes of CO_2 by spreading 25,000 tonnes of basalt in the UK [237]. In 2024, agreed with power station operator Ørsted to capture 430,000 tonnes of biogenic CO2 annually and store it in a deep reservoir in Norway's North Sea [238], and agreed to buy 500,000 metric tonnes of carbon removal credits from a 1PointFive (a subsidiary of Occidental Petroleum) direct air capture (DAC) facility at its Stratos plant in Texas, the largest purchase on record [239]

(continued)

Table 6.5 (continued)

Company	Investments in carbon-removal and environmental restoration
Salesforce *Salesforce Climate Action Plan* [240]	Uses credits to match the gaps in its targets, primarily from nature-based projects and carbon removal technologies; operates climate funding mechanisms such as the Sustainable Bond Framework [241], to attract investment in nature-based and carbon removal projects, and the Salesforce Venture Impact Fund [242], to promote ESG programmes, including ways to reduce carbon emissions. As a member of Symbiosis, has pledged USD25 million to the coalition's pledge of USD1 billion by 2030 for carbon removal purchases. This is part of a commitment by Salesforce in 2022 to spend USD100 million on carbon removal contracts by 2030 [243] when it joined the WEF's buyers' group, the First Movers Coalition [244]
Symbiosis	Symbiosis was created in 2024 as a 'buyer's club' [245] created by Google, Meta, Microsoft, and Salesforce as an Advance Market Commitment (AMC) [246] which is an agreement to invest in, or to purchase carbon credits from nature-based projects in collaboration with the LEAF Coalition [247]. It is coordinated by Emergent [248] which signs Emissions Reductions Purchase Agreements (ERPAs) with countries such as Costa Rica and Ghana [249], or through direct carbon removal projects in collaboration with Frontier Climate [250]. By late 2024 between them they had pledged to contract up to 20M tonnes of 'high-quality nature restoration carbon credits' by 2030
Use of nuclear	Small modular reactors—Amazon [251], Google [252], Meta [253] Three Mile Island nuclear plant and others [254]—MSFT [255] Buy from the Grid—Apple only uses nuclear indirectly through the Grid [256]

Summary of Funding Estimates

The story of climate-finance and biodiversity conservation is a sadly familiar one of insufficiency, bordering on the 'too little, too late'. A case in point is the difficult to reverse drying up of the Caspian Sea, the world's largest inland body of water in Central Asia, together with its horrendous pollution and the threatened extermination of the unique Caspian seals [198]. The prospects for the Aral Sea, also in Central Asia, which did completely dry up, are only slightly more optimistic as conservation efforts have restored some water and some fish [199]. Table 6.4 provides a brief summary of financial flows estimated as of 2024 as necessary to make up the shortfalls.

An optimistic, but politically hardly realistic, view of filling the gap between a degraded biosphere and a restoration and conservation of biodiversity worldwide is the WEF's estimate of *an annual* global investment of USD2.7 trillion to 2030—see Fig. 6.6. The WEF cites the response of the US Government to COVID-19 as an example of what can be done as needed when it announced a programme of USD2.2 trillion in 2020. Since 2025 in the USA, the prospects look decided less optimistic.

Table 6.4 Biodiversity funding sources and outcomes

Source	Biodiversity funding projections and outcomes
Paulson Institute [1], The Nature Conservancy [200], Cornell Atkinson Center for Sustainability	Financial flows into conservation 2012–2019 tripled to upwards USD140 billion, but subsides on activities that degrade agriculture, forestry, and fisheries were at least × 2 to × 4 greater, excluding subsidies for fossil fuels. Carbon sinks annually boost GDP by well over USD100 trillion (at USD600 per tonne of CO2 captured) and by WEF estimates, around 50% global GDP depends upon natural capital, but to reach zero biodiversity loss a further USD600–800 billion is needed annually
World Economic Forum (2024) [201] International Finance Corporation [202]	USD7 trillion invested globally annually in activities that have a direct negative impact on nature from both public and private sources; USD2.7 trillion average annual investment required to unlock a 'nature positive' economy by 2030; USD711 billion average annual global financing gap specifically required for the protection, restoration, and enhancement of nature
Andrew Seidl et al. (2021) 'The effectiveness of national biodiversity investments to protect the wealth of nature' *Nature, Ecology and Evolution Journal* [203]	Funds to biodiversity are associated with reduced rates of threatened species and biodiversity loss of 1% per year. Each USD1 billion investment in biodiversity is associated with an annual reduction in the proportion of threatened to total species of about 0.57%. Population growth correlates with lower financial support for biodiversity and an increase in the proportion of threatened to total species in a country
Forest and Finance *Banking on Biodiversity Collapse: Tracking the Banks and Investors Driving Tropical Forest Destruction* (2023) [140, 204]	Sounding the alarm, USD154 billion of credit has been extended to 'forest-risk commodity companies' in the three tropical forest regions of South America, Southeast Asia, and Central and West Africa since the Paris Agreement
UNEP [205]	*State of Finance for Nature* (2023) identifies a USD4.1 trillion financing gap in nature by 2050. USD200 billion spend on Nature-based Solutions (NbS) are massively outweighed by USD7 trillion finance flows with direct negative impacts on nature. Private finance flows have USD5 trillion direct negative impact on nature 'which is 140 times larger than private investments into NbS'

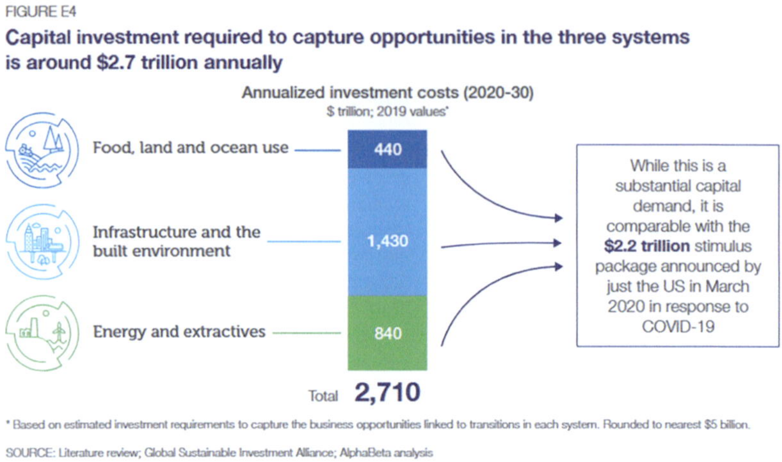

Fig. 6.6 WEF estimates of investment requirements to generate equivalent $2.2 trillion by 2030. *Source* Paulson Institute, The Nature Conservancy, Cornell Atkinson Center for Sustainability *The Future of Nature and Business* (2024) [206]

USD2.2 trillion is the scale of state-sponsored investment the WEF sees as necessary to trigger USD10 trillion worth of new business opportunities per year across 15 transitions of three systems: environment, infrastructure, and energy—see Fig. 6.7 (Fig. 6.8).

As illustrated in Fig. 6.7, the optimism of the WEF, which strenuously promotes green environmental policies and a transition to clean energy, reckons just three socio-economic systems alone could generate USD10 trillion business opportunities by 2030 and up to 400 million jobs. The problem with all such projections is that the only component that could, in theory at least, be pinned down is what governments commit to, although in reality that is not assured. The major component is what would be forthcoming from private investments, and even the private sector has no way of knowing. Changes in global trade, tariffs, currency movements, technologies, markets, governments, etc. are unpredictable. There seem to be no estimates of how much the private sector at a national or global level is investing in such low-carbon or no-carbon production and distribution processes (Scopes 1, 2, and 3), and even if every enterprise bought carbon offsets in projects to the tune of USD10 trillion, this would not be a true reflection of funds going towards emissions reductions or of environmental restoration and conservancy, because carbon trading prices are so low. They do not reflect the social costs involved.

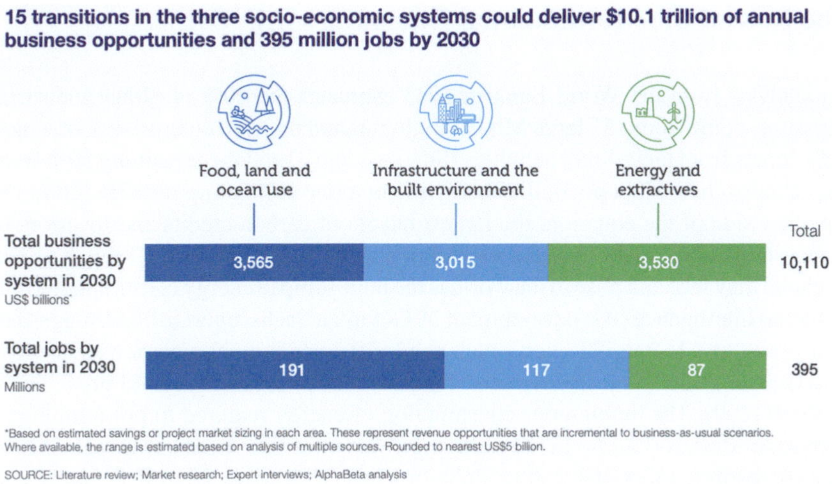

Fig. 6.7 Three sectors the WEF sees as creating $10 trillion returns and 395 million jobs by 2030. WEF projections of a 1:5 ratio of state investment to business investment. *Source* WEF The Future of Nature and Business (2020) [207]

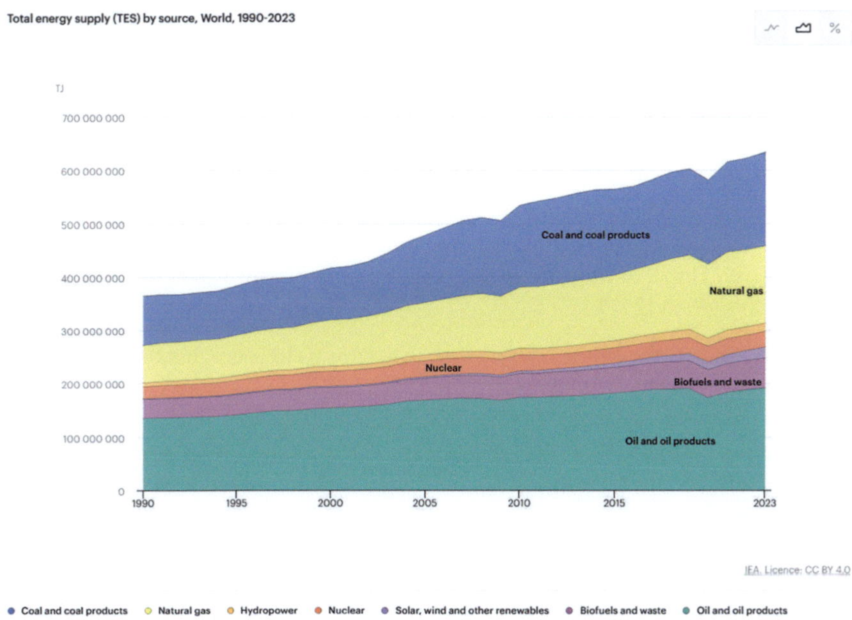

Fig. 6.8 Total energy supply (TES) by source, World, 1990–2022. *Source* IEA World Energy Statistics and Balances [259]

The Role of the Multinational Companies (MNCs)

An analysis from the World Bank in 2023 estimates the 60% of global industrial emissions come from 157 large MNEs (multinational national enterprises), of which 10% comes from their direct activities (Scopes 1 and 2) and the remaining 50% from their supply chains (Scope 3); the energy sector alone accounting for 37% [208]. On the other side of the equation, the largest buyers of carbon credits and investors in carbon-reduction projects are the multinational Hi-Tech giants—see Table 6.5. The dilemma they face and pose to the world is the huge jump in energy consumption from Artificial Intelligence (AI) development. A Goldman Sachs report in 2024 suggested, on average, an AI ChatGPT query needs nearly 10 times as much electricity to process as a Google search, and that demand for data centre electricity would grow 160% by 2030 [209]. The humungous consumption of energy required to pre-train Large Language Models (LLMs) [210] is driving companies like Amazon, Google, Meta, and Microsoft to invest in small modular (nuclear) reactors to avoid fossil fuels, and increasing their search for carbon-offsets that have credible levels of integrity—see Box 6.6.

Box 6.6 The *Other* Amazon and its Neighbour

The region of the Amazon rainforest is an area of biodiversity that acts as a critical lung and carbon sink for the planet. The importance of its survival from human degradation cannot be overstated. Microsoft, despite (or rather because of) its own carbon footprint, which at 17 million tonnes of carbon dioxide in 2024 was a 40% increase over 2020 due primarily to its investment in AI, has become one of the important multinationals contributing to the region's natural survival. For example, in 2025 it purchased 3.5 million carbon credits over 2 years from Brazil startup Re.green, which is likely to translate into around USD200 million [211].

Also at stake is the biodiversity of the Amazon's close neighbour in Brazil, the Cerrado savannah, which since the 1950s has lost almost half of its native vegetation to unsustainable agricultural commodities: beef, soy, and palm oil production [212]. In 2024, the Latin America reforestation strategy of the BTG Pactual Timberland Investment Group (TIG), and the Timberland Investment Management Organization (TIMO) [213] received a commitment from the International Finance Corporation (IFC) for USD50 million of carbon credits as a means to mobilise an additional USD1 billion [214]. The strategy aims to restore 135,000 hectares of natural forests and develop sustainable commercial tree farms on an additional 135,000 hectares. This is the conditional public–private catalytic model in practice. Progress was made in September 2024 when Meta agreed to purchase 1.3 million carbon removal credits, with options for delivery of an additional 2.6 million through to 2038 [215].

The buying of carbon credits by MNCs as offsets has been controversial. Taken on its own, it would be a hands-off approach compared with direct investments and management of nature-based projects which are not the skill sets of Hi-Tech companies. The apparent reason behind Jeff Bezos (Amazon) withdrawing his USD10 billion *Earth Fund* support for the SBTi (Science-based Targets initiative) in February 2025 was the SBTi's decision in 2024 to exclude carbon credits from its approved list (see Chap. 5), although it does coincide with an obsequious rallying of oligarchs around the new occupant of the White House, several of whom have other commercial interests at play [216]. One source describes the purchase of credits for carbon dioxide removal (CDR) as 'an arms race, with technology giants competing for access to offsets in what is a supply-constricted market' [217]. For example, Google, whose emissions were 48% higher than its 2019 benchmark year due to its AI development, started 2025 buying 100,000 tonnes of biochar, the largest purchase up to that time, but Microsoft remained the leader buying 4.5 million tonnes in 2024. These purchases use BECCS (bioenergy with carbon capture and storage), less expensive than DAC (Direct Air Capture) technologies which mostly remain in their trial and development stage. In his book *How to Avoid a Climate Disaster* [218], Bill Gates argues strongly for companies like Microsoft to take a lead in investing in start-up CDR technologies despite at the risk of failure.

There are, and will be, ways to reduce energy consumption through better use of the technologies involved, including AI [219], which is especially important if AI agents can contribute to reducing carbon emissions. In 2024, the US Department of Energy (DOE) issued its *Recommendations on Powering Artificial Intelligence and Data Center Infrastructure* [220] which includes an AI test-bed in the DOE. Yet for as long as societies becoming more dependent upon digital apps remains unabated, energy consumption will grow. Table 6.5 summarises the ways in which several Hi-Tech giants *present* their own carbon-abatement strategies.

But Table 6.5 comes with Health Warnings. First, although several of these Hi-Tech giants have plans for their own sources of renewable energy, such as small modular reactors (SMRs) to support their AI data centres, and buy as much as they can globally from renewable sources, such as solar farms, windfarms, and hydroelectric generators, they are all highly dependent upon local and national grids, as are their supply chains. But the grids rely more heavily upon fossil fuels than they do on renewables. As Fig. 6.4 from the IEA shows, hydro (just above nuclear) plus solar and wind power (just below nuclear) plus biofuels and waste in total constituted just slightly above 14% of global energy supplies by 2022. Nuclear was close to a further 5%. And, unfortunately, because renewable sources are not sufficiently reliable to guarantee grid electricity, it will take huge additional investments to reach a point of sufficiency, supported by additional investments in battery storage facilities. No grid in the world relies upon 100% renewables, so any claims that data centres are sourced from 100% renewable energy are false, and invariably imply that offsets are included as an equivalence of the shortfall [257]. Even in the USA, 60% of energy production comes from fossil fuels [258].

Second, the investments shown in Table 6.5 are therefore offsets. Data on reducing Scope 1 emissions by modifying industrial processes is not available unless specified in annual reports. Some companies report to the Climate Disclosure Project [260], but it requires licences to access data on emissions reductions, and there seems to be no estimate of actual total tonnage of C02e avoided. So, besides the evidence of investments in third-party projects, making assessments of MNCs efforts to reduce Scopes 1, 2 and 3 is reliant upon whatever disclosures they choose to make.

The Role of Banks and Scope 3 Emissions

Since the coming of President Trump to the White House, financial institutions in the USA have been keeping their heads below the parapets, but this need not apply outside the USA. Yet a more ongoing problem is the lack of reliable data and lack of bank transparency on their funding activities, something which is very traditional, protecting them from overt scrutiny, whether it is to do with protecting clients, money laundering, tax evasion, financing fossil-fuel extraction, liquidity status, or other reasons.

A study by the Transition Pathway Initiative (TPI) of the Granthem Research Institute on Climate Change and the Environment, established at the London School of Economics (LSE) in 2022 [261], found that banks continually failed to disclose all on-and-off balance sheet activities, meaning that 'banks could continue to finance high-emitting activities in the long term' due to the lack of full data disclosure and standardised reporting. The study included 26 leading global banks using a 10-factor assessment framework was published in July 2023 [262]. Another study in 2024 by the Climate Policy Initiative (CPI) of over 600 financial institutions [263] showed they prioritised high-emitting sectors within their portfolios even 'where clear decarbonisation pathways exist', finding that 'direct finance to new fossil-fuel plants outstripped investments in renewables, reaching USD33.4 billion in 2023'. It also highlighted the challenge of data availability, comparability and lack of 'independent validation on Scope 3 financed emissions'. In the *GHG Accounting and Reporting Standard* [264], the Scope 3 standard for financial institutions, as noted by the Grantham Institute, also includes the Scope 1 and 2 emissions of their clients to the extent they were invested in them.

The regulatory burden of these reporting standards may be high, but the profitability of banks, which has been exceptional over recent years [265], should easily accommodate them, and the push for data clarity is central to climate change policy-making. Further, using a spreadsheet template or an AI agent for ease of updating the data annually should not be over-complicated. Banks like most commercial companies push back against regulations when they would be better occupied upgrading their IT systems to simplify the compliance procedures.

An AI Large Language Model (LLM) was used in 2024 by Climate X [266], to scan a database of the world's 50 largest commercial banks (excluding China). It showed 86% of them achieved less than half of '17 qualitative indicators, including alignment with international adaptation goals, physical risk data collection, and client engagement'. To avoid AI 'hallucination' (false facts) a Retrieved Augmented Generation (RAG) [267] technique was used which grounds the answers to specific questions in a defined knowledge base [268]. The lack of achievement by these banks will be a challenge for the EU's 2023 *Corporate Sustainability Reporting Directive* which requires comprehensive reporting of emissions across company value chains. Compliance could, as pointed out by the Grantham institute, be a basis for a carbon tax on financial institutions, a source of much needed climate-project funding [269]. In the US the closest equivalent was the climate-related disclosure rules issued by the Securities and Exchange Commission (SEC) which require registrants (companies filing for exchange listings) to identify (for investor information) climate-related risks to their business and what steps they are taking in regards Scopes 1, 2, and 3 [270]. However, in the pushback from Republicans, the courts have been asked to rule that this information is 'not financially material' [271]. And in March 2025, the SEC announced the rule would no longer operate [272].

In response to the continued poor showing of commercial banks in promoting climate finance and their continuing to finance fossil-fuel projects, the WWF (World Wildlife Fund) has called upon banking regulators to consider stricter rules governing bank assets that are accepted as collateral for financial support from central banks, such as restricting highly carbon-intensive assets banks can use, and making it less attractive for banks to support investments and loans to companies pursing policies that are not science-based transition pathways [273]. In 2020, the European Union (EU) developed a specific taxonomy for what is and is not 'sustainability' [274], including a *Platform on Sustainable Finance* [275]. Within this taxonomy, Nature-based Solutions (NbS) are included, but carbon credit offsets are absolutely not. In 2024, the Swedish-based Sustainable Finance Lab in a report *A Green Interest Rate for the Eurozone* [276] views EU taxonomy as an opportunity for the European Central Bank (ECB) to use TLTROs (Targeted Longer-Term Refinancing Operations) [277] for 'refinancing operations where the interest rate that banks pay depends on their volume of lending that complies with the EU's green taxonomy' [278], and in that way to offer banks incentives to scale up green financing.

Financing Nature-Based Projects by Debt Swaps

Debt swaps are a way to finance carbon and biodiversity projects in indebted EMDEs. A national debt in the shape of bonds that are unlikely ever to be paid back in full, in effect non-performing assets (NPAs), can in principle be swapped at a major discount in exchange for dollars-converted-into-local currency, namely, the redemption rate. The investment comes from development banks, aid agencies, NGOs and private banks, and other investors supporting the debt-for-nature swap. According to one view, debt-swaps are an effective way to help EMDEs finance long-term avoidance and mitigation projects [279], while another challenges their effectiveness in reducing debt and sees them as modern-day colonialism, as a way for a poor country to lose control of its natural assets to a richer country [280].

The arrangement was first introduced as an idea by the WWF in the 1980s, followed by swaps occurring in some Latin American economies, in the Philippines, and subsequently in Africa [281]. Where swaps did not materialise, in several cases local conservation funds were established with the help of concessionary finance. Ironically, as the indebted status of low-income countries improved during the early years of globalisation in the late 1990s onwards, their debt discounts fell undermining the incentive for private banks and investors to participate. The reemergence of poor country indebtedness since the Great Recession of 2008, and then COVID-19, has reignited an interest in nature-debt swaps. It is basically a way to mortgage the land, whether to domestic or overseas commercial financial interests. It would be aid only if future revenues streams were shared as collateral with the financial institutions but not the land ownership; in other words, the financial institutions would bear the risk, but they could also contractually insist upon the transparent management of the land.

The International Institute for Environment and Development (IIED) using IMF/World Bank data for 2022 calculates the total debt of the 49 countries most at risk of defaulting on their external debts, to be USD431 billion. Using a methodology derived from previous international debt-reduction schemes, IIED estimates that USD103.4 billion of that total could again be freed up for swaps [282]. Yet another source, Debt Justice [283], suggests 54 countries were facing a severe debt crisis in 2022, with a substantial additional number at risk, which widens the potential for swaps. This follows what the WEF (World Economic Forum) [284]calls a landmark deal by Ecuador in May 2023 that will help protect the endangered ecosystem of the Galapagos Islands through the sale of a blue bond that matures in 2041. In October 2024, El Salvador completed a US-backed loan for USD1 billion from JPMorgan bank to buy back sovereign bonds and use the debt-servicing savings of USD350 million to fund the restoration of one of the country's longest rivers [285].

Debt swaps are one way to engage private capital in financing such nature-based projects. They are certainly not decisive in meeting the Paris goals, nevertheless beyond the Paris goals conserving nature as carbon sinks and biodiversity is the longer run road to the sustainability of life on Earth.

Concluding Remarks

The yawning gap between the funds made available to mitigate climate change, or avoid it altogether, and the resources required to do the job, especially in poorer countries, is not and will not be closed under prevailing policies. Those policies are based upon the premise that public finances can catalyse private finances, either by reducing or sharing the risks of climate change and adaptation project investments, or by creating investment opportunities for the private markets, often associated with grants, loans, and investments by multilateral agencies to EMDEs, or by creating a regulatory environment to encourage private investments, such as tax incentives.

This has been the clear outcome of all the COPs, and especially at the 'finance COP-29'. The nail in the coffin of an internationally agreed approach has been the withdrawal, under President Trump, of the USA from both the Paris Agreement and the UN Sustainable Development Goals. Even before that stab-in-the back to internationalism, the Global North countries have been unwilling to devote sufficient funds to meet the Paris goals. Equally, the Global South has been unable or unwilling to change policies, such as tax reforms and carbon taxes, to sufficiently finance domestic GHG mitigation and avoidance. The rules of capitalist markets work across all countries, North and South. Even China is not totally immune.

The problem is that with few exceptions, carbon-abatement projects are not sufficiently profitable compared with the risks associated with shortfalls. This is especially so in the case of the voluntary market for carbon credits. In 2021, the TSVCM observed that if a target 2Gt CO2 were to be removed by 2030 relying solely on carbon credits, often regarded as a 'get-out-of-jail card' for companies unwilling (or unable) to invest in carbon-reducing production processes of their own, this would require a '15-fold scale up of voluntary offsetting in 2030 versus 2019' [99]. This is just not going to happen.

It is a sad fact that globally, subsidies by governments and banks to fossil-fuel energy operations are not decreasing, because renewable energy supplies cannot meet the ever-growing demand for energy. That leads to 'energy security' challenges. This is exacerbated by the rapid development of AI LLMs, and will grow even further with the deployment of AI agents across economies. The rivalry between the AI giants, such as Amazon, Google, Meta, Microsoft, Xai, and others, now including Chinese companies, is so intense that energy conservation is being traded-off against competitive advantage. At the upstream level, AI's net contribution to net negative is itself net negative.

But the biggest game-changer is the withdrawal for the second time of the USA from the Paris Agreement. It changes all the parameters of global efforts. The diminishing of GFANZ rips holes in the entire fabric of the UN's efforts to use public finances to mobilise private capital to fund climate change mitigation and adaptation projects in the EMDEs. This top-down approach to what is *already* an existential problem for many on the planet, while not totally dead-in-the-water, is sinking because the scale of the funds involved (trillions of dollars) is only available through these public–private-partnerships, and there are no other sources available *unless* there was a global agreement to raise carbon taxes and implement the 15% Global Minimum Tax on multinational company profits [286]. In other words, existing financial policies (green finances) are always going to be insufficient to combat climate change, and need to be replaced by the only thing left, the direct, active, and sustained intervention of governments.

Another Intelligent Approach

More radical, direct, active, and sustained interventionist policies by governments are available, but are by definition a challenge to unfettered capitalism. This should be a green agenda involving a combination of pro-active top-down and bottom-up policies that it is easy to outline, but if put into practice will be relentlessly attached by the fossil-fuel industry and those in its pay. Further, capital flight will always be an Achilles' heel to radical economic and social policies-even to radical Right-wing policies as shown in the bond market's reaction to Donald Trump's anachronistic tariffing policy-and likely only active bottom-up campaigning by civil society will be able to move the political dial.

The top-down strategy is to change the direction of economic growth to one that starts with a redistribution of national wealth towards civil society. This should include policies that incentivise businesses to adopt carbon avoidance or carbon reduction technologies and products. This is necessary for two reasons. First, without a redistribution, civil society will see nothing new to support, and therefore will be more tempted by the vacuous but emotional, mostly xenophobic, plays of Right-wing popularism. Second, redistribution is not only about anti-poverty measures such as a basic minimum income and vital social services, but also and crucially about job creation involving a series of green industries, and the clean air and water that follow from green processes of production and distribution. It needs consist of infrastructure projects which are typically dependent upon public finances, such as (1) an upgraded national electricity grid fully equipped with transformers and renewable energy battery storage facilities together with AI-supported control and monitoring centres; a similar infrastructure for electric vehicle recharging points and hydrogen storage and refueling stations; (2) upstream manufacturing of solar panels, storage batteries, wind turbines and blades, components and parts to be outsourced to industry; and (3) downstream climate-friendly designs for new buildings using environmentally friendly materials, installation of insulation, heat pumps and cooling systems for retrofitting existing buildings and homes; trade-in subsidies to replace internal combustion engine vehicles with electric and hydrogen vehicles, and so on. The list is endless, but at each level job creation, apprenticeships, and digital skills training would need to be in parallel with public-sector-funded initiatives and industry engagement to adopt carbon avoidance and reduction. Across this spectrum of activities, greater and greater economic surpluses can be created over time in each of the public–private-NGO sectors which will bring in tax revenues long term.

Overall, the private sector would be executing much of the green agenda, but a response to charges of crowding-out the private sector in non-green parts of the economy would be to offer them green investment bonds. Green bonds could also be used to fund an International Green Bank (IGB) - see Chap. 7.

The bottom-up approach involves engaging local communities, civil society more broadly, and especially housing associations, parental groups, trade unions, workers, and staff associations in campaigning for the replacement of dirty fuels with clean energy, cleaner production and distribution processes, and clean air for children to breathe. Ways to institutionalise this civic activism should be found so that a new *sustainable* layer to social democracy emerges. The key word is 'activism' meaning that this intervention from below needs to be more than just a source of criticism of policies, but with campaigning energy, a constant source of creative ideas, similar to the creative activism of entrepreneurs behind start-ups in industry incubators, accelerators, and science parks. Such a movement is bound to be opposed and attacked by the fossil-fuel sector and by the Right who will see it as eating into the managerial privileges of the owners of capital, which it will be, because it will be offering an alternative way to manage the interests of society-at-large, in the same way as collective bargaining does. While there is no shortage of collectives, cooperatives and unions already contesting climate change and environmental issues [287], and despite green policy advocacy by organisations such as the International Labour Organisation (ILO) [288], research and surveys suggest that, on the one hand, recognition of the need for climate change justice is growing, especially in countries and among communities where severe weather events have been frequent, but, on the other, there is little understanding of what climate change justice really means [289].

The question then arises: can such national policies work. Policies that strongly redirected national resources were accepted as the norm during two world wars, and even to an extent during COVID-19, so there is nothing in principle to suggest they cannot be implemented. The policies have to start with the national budget which identifies taxes and expenditures. A government that proposes a green agenda as defined above faces expenditures in the immediate future before the tax returns arise from green growth. Government debt is therefore inevitable to launch such a programme. A combination of measures to redistribute wealth, including creating green jobs, will be opposed by sections of capital, especially those that do not stand to enjoy immediate gains through public works contracts. The worst-case scenario is that capital flight may ensue. That would require interest rates to rise to stem the outflow and attract an inflow to support the value of the currency. A high interest rate increases the cost of borrowing and servicing government debt. But that may still be sustainable if the green agenda gathers steam early on. Debt itself is not a big problem, since the central bank will always cover the debt as long as it is lawful to do so, meaning if the legislature has agreed the budget and that the debt is kept within the limits of the resources available to prevent runaway inflation. So an act of political will is required to ride out the inevitable storm, and if civil society is sufficiently supportive, the political will should be there. The alternative, massive doses of austerity could reduce consumption, but that is not socially acceptable, and would not work anyway as consumption by the rich would continue, as would the inexorable rise of demand for energy.

Putting it in these terms makes it expressly clear that green finance is only an answer if the political will exists to make it the answer, which is not the case currently. Green finance as it exists makes its contribution, but it is just insufficient. This is true even when green finance includes private sector business investment in cleaner, greener processes of production because the costs are financially internal but the benefits are mostly socially external. There is no other source of green finance at scale than government, but that source will not be mobilised at scale without an irresistible push from civil society.

Post-Script

Climate Policy Initiative (CPI) updated the 2021–2022 data in Fig. 6.1 with the following data for year-ended 2023, Fig. 6.9 [290].

By 2023 there was a slight increase in public finances committed to green funds, overwhelmingly to project-level market-rated debt, mostly by governments, multilateral DFIs, and state-owned FIs, and a more substantial increase in private-sector funds from the commercial sectors but principally from a 150% increase in private/household spending on heat pumps, home insulation, electric vehicles, etc. The proportion of funds financing mitigation projects rose from nearly 90% of the total to close to 95%, while overall finance to mitigation and adaptation rose from USD1.3 trillion to USD1.9 trillion. This was before the election of Donald Trump as President of the USA. As graphed, the 2023 data shows the allocation of finances to sectors, but not to regions nor by development status such as EMDEs. The report does highlight the yawning gap between the mitigation needs and capital flows to EMDEs, however CPI continues to focus on the role of the catalytic model: ‘EMDEs still need more catalytic capital to further scale climate flows across all regions’. The workability of this model is challenged above and in Chap. 7.

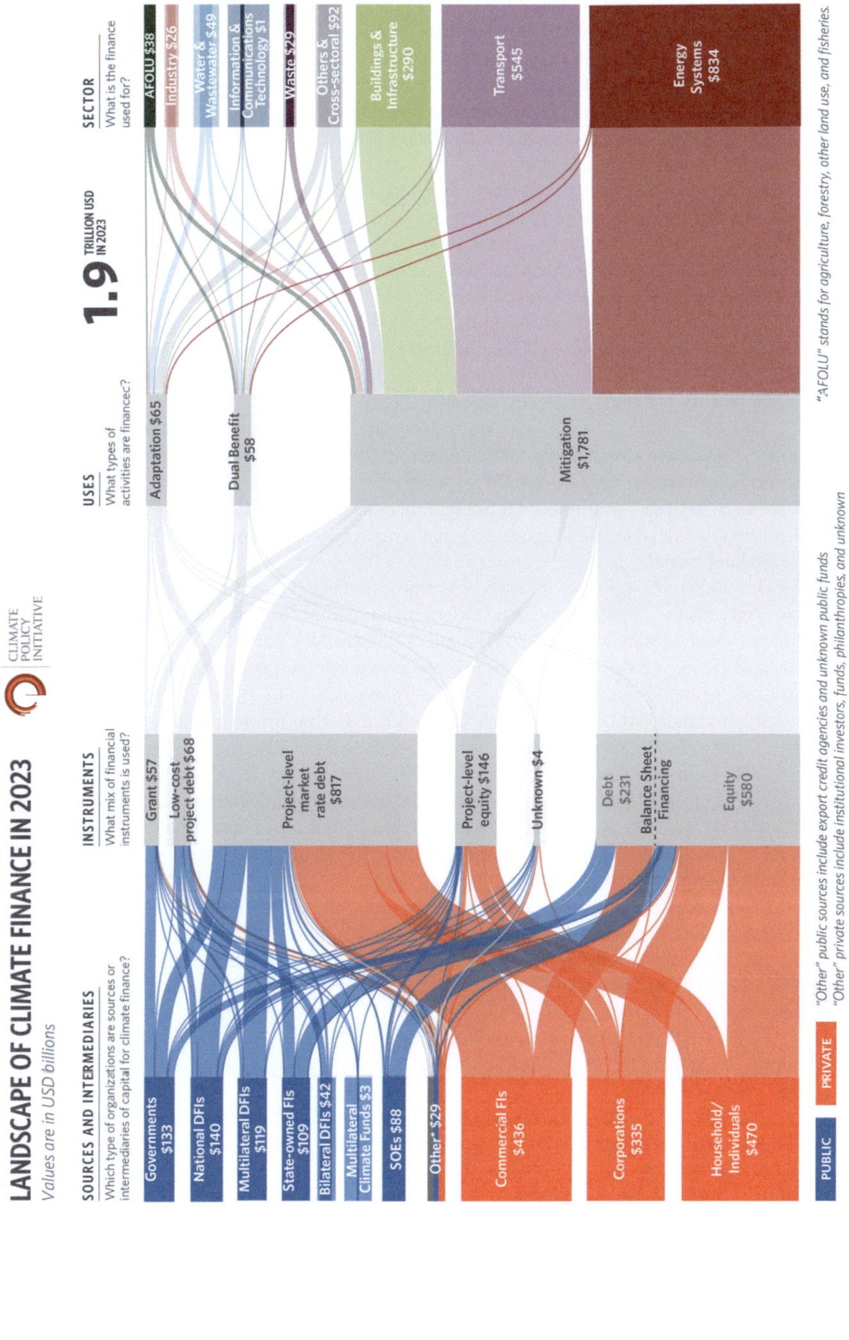

Fig. 6.9 The Climate Financial Landscape 2023

References

1. Paulson Institute, The Nature Conservancy, Cornell Atkinson Center for Sustainability (2024) Financing Nature: Closing the Global Biodiversity Financing Gap https://www.paulsoninstitute.org/wp-content/uploads/2020/10/Updated-10.23.20-FINANCING-NATURE_Exec.-Summary_Final-with-endorsements_101420.pdf
2. World-nuclear.org (May 2024) Financing Nuclear Energy https://world-nuclear.org/information-library/economic-aspects/financing-nuclear-energy
3. BloombergNEF (2025) The definitive resource for clean energy investment https://about.bnef.com/energy-transition-investment/
4. CPI (November 2023) Global Landscape of Climate Finance 2023 https://www.climatepolicyinitiative.org/publication/global-landscape-of-climate-finance-2023/
5. NGFS (2022) NGFS Climate Scenarios for central banks and supervisors https://www.ngfs.net/en/ngfs-climate-scenarios-central-banks-and-supervisors-september-2022
6. UNDP (2022) What is just transition? And why is it important? https://climatepromise.undp.org/news-and-stories/what-just-transition-and-why-it-important#:~:text=The%20concept%20of%20just%20transition,their%20economies%2C%20and%20build%20resilience
7. UNEP (2024) Green Financing https://www.unep.org/regions/asia-and-pacific/regional-initiatives/supporting-resource-efficiency/green-financing
8. LSE (2024) What is the just transition and what does it mean for climate action? https://www.lse.ac.uk/granthaminstitute/explainers/what-is-the-just-transition-and-what-does-it-mean-for-climate-action/
9. TUDCN (2019) The Contribution of Social Dialogue to the 2030 Agenda: Promoting a Just Transition towards sustainable economies and societies for all https://www.ituc-csi.org/IMG/pdf/sdjt-final-en.pdf
10. IHLEG (2023) A climate finance framework: decisive action to deliver on the Paris Agreement https://www.lse.ac.uk/granthaminstitute/wp-content/uploads/2023/11/A-Climate-Finance-Framework-IHLEG-Report-2-SUMMARY.pdf
11. Climateworks.org (2025) Just Transition https://www.climateworks.org/programs/just-transition/
12. Amnesty International (2024) COP29: Climate negotiation drafts are an affront to human rights https://www.amnesty.org/en/latest/news/2024/11/cop29-climate-negotiation-drafts-are-an-affront-to-human-rights/#:~:text=COP29%20negotiation%20drafts%20are%20an%20affront%20to%20human%20rights%20%2D%20Amnesty%20International.
13. CSIS (2024) Experts React: COP29 Highlights, Challenges, and Emerging Trends in Global Climate Policy https://www.csis.org/analysis/experts-react-cop29-highlights-challenges-and-emerging-trends-global-climate-policy
14. World Resources Institute (2025) When e Do Emissions Come From? 4 Charts Explain Greenhouse Gas Emissions by Sector https://www.wri.org/insights/4-charts-explain-greenhouse-gas-emissions-countries-and-sectors
15. UNEP (2024) Emissions Gap Report 2024 https://www.unep.org/resources/emissions-gap-report-2024
16. JRC (2023) GHG emissions of all world countries - JRC/IEA 2023 Report https://edgar.jrc.ec.europa.eu/report_2023#:~:text=Even%20in%202022%2C%20global%20GHG,F%2Dgases%20(2.6%25)
17. UNCTAD (2024) Classifications https://unctadstat.unctad.org/EN/Classifications.html
18. LSE (2022) How is China tackling climate change? https://www.lse.ac.uk/granthaminstitute/explainers/how-is-china-tackling-climate-change/
19. Politico (2024) China must now lead global warming fight, UN climate chief says https://www.politico.eu/article/china-lead-global-climate-fight-un-climate-chief-simon-stiell-cop-azerbaijan-clean-energy/

20. Cgdev.org (October 2024) Has China Really Provided More Climate Finance to Developing Countries than the US? https://www.cgdev.org/blog/has-china-really-provided-more-climate-finance-developing-countries-us#:~:text=Accounting%20for%20China%27s%20climate%20finance%20under%20the,the%20most%20recent%20five%2Dyear%20period%20since%202017).
21. cfr.org (11 March 2021) The Climate Challenge and China's Belt and Road Initiative https://www.cfr.org/blog/climate-challenge-and-chinas-belt-and-road-initiative
22. NDB NDB's General Strategy: 2017 – 2021 https://www.ndb.int/wp-content/uploads/2017/08/NDB-Strategy.pdf
23. White & Case (2025) US Withdrawal from the Paris Agreement: Impact and Next Steps https://www.whitecase.com/insight-alert/us-withdrawal-paris-agreement-impact-and-next-steps
24. CPI (2024) Global Landscape of Climate Finance 2024: Insights for COP29 https://www.climatepolicyinitiative.org/wp-content/uploads/2024/10/Global-Landscape-of-Climate-Finance-2024.pdf
25. CPI (2025) About Us https://www.climatepolicyinitiative.org/about-cpi/team/
26. IRENA (2023) Global Landscape of Renewable Energy Finance https://www.irena.org/Energy-Transition/Finance-and-investment/Investment
27. New York Times Events (2024) Michael Oppenheimer and Stefan Rahmstorf | How Close is Climate to a Tipping Point? YouTube https://www.youtube.com/watch?v=cJ1GVMWzWIE
28. Climate.copernicus.eu (January 2025) 2024 - a second record-breaking year, following the exceptional 2023 https://climate.copernicus.eu/global-climate-highlights-2024
29. Bankrate.com (2024) The impact of natural disasters on insurance rates in 2024 https://www.bankrate.com/insurance/homeowners-insurance/natural-disaster-costs/#billion-dollar-disasters
30. UNDRR (2022) Disaster Risk Reduction in the Republic of Vanuatu https://www.undrr.org/media/83389/download?startDownload=20250215
31. World Resources Institute (November 2024) What Is 'Loss and Damage' from Climate Change? 8 Key Questions, Answered https://www.wri.org/insights/loss-damage-climate-change
32. UNEP (2024) Climate action https://www.unep.org/topics/climate-action
33. UNEP (2022) Emissions Gap Report 2022 https://www.unep.org/resources/emissions-gap-report-2022
34. World Resources Institute (2024) What Is the Paris Agreement's Article 2.1(c) on Climate Finance, and Why Does it Matter? Key Questions, Answered https://www.wri.org/insights/article-2-1-c-paris-agreement-explained#:~:text=Article%202.1(c)%20of%20the%20international%20Paris%20Agreement%20on%20climate,on%20what%20exactly%20it%20entails.
35. UNFCCC (2024) UNFCCC Standing Committee on Finance Sixth Biennial Assessment and Overview of Climate Finance Flow https://unfccc.int/sites/default/files/resource/UNFCCC_BA6_Report_Web_FINAL.pdf
36. NRDC (2024) COP29: Key Outcomes Signal Need to Deliver Greater Climate Action https://www.nrdc.org/bio/jake-schmidt/cop29-key-outcomes-signal-need-deliver-greater-climate-action
37. G77.Org About the Group of 77 https://www.g77.org/doc/
38. G77.Org (February 2024) Statement on Behalf of the Group of 77 and China by Ms. Celia Kafureka Nabeta, Minister Counsellor, Permanent Mission of Uganda to the United Nations, on Chapter One of the Pact for the Future https://www.g77.org/statement/getstatement.php?id=240206b
39. Downtoearth.com (12 November 2024) CSE-DTE at COP29: G77 and developing country allies reject first draft of climate finance text https://www.downtoearth.org.in/climate-change/cse-dte-at-cop29-g77-and-developing-country-allies-reject-first-draft-of-climate-finance-text
40. UNFCC (2024) COP29 UN Climate Conference Agrees to Triple Finance to Developing Countries, Protecting Lives and Livelihoods https://unfccc.int/news/cop29-un-climate-conference-agrees-to-triple-finance-to-developing-countries-protecting-lives-and

41. UNFCCC (2024) New Collective Quantified Goal on Climate Finance https://unfccc.int/NCQG
42. UNFCC (2024) New collective quantified goal on climate finance https://unfccc.int/sites/default/files/resource/cma2024_L22_adv.pdf
43. LSE.AC.UK (2022) Finance for climate action: scaling up investment for climate and development https://www.lse.ac.uk/granthaminstitute/wp-content/uploads/2022/11/IHLEG-Finance-for-Climate-Action-1.pdf
44. Grantham Institute (November 2023) Second report of the Independent High-Level Expert Group on Climate Finance https://www.lse.ac.uk/granthaminstitute/wp-content/uploads/2023/11/A-Climate-Finance-Framework-IHLEG-Report-2-SUMMARY.pdf
45. France 24 (24 November 2024) Developing nations blast 'paltry' $300 billion deal approved at COP29 UN climate summit https://www.france24.com/en/europe/20241124-developing-nations-blast-paltry-300-billion-deal-approved-at-un-climate-summit
46. Kleinman Center for Energy Policy (November 2024) China and Global Climate Governance at COP29 and Beyond https://kleinmanenergy.upenn.edu/commentary/blog/china-and-global-climate-governance-at-cop29-and-beyond/#:~:text=At%20COP29%2C%20China%20faced%20increased%20pressure%20to,could%20reshape%20global%20climate%20policy%20moving%20forward.
47. UNFCCC (2024) Second report on the determination of the needs of developing country Parties related to implementing the Convention and the Paris Agreement https://unfccc.int/sites/default/files/resource/UNFCCC_NDR2_ES_Web_Final.pdf
48. UNFCCC (December 2023) LT-LEDS Synthesis Report https://unfccc.int/lt-leds-synthesis-report
49. bcg.com (2024)To Make Adaptation Projects Bankable, Reimagine Returns https://www.bcg.com/capabilities/climate-change-sustainability/expert-insights/adham-abouzied
50. Science Media Centre (November 2024) Expert reaction to closing stages of COP29 https://www.sciencemediacentre.org/expert-reaction-to-closing-stages-of-cop29/
51. UNFCCC (2024) COP29 UN Climate Conference Agrees to Triple Finance to Developing Countries, Protecting Lives and Livelihoods https://unfccc.int/news/cop29-un-climate-conference-agrees-to-triple-finance-to-developing-countries-protecting-lives-and
52. Bloomberg Live (November 2024) 'COP29: What We're Watching in Baku' YouTube video https://www.youtube.com/watch?v=INKGbBCThkU
53. President Babayev (11th November 2024) 'COP29 President Mukhtar Babayev Opening Speech' YouTube video https://www.google.com/search?client=firefox-b-1-d&q=opening+remaks+by+the+President+at+COP29#fpstate=ive&vld=cid:48b8adfc,vid:ktjr9aG65iY,st:255
54. Kurt Vonnegut (1969) Slaughterhouse-Five Delacorte
55. John Ure (2024) Achieving the UN Sustainable Development Goals: Late or Too Late? Emerald Publications.
56. Convergence.finance (2024) Blended Finance https://www.convergence.finance/blended-finance
57. Committees.parliament.uk (2022–2023) Investment for development: The UK's strategy towards Development Finance Institutions https://committees.parliament.uk/publications/41461/documents/203966/default/
58. Reuters (20 November 2024) Brazil development bank gets $690 million loan from China https://www.reuters.com/business/finance/brazil-development-bank-gets-690-million-loan-china-2024-11-20/
59. Pilot.com What is Balance Sheet Financing? https://pilot.com/glossary/balance-sheet-financing#:~:text=Balance%20Sheet%20Financing%20(BSF)%2C,appropriate%20accounting%20standards%20and%20regulations.
60. IDA (September 2023) 2023 Joint Report on Multilateral Development Banks Climate Finance https://publications.iadb.org/en/2023-joint-report-multilateral-development-banks-climate-finance

61. kfw-entwicklungsbank.de (2024) About Us https://www.kfw-entwicklungsbank.de/International-financing/KfW-Development-Bank/Tasks-and-goals/Unsere-Finanzprodukte/
62. DFC.GOV(2024) Our Priorities https://www.dfc.gov/
63. Climatebonds.net (November 2 024) The Role of Development Finance Institutions in Accelerating the Mobilisation of Green Capital https://www.climatebonds.net/files/reports/the_role_of_dfis_in_accelerating_the_mobilisation_of_green_capital.pdf
64. Greenclimate.fund (2024) About GCF https://www.greenclimate.fund/about
65. GEF.Org (2024) Newsroom https://www.thegef.org/newsroom
66. UNFCCC (2024) Doubling Adaptation Finance: Efforts to Respond to the Call of the Glasgow Climate PACT https://unfccc.int/sites/default/files/resource/Doubling%20Adaptation%20Finance_Efforts%20to%20Respond%20to%20the%20Call%20of%20the%20Glasgow%20Climate%20Pact.pdf
67. OECD (2024) Multilateral Development Finance 2024 https://www.oecd.org/en/publications/multilateral-development-finance-2024_8f1e2b9b-en.html
68. WWF (2024) 2024 Living Planet Report: A System in Peril https://wwflpr.awsassets.panda.org/downloads/2024-lpr-executive-summary.pdf
69. Livingplanetindex.org (2024) 2024 Living Planet Index https://www.livingplanetindex.org/
70. IPBES (2019) The global assessment report on Biodiversity and Ecosystem Services file:///C:/Users/urejo/Downloads/ipbes_global_assessment_report_summary_for_policymakers.pdf
71. WNA (January 2025) Plans For New Reactors Worldwide https://world-nuclear.org/information-library/current-and-future-generation/plans-for-new-reactors-worldwide#:~:text=Today%20there%20are%20about%20440,are%20building%2C%20new%20power%20reactors.
72. World-nuclear.org (2024) Six More Countries Endorse the Declaration to Triple Nuclear Energy by 2050 at COP29 https://world-nuclear.org/news-and-media/press-statements/six-more-countries-endorse-the-declaration-to-triple-nuclear-energy-by-2050-at-cop29
73. Wna.origindigital.co (December 2023) Declaration to Triple Nuclear Energy https://wn a.origindigital.co/images/articles/Declaration-to-Triple-Nuclear-Energy-_-COP28-_-28-November-2023-FINAL.pdf
74. World Nuclear Association (April 2024) Storage and Disposal of Radioactive Waste https://world-nuclear.org/information-library/nuclear-fuel-cycle/nuclear-waste/storage-and-disposal-of-radioactive-waste
75. Nature (2016) Policy: Reassess New Mexico's nuclear-waste repository https://www.nature.com/articles/529149a
76. NRC (January 2025) Backgrounder on Seismic Reviews at U.S. Nuclear Power Plants https://www.nrc.gov/reading-rm/doc-collections/fact-sheets/fs-seismic-issues.html
77. OneEarth October 2024) 7 reasons why nuclear energy is not the answer to solve climate change https://www.oneearth.org/the-7-reasons-why-nuclear-energy-is-not-the-answer-to-solve-climate-change/#:~:text=To%20recap%2C%20new%20nuclear%20power%20costs%20about,meltdown%2C%20mining%20lung%20cancer%2C%20and%20waste%20risks
78. NREL.GOV (2022) Examining Supply-Side Options to Achieve 100% Clean Electricity by 2035 https://www.nrel.gov/docs/fy22osti/81644.pdf
79. OneEarth 100% Renewable Energy https://www.oneearth.org/100-renewable-energy/#:~:text=Projecting%20out%20current%20levels%20of,to%20start%20reflecting%20this%20reality.
80. World-nuclear.org (September 2023) Economics of Nuclear Power https://world-nuclear.org/information-library/economic-aspects/economics-of-nuclear-power
81. DoE (September 2024) Pathways to Commercial Liftoff: Advanced Nuclear https://liftoff.energy.gov/wp-content/uploads/2024/10/LIFTOFF_DOE_Advanced-Nuclear_Updated-2.5.25.pdf
82. Superagent.sh (January 2025) The future of LLM costs https://www.superagent.sh/blog/the-future-of-llm-costs
83. HoganLovells.com (November 2024) What's really going on with data centers and nuclear? https://www.hoganlovells.com/en/publications/whats-really-going-on-with-data-centers-and-nuclear

84. Lex Fridman (11th February 2025) Detailed analysis of AI megaclusters from OpenAI, xAI, Meta, Google, Microsoft https://www.youtube.com/watch?v=WY36mAUumDg
85. Accenture.com (November 2924) Only 16% of Largest Companies on Track for Net Zero Goals with Nearly Half Seeing Increased Emissions, Accenture Analysis Finds https://newsroom.accenture.com/news/2024/only-16-of-largest-companies-on-track-for-net-zero-goals-with-nearly-half-seeing-increased-emissions-accenture-analysis-finds#:~:text=NEW%20YORK;%20November%2011%2C%202024,Accenture%20(NYSE:%20ACN).
86. ADB (2023) Greening Financial Systems and the Role of Multilateral Development Banks https://www.adb.org/news/events/greening-financial-systems-and-role-multilateral-development-banks
87. Brookings (August 2024) Reforms for a 21st century global financial architecture https://www.brookings.edu/articles/reforms-for-a-21st-century-global-financial-architecture/
88. Clubgestionriesgos.org (April, 2023) Boosting MDBs Investing Capacity https://www.clubgestionriesgos.org/wp-content/uploads/2023-MDBs-Capital-Adequacy-Frameworks-CGR.pdf
89. World Bank Group Joint MDB Methodological Principles for Assessment of Paris Agreement Alignment https://www.worldbank.org/en/publication/paris-alignment/joint-mdb-paris-alignment-approach
90. WRI (2024) Development Banks Are Starting to Spark Climate Action. Will They Complete the Task? https://www.wri.org/insights/mdb-reform-climate-action?apcid=0065a83d31d8c2c5be4fb000&utm_campaign=fin-bigger-in-baku-webinar&utm_content=fin-bigger-in-baku-webinar&utm_medium=email&utm_source=ortto
91. The World Bank Sustainable Finance and Environmental, Social, and Governance (ESG) Advisory https://thedocs.worldbank.org/en/doc/a988d4ca240396268b0194b2afaa5826-0340012023/original/BK-Product-Note-Sustainable-Finance-and-ESG-Advisory-Services.pdf
92. UNEP (2 024) Green Financing https://www.unep.org/regions/asia-and-pacific/regional-initiatives/supporting-resource-efficiency/green-financing
93. Research.UN.Org (2024) UN Documentation: Environment https://research.un.org/en/docs/environment/bodies
94. WEI (2022) World Energy Investment 2022: Methodology Annex https://iea.blob.core.windows.net/assets/79c65e95-baf6-4cc3-b16f-9ea7c7e6be42/WorldEnergyInvestment2022MethodologyAnnex.pdf
95. ICMA (June 2022) Green Bond Principles https://www.ifc.org/content/dam/ifc/doclink/2022/the-green-bond-principles-202206.pdf
96. Climate Bonds (2024) The Climate Bonds Standard https://www.climatebonds.net/standard/the-standard
97. RBC What are ESG ETFs? https://www.rbcgam.com/en/ca/learn-plan/types-of-investments/what-are-esg-etfs/detail
98. VCMI (2024) Scope 3 Claim https://vcmintegrity.org/wp-content/uploads/2024/09/VCMI-Scope-3-1.3_pages-Final.pdf?utm_source=vcmi_platform&utm_medium=referral&utm_campaign=vcmi_platform
99. IIF (2020) TSVCM https://www.iif.com/tsvcm
100. TSVCM (2021) Taskforce on Scaling Voluntary Carbon Markets: Final Report https://www.iif.com/Portals/1/Files/TSVCM_Report.pdf
101. FSB-TCFD (2024) About https://www.fsb-tcfd.org/about/
102. TCFD (2017) Final Report: Recommendations https://assets.bbhub.io/company/sites/60/2021/10/FINAL-2017-TCFD-Report.pdf
103. IFRS (2024) About Us https://www.ifrs.org/
104. CDSB (2022) About CDSB https://www.cdsb.net/our-story
105. IFRS (2024) International Sustainability Standards Board https://www.ifrs.org/groups/international-sustainability-standards-board/
106. IFRS (2023) ISSB issues inaugural global sustainability disclosure standards https://www.ifrs.org/news-and-events/news/2023/06/issb-issues-ifrs-s1-ifrs-s2/

107. UN Climate Change COP26 Outcomes: Finance for Climate Adaptation https://unfccc.int/process-and-meetings/the-paris-agreement/the-glasgow-climate-pact/cop26-outcomes-finance-for-climate-adaptation#What-specific-financial-pledges-were-made-at-COP26
108. UFCCC SCF (2020) Fourth (2020) Biennial Assessment and Overview of Climate Finance Flows https://unfccc.int/sites/default/files/resource/54307_1%20-%20UNFCCC%20BA%202020%20-%20Report%20-%20V4.pdf
109. Climate Champions (2022) COP27 Presidency launches Adaptation Agenda to build climate resilience for 4 billion by 2030 https://climatechampions.unfccc.int/cop27-presidency-announces-ambitous-climate-resilience-agenda/
110. GBAC/IFC (2022) COP27: Implications for Financial Institutions https://ecagbac.org/cop27-implications-for-financial-institutions/
111. ECGBAC@IFC.Org (2025) COP27: Implications for Financial Institutions https://ecagbac.org/cop27-implications-for-financial-institutions/
112. UNFCCC Overseas Development Institute (ODI) https://unfccc.int/topics/adaptation-and-resilience/resources/santiago-network/overseas-development-institute-odi
113. Media.odi.org (April 2019) Blended finance in the poorest countries: The need for a better approach https://media.odi.org/documents/12666.pdf
114. UN-convened Net Zero Asset Owner Alliance (September 2022) Call on Policymakers to Support Scaling Blended Finance https://www.unepfi.org/wordpress/wp-content/uploads/2022/09/NZAOA_Scaling-Blended-Finance.pdf#:~:text=Blended%20finance%20is%20a%20finance%20structure
115. NGFS (2024) Origin and Purpose https://www.ngfs.net/en
116. NGFS Membership https://www.ngfs.net/en/about-us/membership
117. IISD (2024) Mission and Goals https://www.iisd.org/mission-and-goals
118. White House (November 2022) Indonesia and International Partners Secure Groundbreaking Climate Targets and Associated Financing https://www.whitehouse.gov/briefing-room/statements-releases/2022/11/15/indonesia-and-international-partners-secure-groundbreaking-climate-targets-and-associated-financing/
119. Politico (14 February 2023) Bankers strike back on ESG https://www.politico.com/newsletters/the-long-game/2023/02/14/bankers-strike-back-on-esg-00082741
120. Financial Services Committee (September 2024) McHenry, Barr Request GAO Probe on Biden-Harris Financial Regulators' Ties to Network for Greening the Financial System https://financialservices.house.gov/news/documentsingle.aspx?DocumentID=409373
121. Trellis.Net (14th January 2025) BlackRock's exit from once-heralded net-zero group creates climate finance chaos https://trellis.net/article/blackrocks-exit-from-once-heralded-net-zero-group-creates-climate-finance-chaos/
122. Mayerbrown.com (22nd January 2025) Federal Reserve and FDIC Withdraw From the Network for Greening the Financial System https://www.mayerbrown.com/en/insights/publications/2025/01/federal-reserve-and-fdic-withdraw-from-the-network-for-greening-the-financial-system
123. US Treasury.Gov (30th January 2025) U.S. Department of the Treasury Withdraws from the Network of Central Banks and Supervisors for Greening the Financial System (NGFS) https://home.treasury.gov/news/press-releases/sb0003
124. ITA (2024) Our Vision https://ita.missionpossiblepartnership.org/about-us/
125. Gfanzero.com (January 2025) 2025: New Year update from Gfanz Secretariat https://www.gfanzero.com/press/2025-new-year-update-from-gfanz-secretariat/
126. NGFS (August 2024) Acute physical impacts from climate change and monetary policy https://www.ngfs.net/sites/default/files/medias/documents/ngfs_acute_physical_impacts_from_climate_change_and_monetary_policy.pdf
127. NGFS (2024) Net-Zero Banking Alliance https://www.unepfi.org/net-zero-banking/
128. Transition Pathway Initiative (October 2024) State of transition in the banking sector https://www.transitionpathwayinitiative.org/publications/uploads/2024-state-of-transition-in-the-banking-sector-report-2024.pdf

129. GFANZ (2024) GFANZ delivers on the year of the transition plan with continued growth and progress to close key gaps in the global financial system and accelerate climate investment https://www.gfanzero.com/press/gfanz-delivers-on-the-year-of-the-transition-plan-with-continued-growth-and-progress-to-close-key-gaps-in-the-global-financial-system-and-accelerate-climate-investment/#:~:text=GFANZ%20has%20developed%20a%20draft%20framework
130. GFANZ (2023) Mobilizing Capital in and to Emerging Markets https://assets.bbhub.io/professional/sites/24/Mobilizing-Capital-in-and-to-EMDEs.pdf
131. Reuters (September 2024) Scant green premium stalls over 450 low-carbon projects, analysis shows https://www.reuters.com/sustainability/climate-energy/scant-green-premium-stalls-over-450-low-carbon-projects-analysis-shows-2024-09-20/
132. Bloomberg/CFLI (2024) About https://www.bloomberg.com/cfli/about/
133. GEAPP (2024) GEAPP works to accelerate the green energy transition to end energy poverty across emerging economies https://energyalliance.org/
134. GEAPP (2024) Powering People and Planet 2023 https://energyalliance.org/powering-people-planet-2023/
135. IFACC (2024) Innovative Finance for the Amazon, Cerrado and Chaco https://www.ifacc-initiative.org/home
136. CPI (2024) FAST-Infra https://www.climatepolicyinitiative.org/fast-infra/#:~:text=GREG%20(Global%20Revenue%20Guarantee):%20Guaranteeing%20timely#:~:text=GREG%20(Global%20Revenue%20Guarantee):%20Guaranteeing%20timely
137. MOBILIST Mobilising Capital to meet the Global Goals https://www.mobilistglobal.com/about/
138. Forests & Finance (2024) About Us https://forestsandfinance.org/about-us/
139. Forests & Finance (2023) Banking on Biodiversity Collapse: Tracking the Banks and Investors Driving Tropical Forest Destruction https://forestsandfinance.org/wp-content/uploads/2023/12/BOBC_2023_vF.pdf
140. Forests & Finance (2024) Data Deep Dive https://forestsandfinance.org/data/
141. GCBC (2024) About Us https://capacity-building.org/
142. GCBC (2024) Building Capacity, Unlocking Capital https://pub-1de91bfb7f524dbb8f3048992be81d54.r2.dev/document/GCBC_bch_FNL_DIG__2___1_.pdf
143. WEF (2024) Unlocking climate solutions at scale through blended finance https://www.weforum.org/stories/2024/09/blended-finance-drive-climate-solutions/
144. Partha Dasgupta (2021), The Economics of Biodiversity: The Dasgupta Review. (London: HM Treasury) https://assets.publishing.service.gov.uk/media/602e92b2e90e07660f807b47/The_Economics_of_Biodiversity_The_Dasgupta_Review_Full_Report.pdf
145. World Bank blogs (2024) Harnessing the power of capital markets to conserve and restore global biodiversity through 'Natural Asset Companies' https://blogs.worldbank.org/en/psd/harnessing-power-capital-markets-conserve-and-restore-global-biodiversity-through-natural-asset
146. Tax Foundation (2024) Pigouvian Tax https://taxfoundation.org/taxedu/glossary/pigouvian-tax/
147. Investopedia Shadow Pricing: Definition, How It Works, Uses, and Example https://www.investopedia.com/terms/s/shadowpricing.asp
148. William D. Nordhaus (2021) The Spirit of Green Princeton University Press
149. IMF Externalities: Prices Do Not Capture All Costs https://www.imf.org/external/pubs/ft/fandd/basics/38-externalities.htm
150. DelawareInc.Com (2024) Non-Profit Corporation vs Public Benefit Corporation https://www.delawareinc.com/blog/non-profit-corporation-vs-public-benefit-corporation/#:~:text=Public%20Benefit%20Corporation%20vs%20Non-profits&text=There%20are%20no%20owners%20or,as%20defined%20by%20the%20IRS
151. HEC (Paris) What is the "Triple Bottom Line"? https://www.hec.edu/en/faculty-research/centers/sustainability-organizations-institute/think/so-institute-executive-factsheets/what-triple-bottom-line

152. Capital Services (2024) Benefit Corporations and Beyond https://www.capitolservices.com/benefit-corporations-and-beyond/#:~:text=Making%20an%20initial%20appearance%20in,as%20D.C.%20and%20Puerto%20Rico
153. Bcorporation.Net (2024) B Lab is the nonprofit network transforming the global economy to benefit all people, communities, and the planet https://www.bcorporation.net/en-us/movement/about-b-lab/
154. Impact Entrepreneur (January 2022) Natural Asset Companies (NACs) https://impactentrepreneur.com/natural-asset-companies-nacs/
155. IEG (2024) How NACs Work https://www.intrinsicexchange.com/nacs
156. WEF (2024) The Future of Nature and Business https://www3.weforum.org/docs/WEF_The_Future_Of_Nature_And_Business_2020.pdf
157. LandTrustAlliance (February 2024) Natural Assets Companies, natural asset credits and land trusts https://landtrustalliance.org/resources/learn/explore/defending-voluntary-private-land-conservation-community-toolkit/natural-assets-companies-natural-asset-credits-and-land-trusts
158. CPAC (December 2023) The Troubling Rise of Natural Asset Companies https://www.cpac.org/post/the-troubling-rise-of-natural-asset-companies#:~:text=The%20New%20York%20Stock%20Exchange%20(NYSE)%20and%20the%20Intrinsic
159. National Ocean Service What is lidar? https://oceanservice.noaa.gov/facts/lidar.html
160. Thomas J. Snyder and Randall Haas (2024) 'Unstructured satellite survey detects up to 20% of archaeological sites in coastal valleys of southern Peru' PLoS ONE vol.19(2) https://journals.plos.org/plosone/article?id= https://doi.org/10.1371/journal.pone.0292272
161. Geographyrealm (2018) Creating Ground-level Views from Satellite Imagery https://www.geographyrealm.com/creating-ground-level-views-satellite-imagery/
162. BeZero (2024) Buyers pay more for carbon credits with SDG claims https://bezerocarbon.com/insights/buyers-pay-more-for-carbon-credits-with-sdg-claims?utm_campaign=Marketing%20Newsletter&utm_medium=email&_hsenc=p2ANqtz-8ytAwYF5mHvdyWMN1lapm3oqnyhCsmAZArQ7S866YrfmPWtqw5RuwsKIq1F-NqvSZA8YLNdpbRz590NtZhLoHbArWoyA&_hsmi=326938042&utm_content=326938042&utm_source=hs_email
163. Utility Dive (2024) Banks endorse goal of tripling nuclear capacity as Climate Week NYC begins https://www.utilitydive.com/news/climate-week-nyc-2024-roundup-green-bank-carbon-removal-nuclear-finance-gb50-frontier-climeworks/728005/#:~:text=Kirsten%20Stasio%20(speaking),%20CEO%20of%20Nevada
164. USGreenbanks.Org (2024) The U.S. Green Bank 50 https://www.usgreenbanks.org/
165. EPA (2024) Greenhouse Gas Reduction Fund https://www.epa.gov/greenhouse-gas-reduction-fund#:~:text=The%20Fund%20will%20provide%20competitive%20grants
166. EENews.Net (2024) Green banks launch coalition to coordinate IRA funding https://www.eenews.net/articles/green-banks-launch-coalition-to-coordinate-ira-funding/#:~:text=Green%20banks%20across%20the%20country%20launched
167. Mother Joines (March/April 2025) Can the Rule of Law Save the IRA? https://www.motherjones.com/politics/2025/02/targeted-package-biden-trump-ira-green-energy-investment/
168. Wikipedia Convention on Biological Diversity https://en.wikipedia.org/wiki/Convention_on_Biological_Diversity#:~:text=The%20Convention%20on%20Biological%20Diversity%20%28CBD%29%2C%20known%20informally,equitable%20sharing%20of%20benefits%20arising%20from%20genetic%20resources
169. Mongabay (20 December 2022) Nations adopt Kunming-Montreal Global Biodiversity Framework https://news.mongabay.com/2022/12/nations-adopt-kunming-montreal-global-biodiversity-framework/#:~:text=The%20United%20States%20is%20the%20only%20major,nation%20not%20a%20party%20to%20the%20agreement.
170. BIOFIN (2024) Financial Resources (FIRE) for Biodiversity https://fire.biofin.org/
171. UNEP Strategic Plan for Biodiversity 2011–2020 https://www.cbd.int/undb/media/factsheets/undb-factsheet-sp-en.pdf
172. CBD.Int (2022) Kunming-Montreal Global Biodiversity Framework https://www.cbd.int/doc/decisions/cop-15/cop-15-dec-04-en.pdf

173. CBD.Int (2022) Global Diversity Framework https://www.cbd.int/gbf#:~:text=Among%20the%20Framework%E2%80%99s%20key%20elements%20are%204%20goals%20for%202050
174. Friends of the Earth Europe (2016) The Cancun Agreement: Not Worth Cheering https://friendsoftheearth.eu/wp-content/uploads/2012/02/foee_cancun_agreement_1210.pdf
175. Wikipedia (2024) Green Climate Fund https://en.wikipedia.org/wiki/Green_Climate_Fund#:~:text=The%20Green%20Climate%20Fund%20(GCF)%20is%20a%20fund%20for%20climate
176. Wikipedia (2024) Global Environment Facility https://en.wikipedia.org/wiki/Global_Environment_Facility#:~:text=Since%201994,%20the%20World%20Bank%20has%20served%20as%20the%20Trustee
177. GEF (2024) Global Biodiversity Framework Fund https://www.thegef.org/what-we-do/topics/global-biodiversity-framework-fund#:~:text=The%20GBFF%20aims%20to%20help%20countries%20achieve%20the%20Kunming-Montreal%20Global
178. GEF (2024) Least Developed Countries Fund – LDC: Main Issue https://www.thegef.org/what-we-do/topics/least-developed-countries-fund-ldcf#:~:text=In%202001%2C%20the%20194%20parties%20to%20the%20United,helping%20these%20countries%20adapt%20to%20new%20climate%20realities.
179. GEF (2024) Special Climate Change Fund – SCCF: Main Issue https://www.thegef.org/what-we-do/topics/special-climate-change-fund-sccf#:~:text=The%20Special%20Climate%20Change%20Fund,%20one%20of%20the%20world%E2%80%99s%20first
180. IISD (2012) GEF Approves First Access and Benefit Sharing Project Under Nagoya Protocol Fund https://sdg.iisd.org/news/gef-approves-first-access-and-benefit-sharing-project-under-nagoya-protocol-fund/#:~:text=The%20NPIF%20is%20a%20multi-donor%20trust%20fund%20administered,by%20the%20Governments%20of%20Japan%2C%20Switzerland%20and%20Norway.
181. IISD (2016) CBIT Trust Fund Supporting Transparency under Paris Agreement Operationalized https://sdg.iisd.org/news/cbit-trust-fund-supporting-transparency-under-paris-agreement-operationalized/#:~:text=The%20CBIT%20trust%20fund%20was%20established%20by%20the,1%2FCP.21%20adopting%20the%20Paris%20Agreement%20on%20climate%20change.
182. Manifestclimate.Com (March 2023) What is the TNFD and how does it Fit into Climate Reporting? https://www.manifestclimate.com/blog/tnfd-climate-reporting/#:~:text=For%20example%2C%20the%20TCFD%20is,develop%20appropriate%20risk%20management%20processes.
183. TNFD (2024) Recommendations of the TNFD https://tnfd.global/recommendations-of-the-tnfd/#overview
184. TNFD (October 2023) Guidance on the identification and assessment of nature-related issues: the LEAP approach https://tnfd.global/publication/additional-guidance-on-assessment-of-nature-related-issues-the-leap-approach/
185. Leap.sei.org (2024) LEAP https://leap.sei.org/default.asp?action=introduction
186. TNFD.Global (June 2024) TNFD adoption now over 400 organisations and new sector guidance released https://tnfd.global/tnfd-adoption-now-over-400-organisations-and-new-sector-guidance-released/
187. Ran.org (2022) Keep Forests Standing 2022 https://www.ran.org/wp-content/uploads/2022/06/KFS-Scorecard-Brief-2022-_WEB.pdf
188. Ran.org (September 2023) Is TNFD an answer to the biodiversity crisis? https://www.ran.org/the-understory/is-tnfd-an-answer-to-the-biodiversity-crisis/#:~:text=TNFD%20not%20only%20fails%20to,avoiding%20accountability%20to%20frontline%20communities
189. Green Finance Institute Hive (2024) ASN Biodiversity Fund https://hive.greenfinanceinstitute.com/gfihive/case-studies/asn-biodiversity-fund/#
190. Finance for Biodiversity Foundation (2024) Finance for Biodiversity Pledge https://www.financeforbiodiversity.org/about-the-pledge/

191. Finance for Biodiversity Foundation (2024) Finance for Biodiversity Guide on biodiversity measurement approaches https://www.financeforbiodiversity.org/wp-content/uploads/Finance-for-Biodiversity_Guide-on-biodiversity-measurement-approaches_3rd-edition-1.pdf
192. IMF.Org (October 2024) Embedded in Nature: Nature-Related Economic and Financial Risks and Policy Considerations https://www.imf.org/en/Publications/staff-climate-notes/Issues/2024/10/01/Embedded-in-Nature-Nature-Related-Economic-and-Financial-Risks-and-Policy-Considerations-555072
193. Naturepositive.org (2024) A Global Goal for Nature: Nature Positive by 2030 https://www.naturepositive.org/resources/
194. The Oxford Nature Positive Hub (2023) What is Nature Positive? https://naturepositive.web.ox.ac.uk/nature-positive-oxford#:~:text=In%202019%2C%20a%20group%20of%20environmental%20organisations%20and,as%20a%20proposed%20headline%20global%20goal%20for%20nature
195. NPI (2023) The Definition of Nature Positive https://www.naturepositive.org/app/uploads/2024/02/The-Definition-of-Nature-Positive.pdf
196. CBD.Int (2022) Convention on Biological Diversity https://www.cbd.int/doc/decisions/cop-15/cop-15-dec-06-en.pdf
197. CBD.Int (2024) National Biodiversity Strategies and Action Plans (NBSAPs) https://www.cbd.int/nbsap
198. The Moscow Times (September 2024) Is Russia Running Out of Time to Save the Caspian Seal From Extinction? https://www.themoscowtimes.com/2024/09/17/is-russia-running-out-of-time-to-save-the-caspian-seal-from-extinction-a85827
199. National Geographic (2018) Once Written Off for Dead, the Aral Sea Is Now Full of Life https://blog.education.nationalgeographic.org/2018/03/21/once-written-off-for-dead-the-aral-sea-is-now-full-of-life/
200. The Nature Conservancy (2024) Financing Nature: Closing the Global Biodiversity Financing Gap https://www.nature.org/content/dam/tnc/nature/en/documents/FINANCINGNATURE_FullReport_091520.pdf
201. WEF (2024) Financing for Nature https://initiatives.weforum.org/financing-for-nature/home
202. IFC (2023) Biodiversity Finance: Protect the Planet, Strengthen Livelihoods https://www.ifc.org/en/stories/2023/biodiversity-finance-interview-with-irina-likhachova
203. Andrew Seidl et al (2021) Nature, Ecology and Evolution Journal v.5, pp.530–539 (2021) 'The effectiveness of national biodiversity investments to protect the wealth of nature' https://www.nature.com/articles/s41559-020-01372-1#:~:text=The%20Biodiversity%20Finance%20Initiative%20provides%20a%20means%20for%20countries%20to
204. Forests & Finance (2023) Banking On Biodiversity Collap$E https://forestsandfinance.org/wp-content/uploads/2023/12/BOBC_2023_vF.pdf
205. UNEP (2023) State of Finance for Nature the Big Nature Turnaround: Repurposing $7 trillion to combat nature loss file:///C:/Users/urejo/Downloads/state_finance_nature_2023%20(6).pdf
206. WEF (2020) The Future of Nature and Business https://www3.weforum.org/docs/WEF_The_Future_Of_Nature_And_Business_2020.pdf
207. WEF (June 2024) How to spot a nature crisis – and why we should use natural capital to combat it https://www.weforum.org/agenda/2024/06/3-traits-nature-crisis-natural-capital/
208. World Bank (2023) The Effect of Multinational Enterprises on Climate Change Supply Chain Emissions, Green Technology Transfers, and Corporate Commitments https://openknowledge.worldbank.org/server/api/core/bitstreams/def956de-eac5-4ede-a366-e01331481c99/content
209. GoldmanSachs.Com (May 2024) AI is poised to drive 160% increase in data center power demand https://www.goldmansachs.com/insights/articles/AI-poised-to-drive-160-increase-in-power-demand

210. Investopedia (6th February 2025) Amazon Follows Google, Meta, and Microsoft With Plans To Boost Spending on AI https://www.investopedia.com/amazon-follows-google-meta-and-microsoft-with-plans-to-boost-spending-on-ai-8787507
211. Financial Times (21st January 2025) Microsoft secures deal to restore Amazon rainforest and offset AI emissions https://www.ft.com/content/3ac7e92f-bea5-4460-8791-23a2697d8b80
212. WWF (2024) The Corrado is disappearing https://www.worldwildlife.org/pages/save-the-cerrado-our-climate-depends-on-it
213. TIG (2024) Sustainably Managing the World's Timberland https://timberlandinvestmentgroup.com/
214. IFC (July 2024) IFC commits $50 million to BTG Pactual TIG's Latin American reforestation strategy https://www.ifc.org/en/pressroom/2024/ifc-commits-50-million-to-btg-pactual-tigs-latin-american-refore
215. TIG (18th September 2024) BTG Pactual Timberland Investment Group and Meta announce long-term contract for delivery of 1.3 million nature-based carbon removal credits, with options for delivery of additional 2.6 million credits https://timberlandinvestmentgroup.com/btg-pactual-timberland-investment-group-and-meta-announce-long-term-contract-for-delivery-of-1-3-million-nature-based-carbon-removal-credits-with-options-for-delivery-of-additional-2-6-million-credit/
216. Financial Times (4th February 2025) Jeff Bezos's $10bn Earth Fund cuts ties with climate group https://www.ft.com/content/c3bf3f5f-afa2-44ce-8314-18ad321cd983
217. Verdantiz (2025) The Carbon Removal Arms Race Continues With Google's First 2025 Offtake Agreements https://www.verdantix.com/insights/blogs/the-carbon-removal-arms-race-continues-with-google-s-first-2025-offtake-agreements
218. 226. Bill Gates (2021) How to Avoid a Climate Disaster Penguin, Random House, UK
219. MIT News (October 2023) New tools are available to help reduce the energy that AI models devour https://news.mit.edu/2023/new-tools-available-reduce-energy-that-ai-models-devour-1005#:~:text=Training%20is%20just%20one%20part,any%20better%2C%22%20Gadepally%20says.
220. Energy.gov (2024) Recommendations on Powering Artificial Intelligence and Data Center Infrastructure https://www.energy.gov/sites/default/files/2024-08/Powering%20AI%20and%20Data%20Center%20Infrastructure%20Recommendations%20July%202024.pdf
221. Apple (2024) Carbon Removal Strategy https://www.apple.com/environment/pdf/Apples_Carbon_Removal_Strategy_White_Paper.pdf
222. CAM.Com (2023) Climate Asset Management will manage a new investment fund for Apple as Apple doubles the size of its Restore Fund https://climateassetmanagement.com/insight/climate-asset-management-will-manage-a-new-investment-fund-for-apple-as-apple-doubles-the-size-of-its-restore-fund/
223. Amazon.com (2023) Amazon Sustainability Report (2023) https://sustainability.aboutamazon.com/2023-amazon-sustainability-report.pdf
224. Spglobal.com (July 2022) Amazon seeks to restore trust in voluntary carbon market with new credit designation https://www.spglobal.com/commodityinsights/en/market-insights/latest-news/energy-transition/070722-amazon-seeks-to-restore-trust-in-voluntary-carbon-market-with-new-credit-designation
225. Leafcoalition.org What is the LEAF Coalition? https://www.leafcoalition.org/home
226. Reuters (September 2024) Firms including Amazon to buy $180 million in carbon credits from namesake rainforest https://www.reuters.com/sustainability/firms-including-amazon-buy-180-million-carbon-credits-namesake-rainforest-2024-09-24/
227. Google (2024) Google Environmental Report, 2024 https://www.gstatic.com/gumdrop/sustainability/google-2024-environmental-report.pdf
228. DCD.com (October 2024) Meta joins Google in DOE carbon removal investment scheme https://www.datacenterdynamics.com/en/news/meta-joins-google-in-doe-carbon-removal-investment-scheme/
229. Meta (2024) For a better reality https://sustainability.atmeta.com/wp-content/uploads/2024/08/Meta-2024-Sustainability-Report.pdf

230. Green.earth.com (June 2023) Meta signs major carbon credits deal with Aspiration for nature-based solutions https://www.green.earth/news/meta-signs-major-carbon-credits-deal-with-aspiration-for-nature-based-solutions#:~:text=In%20a%20significant%20move%20towards%20sustainability%2C%20Meta%2C%20the,Meta%27s%20quest%20to%20achieve%20net-zero%20emissions%20by%202030.
231. ESGDive.com (October 2024) Dive Brief: Meta joins DOE initiative, pledges $35M for carbon removal projects https://www.esgdive.com/news/meta-joins-doe-initiative-commits-35m-carbonremoval-projects/730044/
232. Microsoft.com (2023) How can we advance sustainability? https://query.prod.cms.rt.microsoft.com/cms/api/am/binary/RW1lMjE#page=10
233. Edie.net (2024) Microsoft signs major carbon credits agreement https://www.edie.net/microsoft-signs-major-carbon-credits-agreement/
234. Reuters (18 June 2024) Microsoft to buy 8 million carbon credits from BTG Pactual in largest-ever sale https://www.reuters.com/sustainability/climate-energy/microsoft-buy-8-million-carbon-credits-btg-pactual-largest-ever-sale-2024-06-18/
235. Indigo Ag (2024) The Leader in Sustainable Agriculture Solutions https://www.indigoag.com/
236. Carboncredits (September 2023) Microsoft's $200M Carbon Removal Deal Advances Heirloom's DAC Solution https://carboncredits.com/microsofts-200m-carbon-removal-deal-supercharges-heirlooms-dac-solution/
237. Carboncredits.Com (2023) Microsoft and UNDO Partner for 15,000 Tons of Carbon Removal Using Enhanced Rock Weathering! https://carboncredits.com/microsoft-and-undo-partner-for-15000-tons-of-carbon-removal-using-enhanced-rock-weathering/
238. Carboncredits.Com (May 2024) Ørsted Secures Major Carbon Removal Deal with Microsoft https://carboncredits.com/orsted-secures-major-carbon-removal-deal-with-microsoft/
239. Geekwire.com (July 2024) Microsoft inks the largest carbon removal deal ever https://www.geekwire.com/2024/microsoft-to-buy-500k-tons-of-carbon-removal-in-largest-deal-ever-made/
240. Salesforce.com Our Net Zero Journey https://www.salesforce.com/content/dam/web/en_us/www/assets/pdf/reports/salesforce-climate-action-plan.pdf
241. Salesforce.com (2021) Sustainable Bond Framework, June 2021 https://s23.q4cdn.com/574569502/files/doc_downloads/2021/06/Salesforce-Sustainable-Bond-Framework.pdf
242. Salesforce.com (2024) Salesforce Venture Impact Fund https://salesforceventures.com/impact-fund/
243. Trellis.net (June 2024) Salesforce commits $25 million to 'steep learning curve' of carbon removal https://trellis.net/article/salesforce-commits-25-million-steep-learning-curve-carbon-removal/
244. Weforum (2024) First Movers Coalition https://initiatives.weforum.org/first-movers-coalition/home
245. Carbondirect.com (2024) Ecological Restoration in the Voluntary Carbon Market https://www.carbon-direct.com/research-and-reports/ecological-restoration-in-the-voluntary-carbon-market
246. Symbiosis (2024) Introducing Symbiosis https://www.symbiosiscoalition.org/perspectives/introductory-blog
247. Leaf Coalition (2024) Uniting to protect tropical forests: For climate, people & nature https://www.leafcoalition.org/home
248. Emergent (2021) What is Emergent? https://emergentclimate.com
249. Emergent (2021) Costa Rica and Ghana Agree Landmark Deals to Supply Forest Carbon Credits to LEAF Coalition Buyers https://emergentclimate.com/costa-rica-and-ghana-agree-landmark-deals-to-supply-forest-carbon-credits-to-leaf-coalition-buyers/
250. Frontier Climate An advance market commitment to accelerate carbon removal https://gaps.frontierclimate.com
251. Thehill.com (October 2024) Amazon signs deal to build, invest in small modular nuclear reactors https://thehill.com/policy/technology/4935308-amazon-invests-nuclear-energy-ai/

252. Blog.google (October 2024) New nuclear clean energy agreement with Kairos Power https://blog.google/outreach-initiatives/sustainability/google-kairos-power-nuclear-energy-agreement/
253. Power-Eng.Com (4th December 2024) Meta seeks up to 4 GW of new nuclear to power its data centers, AI https://www.power-eng.com/nuclear/meta-seeks-up-to-4-gw-of-new-nuclear-to-power-its-data-centers-ai/
254. Fool.com (September 2024) Microsoft Goes Nuclear. Here Are the Biggest Beneficiaries of Its Massive Power Grab https://www.fool.com/investing/2024/09/22/microsoft-goes-nuclear-here-are-the-biggest-benefi/#:~:text=Microsoft%20signed%20a%20power%20purchase%20agreement%20with%20Constellation,deal%20to%20secure%20renewable%20energy%20from%20Brookfield%20Renewable.
255. Gfmag.com (October 2024) Microsoft Taps Nuclear Power To Fuel Growing AI Demand https://gfmag.com/economics-policy-regulation/microsoft-three-mile-island-nuclear-power-ai-demand/
256. Apple (June 2024) Expanding Renewable Electricity Capacity https://www.apple.com/environment/pdf/Apples_Product_Use_Electricity_Strategy.pdf
257. Center for Industrial Progress (2018) The Truth About Apple's '100% Renewable' Energy Usage https://industrialprogress.com/the-truth-about-apples-100-renewable-energy-usage/
258. EIA.G OV (2024) What is U.S. electricity generation by energy source? https://www.eia.gov/tools/faqs/faq.php?id=427&t=3
259. IEA (2025) World Energy Statistics and Balances https://www.iea.org/data-and-statistics
260. CDP (2025) Brand Hub https://www.cdp.net/en/brand-hub
261. Transitionpathwayinitiative.org (September 2023) Banks and the net zero transition https://www.transitionpathwayinitiative.org/publications/uploads/2023-banks-and-the-net-zero-transition-tracking-progress-with-the-tpi-net-zero-banking-assessment-framework.pdf
262. Transitionpathwayinitiative.org (July 2023) Net Zero Banking Assessment Framework https://www.transitionpathwayinitiative.org/publications/uploads/2023-net-zero-banking-assessment-framework.pdf
263. Netzerofinancialtracker (2024) How are private finance institutions progressing on their Paris Agreement and Net Zero goals? https://netzerofinancetracker.climatepolicyinitiative.org/
264. Carbonaccountingfinancials.com (2022) Financed Emissions https://carbonaccountingfinancials.com/files/downloads/PCAF-Global-GHG-Standard.pdf
265. IMF (July 2024) Bank Profitability in Europe: Not Here to Stay https://www.imf.org/en/Publications/WP/Issues/2024/07/09/Bank-Profitability-in-Europe-Not-Here-to-Stay-551129
266. Bankingdive.com (September 2024) Most global banks unprepared for climate change risks: report https://www.bankingdive.com/news/us-banks-unprepared-climate-change-risks-climate-x-bofa-citi-pnc-goldman-capital-one/728047/
267. Google.com What is Retrieval-Augmented Generation (RAG)? https://cloud.google.com/use-cases/retrieval-augmented-generation
268. Climate-X (2024) Top 50 Banks In The World Tackling Adaptation – 2024 https://www.climate-x.com/files/pdfs/climate_x_-_top_50_banks_in_the_world_tackling_climate_adaptation_wp.pdf
269. lse.ac.uk (2023) The case for a Scope 3 carbon tax for financial institutions https://www.lse.ac.uk/granthaminstitute/news/the-case-for-a-scope-3-carbon-tax-for-financial-institutions/
270. SEC (2024) SEC Adopts Rules to Enhance and Standardize Climate-Related Disclosures for Investors https://www.sec.gov/newsroom/press-releases/2024-31
271. Jdsupra.com (13th February 2025) Acting SEC Chair seeks a pause in SEC climate disclosure rule litigation https://www.jdsupra.com/legalnews/acting-sec-chair-seeks-a-pause-in-sec-8226068/
272. ESGDrive.com (28 March 2025) SEC withdraws defense of climate-risk disclosure rule https://www.esgdive.com/news/sec-withdraws-climate-risk-disclosure-rule-defense-eighth-circuit-reactions/743860/?utm_source=Sailthru&utm_medium=email&utm_campaign=Issue:%202025-03-28%20ESG%20Dive%20%5Bissue:71729%5D&utm_term=ESG%20Dive

273. WWF (October 2024) ECB must take responsibility and stop loans that harm nature, say WWF and Positive Money https://www.wwf.eu/?15315891/ECB-must-take-responsibility-and-stop-loans-that-harm-nature-say-WWF-and-Positive-Money
274. Europa.eu (2024) EU taxonomy for sustainable activities https://finance.ec.europa.eu/sustainable-finance/tools-and-standards/eu-taxonomy-sustainable-activities_en
275. Europa.com (2024) Platform on Sustainable Finance https://finance.ec.europa.eu/sustainable-finance/overview-sustainable-finance/platform-sustainable-finance_en
276. Sustainableinancelab.nl (2024) A Green Interest Rate for the Eurozone https://sustainablefinancelab.nl/wp-content/uploads/sites/334/2024/09/A-green-interest-rate-for-Europe.pdf
277. Microhive (2024) How Will TLTROs Affect ECB Monetary Policy in 2022? https://macrohive.com/hive-explainers/what-are-tltros-and-how-will-they-affect-ecb-monetary-policy-in-2022/
278. Green Central Banking (2024) Why the ECB should consider green TLTROs once again https://greencentralbanking.com/2024/10/22/why-ecb-should-consider-green-tltros/
279. Investopedia (November 2025) Debt/Equity Swap: Definition, Purpose, Example https://www.investopedia.com/ask/answers/06/debtequityswap.asp
280. Debtjustice.Org (November 2024) Debt Swaps Won't Save Us: The Urgent Need for Debt Cancellation and Grant-Based Climate Finance https://debtjustice.org.uk/wp-content/uploads/2024/12/Debt-Swaps-Wont-Save-Us-Briefing-2024-v3.pdf
281. FAO (1996) Debt-for-nature swaps: a decade of experience and new directions for the future https://www.fao.org/4/w3247e/w3247e06.htm
282. IIED (2024) Debt swaps could release $100 billion for climate action https://www.iied.org/debt-swaps-could-release-100-billion-for-climate-action
283. Debt Justice (2022) Debt Data Portal https://data.debtjustice.org.uk/
284. Weforum.org (2024) Climate finance: What are debt-for-nature swaps and how can they help countries? https://www.weforum.org/agenda/2024/04/climate-finance-debt-nature-swap/
285. Financial Times (18 October 2024) El Salvador gets $1bn US-backed loan from JPMorgan for debt refinancing
286. Weforum.org (February 2024) What does the OECD global minimum tax mean for global cooperation? https://www.weforum.org/stories/2024/02/oecd-minimum-tax-rate/
287. CCAC Trade Union Group (2021) Climate Jobs: Building a workforce for climate emergency Campaign Against Climate Change, UK
288. ILO (2024) Ensuring safety and health at work in a changing climate https://www.ilo.org/sites/default/files/2024-07/ILO_SafeDay24_Report_r11.pdf
289. Charles A. Ogunbode et at. (2024) 'Climate justice beliefs related to climate action and policy support around the world' Nature Climate Change V.14, pages 1144–1150, 18 October 2024 https://www.nature.com/articles/s41558-024-02168-y
290. CPI (2025) Global Landscape of Climate Finance 2025 https://www.climatepolicyinitiative.org/publication/global-landscape-of-climate-finance-2025/

Chapter 7
Ways Forward

Global Warming and Cooling

Net zero means as much C02 is removed as is emitted from the Earth into the atmosphere. Net negative is when there is an absolute reduction in C02 as a means to bring down average global temperatures. The equal exchange of heat (radiation) between the Sun and the Earth is the Earth's energy budget, and to keep that equality, when the Earth retains more radiation than previously due to excessive GHGs, requires global warming (or 'boiling' as the UN General Secretary calls it) to force additional radiation from the atmosphere back into Outer Space to reduce the budget deficit and restore equilibrium—see Fig. 7.1 and Chap. 1.

Chapters 1–3

The first three chapters of this book were all about modelling. Chapter 1 focused on the models developed by climatologists which are periodically consolidated in the Assessment Reports (AR) of the IPCC [2]. AR6 came out in 2023, and AR7 is due out in 2027, albeit it seems without official American support as the Trump Administration dismissed around 400 climate scientists working on the country's National Climate Assessment [3]. A central feature of these models is to estimate the effects of 'radiation forcing' or the amount by which atmospheric temperatures will rise with each doubling of CO_2e, that is, of C02 or its equivalence in other GHGs, such as methane. All models are based upon the concept of a balance or equilibrium of heat exchange between the Sun and the Earth, but to maintain that equilibrium the Earth's surface temperature has to heat (or cool) according to the conditions of the atmosphere. A mapping of possible scenarios is derived from the models based upon the assumptions made and the quality of the available data, and being scientific models they are all subject to probability assessment. There are no final proofs…

J. Ure, *Climate Change and Carbon Markets: Late or Too Late?*,
https://doi.org/10.1007/978-3-032-07619-9_7

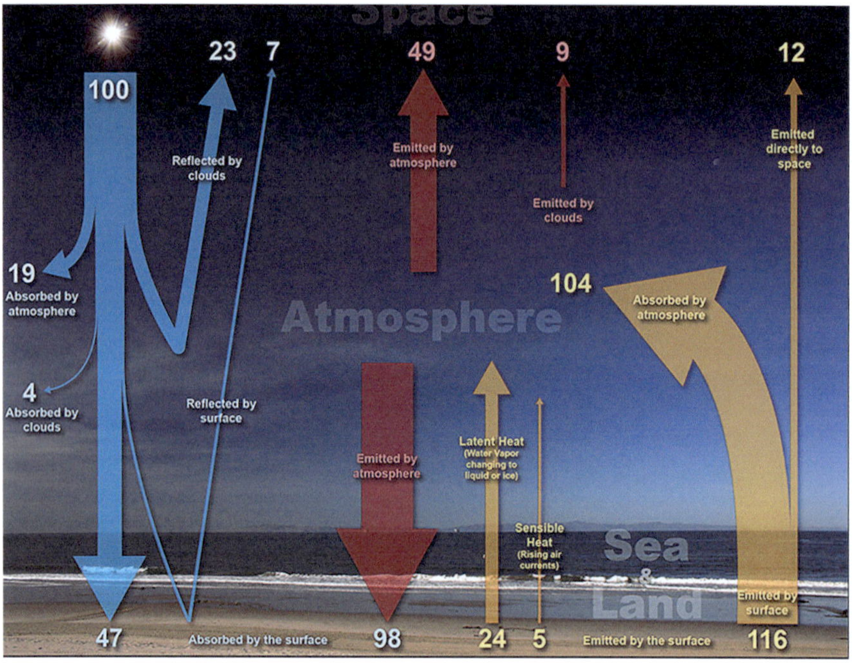

Fig. 7.1 The energy budget. *Source* NOAA https://www.noaa.gov/jetstream/atmosphere/energy

until there are! Their purpose is to assign levels of likelihood of global warming over various timeframes. They alert us all to the possibility of climate change risk. Either we acknowledge the risks and take steps to mitigate them, or we do not.

Chapter 2 focused on the various pathways towards achieving the principal goal of the Paris Agreement of keeping 'the increase in the global average temperature to well below 2°C above pre-industrial levels' and pursue efforts 'to limit the temperature increase to 1.5°C above pre-industrial levels' [4]. The pathways arise from models used by the IEA (GHG emissions) and by IRENA (renewables) to follow through on the range of scenarios of the UNFCCC. Each pathway reaches its target according to the different policy commitments which are assumed to drive them. The main pathways (not forecasts) modelled by the IEA are the Stated Policies Scenario (STEPS), the Announced Pledges Scenario (APS), and the Net-Zero Emissions by 2050 (NZE) Scenario. IRENA has developed two pathway scenarios, the Planned Energy Scenario (PES) and the 1.5°C Scenario. Both stress the importance of energy efficiency, which reduces the energy intensity required for a given task, and energy flexibility for fast reaction to changes in markets, policies and regulations, technologies, etc. It should be noted that improved energy efficiency means little unless total energy demands fall, or until *all* energy is green and renewable, including the energy and materials used in the construction of power plants and transmission grids.

Chapter 3 reviewed the economic models available, especially the so-called 'damage function' of each model as measured by a particular metric, most usually the impact of climate change on GDP growth rates, but alternatively as eco-economists and socio-economists argue for the impact upon nature and the lived-in environment. One feature of this approach is to add nature-based outcomes to investment indices showing the financial implications of gains or losses in nature, as proposed by, among others, The Nature Conservancy (TNC) [5] and adopted by the European Central Bank in 2025 [6]. The divergence of the economic methodologies used is their outstanding feature. The traditional neo-classical-based (neo-liberal) models treat emissions as exogenous variables seemingly independent of the use of labour and capital. More recently, a school of modelling has arisen that place carbon emissions as endogenous within the models, and socio-economic modelling which gives a central role to natural input-outputs rather than market input-outputs. A major difference in the conclusions of these models is that in neoclassical modelling, the damage function is generally small due to a choice of a high discount rate which decreases the present value of future benefits, effectively making them seem less valuable. Other modellers either argue for a low discount rate to acknowledge the right of future generations to inherent a low-carbon economy, or no discount rate at all, either on the grounds that the future is too indeterminate or because it is unethical and meaningless to place a value on the sustainability of nature and the wider environment.

Chapters 4–6

Chapters 4–6 turn attention to what is being done to mitigate climate change: first in the public sphere, looking at the strategies being adopted by governments and the compliance or mandatory carbon markets (MCM), then in the voluntary carbon markets (VCM) sector, and finally in green finance. Running through each of these is the strategy that anticipates and indeed relies upon public capital acting as a catalyst for private capital to play its part. The implicit argument is that governments cannot prioritise climate change project funding to the point where it over-rides all the other claims on public finances, and that therefore the private sector, which is actually responsible for much of the GHGs emitted, has to bridge the gap. An elaborate and sophisticated structure of public–private support mechanisms has been created to realise these objectives, which is detailed in the chapters. By 2023, the mandatory or compliance carbon pricing schemes which have been introduced in many developed economies apply to around one-quarter of global emissions, and over the years have raised more than USD500 billion from polluters. They account for around 50% of the USD100 billion raised annually through carbon taxes and ETS (Emission Trading Schemes) [7]. The funds are retained within country. By contrast, the Voluntary Carbon Market (VCM) funds seek out projects globally, and in theory could be made available to carbon credit transactions in EMDEs (Emerging Market and Developing Economies), for example, for reforestation projects. As recorded in Chap. 5, increasingly the multinational corporations, notably those in the high-tech sector, are by-passing the multifaceted VCMs and investing directly in their chosen

NbPs (Nature-based Projects) where they can exercise more control over project monitoring and evaluation. Despite this, global finance for the climate mitigation and adaptation is proving to be woefully inadequate, and has been severely hampered by the withdrawal of the USA from the Paris Agreement and its funding mechanisms. The failure is particularly true for the EMDEs and especially for the LDCs (Least Developed Countries) who have received only 14% and 2% of global climate finance, respectively [8].

Out of the Bag: Evidence of the Anthropocene

The ultimate test of success or failure to mitigate climate change is what happens to GHG emissions, what happens to average global temperatures, and what happens during signature events such as the melting of ice caps [9]; the rising levels of the oceans [10]; the acidification of the oceans [11]; the reversal of oceanic currents which change forever local and regional weather patterns [12]; a substantial loss of fertile land to droughts or to swamps [13]; a growing frequency of violent climate change-induced weather events [14]; a climate-forced migration of animals [15]; a dramatic loss of animal species [16]; a dramatic loss of biodiversity [17]; and an enforced migration of humans [18]. See Fig. 7.2 from Copernicus [19].

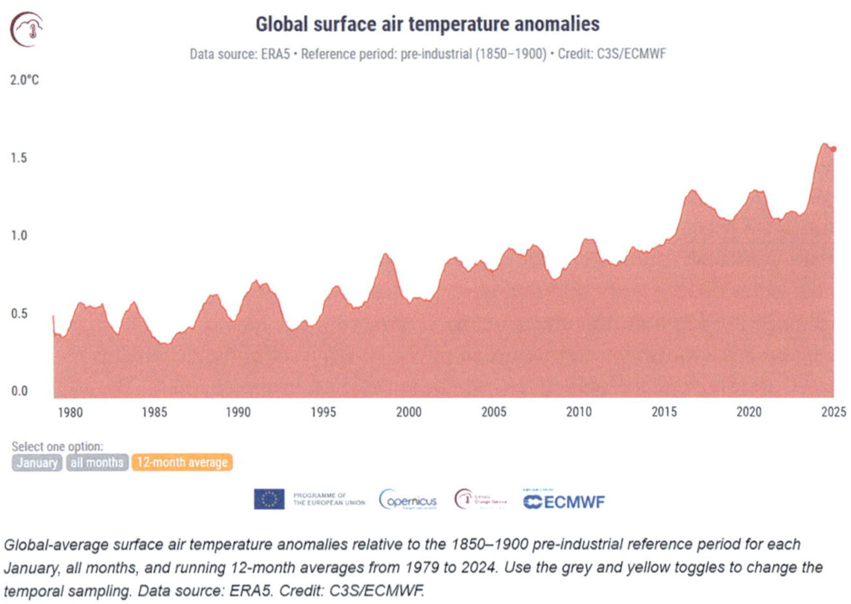

Fig. 7.2 Global surface air temperature anomalies

What is now well established is that (1) the global average temperature anomalies, that is, the difference between current measurements and the average over the pre-industrial long term 1985–1900 has been consistently rising and (2) that since July 2023 the average global temperature has been above the Paris goal of 1.5C in every successive month—see Fig. 7.2 [20].

The detective work of scientists at various research institutions such as NOAA's Global Monitoring Laboratory can identify the carbon isotopes (see Foreword) that make up the GHGs causing climate change, showing that they increasingly originate from human activity of burning fossil fuels [21]. The cat is now well out of the bag.

A Japanese energy economist Yoichi Kaya in 1993 published a simple mathematical model for tracking the drivers of climate change, based upon the role of human activity: F = P x G/P x E/G x F/E, where F is global C02 emissions from human sources, G is global GDP and P is global population, so G/P is global GDP per capita, E is global energy consumption, so G/E is global energy intensity per GDP and F/E is global emissions intensity per capita [22]. Our World in Data [23] has reproduced the tracking up to 2023—see Fig. 7.3.

The data in Fig. 7.3 suggests $C0_2$ and GDP *per capita* remain closely correlated, despite IEA data suggesting the correlation between C02 and GDP growth *in absolute terms* is weakening—see Chap. 3. It should also be noted that as energy intensity (the efficient use of energy) increases (falling graph line) and as the use of clean energy (falling C02 per unit of energy) increases (falling graph line) the implied reduction in energy required per unit of GDP is more than offset by increased GDP activity and a growing demand for energy, the rebound effect [24]. The direct rebound effect, that is, greater demand due to the reduced cost of energy consumption, has been variously estimated to be around 10–30% of possible energy savings, and the macroeconomic rebound from general GDP growth around 30–50% [25].

Increasing the efficient use of energy, in other words, reducing the energy costs of a given level of production, and increasing energy intensity, in other words, increasing energy productivity by increasing output per unit of energy used, are the two directions of travel industrial and service sectors need to go. If the energy is *not* renewable, then these are the ways to reduce GHG emissions and costs, and if they are renewable, then these are the ways to reduce energy costs. While the production and distribution technologies to enable this to happen, such as automated warehousing, are more widely available in OECD countries, they are nevertheless especially relevant for the developing countries of the Global South as they modernise their economies. Leading the pack has been China where, measured as purchasing power parity (PPP), the total energy supply per unit of GDP has fallen 43% over the years 2000–2023 [26].

The question now is how to put an end to further human creation of GHGs, and the simplest answer is to do away with the need for, and use of, fossil fuels. But that is easier said than done, not least because to achieve the capacity level of renewable energy to replace fossil fuels will require levels of investment and social restructuring on a scale difficult to imagine. Using approximate data from IRENA, by 2022 global renewable energy capacity was close to 3,400 GW (3.4 TW), but an *annual* addition of 1,000 GW (1TW) would be required up to 2050 to hit the 2050 target of net zero; that is, around an additional 30,000 GW (30TW) to bring the global total to over

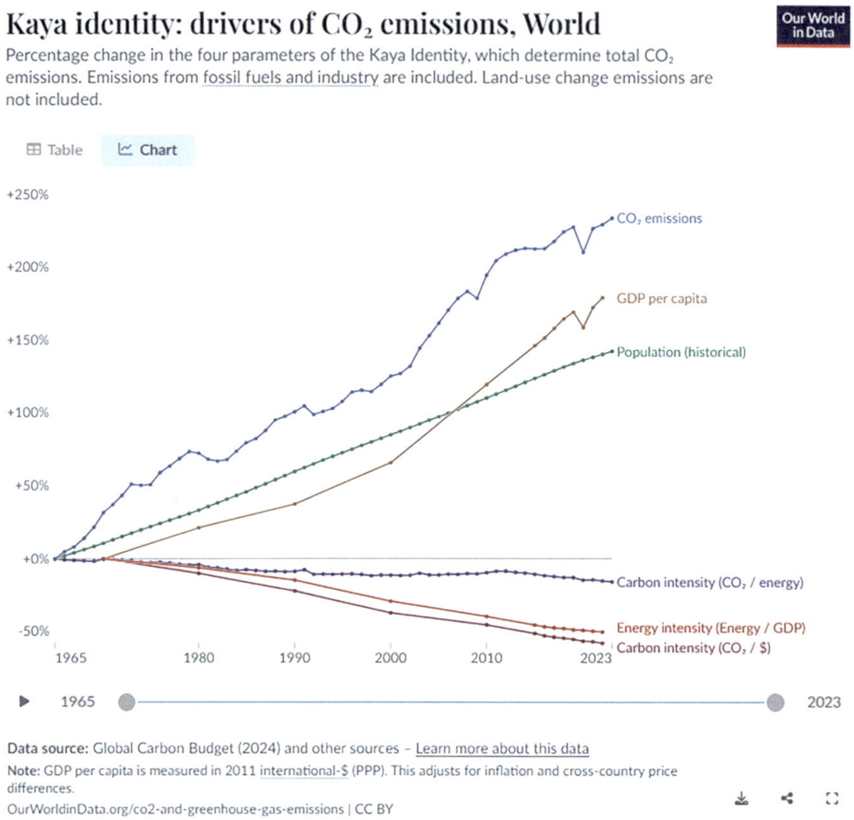

Fig. 7.3 Kaya identity drivers. *Source* Our World in Data

33,000 GW (33TW) [27]. IRENA also assumes that by 2050 renewables will only meet 90% of energy needs. A common calculation of capital investment to build 1TW of extra renewable energy capacity (solar and wind power) using US cost estimates (in China they would be lower) is around USD1 trillion [28].

Yale Environment 360, a journal published by the Yale University School of the Environment, reckons an additional USD4 trillion would be needed to complete the battery power storage infrastructure [29]. Then in addition there would be grid upgrades and buildouts. The countervailing influence will be falling costs as renewables reach scale, notably in China, so using the assumption that the investment costs in renewables fall by an average of 35% over three consecutive decades from 2020 to 2050s, very loosely based upon BloombergNEF projections [30], the cost of an additional 30TW over 30 years could be somewhat over USD5 trillion. While this ignores other costs, such as batteries and grids, it equates to 5 years of the annual defence budget proposed in 2025 for the USA by President Donald Trump [31].

USD5 trillion seems like a good investment for the defence of planet Earth, and a trivial price to pay compared with the costs of cleaning-up GHG-emitting manufacturing—see below. But it is not one that private capital would be willing to make. Infrastructure projects are typically fixed-cost heavy, with negligible marginal costs, so sustainable revenue returns are dependent upon energy prices reflecting average costs (operating costs + fixed-cost mark-ups). A widely accepted practice is to use Contracts of Difference (CfD) as they are known in the UK, which essentially support a development stabilisation fund so when market prices fall below or above the CfD price set by the regulator for the energy producer, the producer is paid from or returns the difference to the fund. Setting the CfD price over time should reflect both falling producer costs and future R&D and repairs and maintenance costs. For a review (critical but government-commissioned) of UK energy pricing policies—see *Cost of Energy Review, 2017* [32]. This is the market-led solution where the wholesaler is likely also the retailer.

The alternative is for the state to directly own and manage the energy supply. The worst solution is to open the retail market to separate competing companies who bid for wholesale prices from the producer and offer their own prices to users. The additional layer of administrative, managerial, and competitive costs such a model adds to a national energy (or any other utility such as water) supply chain creates volatility which messes with investment planning and squanders the advantages of scale, and inevitably ends with many smaller retailers going bankrupt. Market capitalism may claim to offer competitive choices, often erroneously as scale always dominates (think oligopolies or monopolies hidden behind an array of branded products), but the users of energy of all sizes want above all stable and predictable prices.

The Yawning Green Gap in Finances

That is not the end of the story. With the availability of renewables, all users of fossil fuels must make the transition, and that is where the financial challenge comes in. There are insufficient funds, public and private, being made available to make the transition, especially in the EMDEs. The costs of the transition are the costs of carbon mitigation, by replacing agriculture and manufacturing either with carbon avoidance processes, or with carbon reducing processes or with carbon removal processes. The global cost of such a transition is estimated by the Independent Climate Policy Initiative (CPI) at annually USD10 trillion between 2030 and 2050, from public and private sources [33]. The Network of Central Banks and Supervisors for Greening the Financial System (NGFS), also cited by the CPI, estimates a green financial gap of over USD260 trillion exists if a pathway to 1.5C from 2025 to 2050 is to be accomplished, more or less matching the CPI estimate. The CPI further points out, this gap is an under-estimate because it excludes estimates of losses from stranded assets, biodiversity threats, human conflicts, and migration—see Chap. 6. These figures far exceed the USD1.3 trillion of public finances annually agreed at COP-29 in Baku, Azerbaijan, November 2024.

Any suggestion that private capital is able, let alone willing, to fill the financial gap lacks credibility. Private capital is driven by profit, which is tantamount to a fiduciary obligation, and equally lacking credibility are suggestions that environmental sustainability and green growth can guarantee the likelihood of competitive returns on capital. That is not to deny that 'going green' may help reduce certain commercial risks, such as property loss due to violent weather and fires, and work absences and lost productivity due to ill-health related to pollution, but many of these benefits are externalities, shared between all in the community but not directly paid for by individual enterprises other than through rising insurance fees.

The financial model of the UN and its banking and financial institutions is to catalyse private capital through public finances, such as public procurement with private contractors; blended financial investments where risk is borne principally by governments; green bonds issued by governments to the private sector; or commercial bonds with incentives to include ESG (Environment, Social, Governance) commitments; and the encouragement to private banks to step outside their comfort zones, changing their legal requirements if necessary, to issue green loans. But the figures just do not add up. For example, the World Economic Forum (WEF) estimates it would take USD2.2 trillion of public finances to trigger USD10 trillion of private capital into nature-based projects, roughly a 1:4 ratio [34], while the IFC suggests USD1 of public funding has catalysed USD11 of commercial funding, a 1:11 ratio [35]. By contrast, an Overseas Development Institute (ODI) report found that in most cases USD1 of MDB and DFI investment only raised USD0.75 and that 'billions to billions' is more plausible than 'billions to trillions' [36]. There is clearly too much wishful thinking. As the online Green Banking Academy (GBAC) of the IFC concluded, the UN's *Implementation Plan* for a transition to low-carbon economies would need between USD4 and USD6 trillion *annually*, and 'will require a transformation of the financial system and its structures and processes, engaging governments, central banks, commercial banks, institutional investors and other financial actors' [35]. Read 'transformation' as implying 'not fit for purpose'.

In the VCM sector, in 2021, the TSVCM observed that if a target 2Gt CO_2 were to be removed by 2030 relying solely on private capital in the form of carbon credits it would require a '15-fold scale up of voluntary offsetting in 2030 versus 2019' [37]. This is just not going to happen. The simple reality is that present policies based upon the catalytic model of public funding stimulating private funding at the required scale are unrealistic and destined to fail.

The Financial Ecosystem of Tackling Climate Change

The financial ecosystem for tackling climate change is one filled to the brim with acronyms. Unfortunately, to follow the money requires some familiarity with it. Figure 7.4 offers an edited overview for reference.

The UNFCCC stands at the apex and organises the COPs (Conference of the Parties). The Paris Agreement (2015) was decided at COP-21 and Article 6 refers

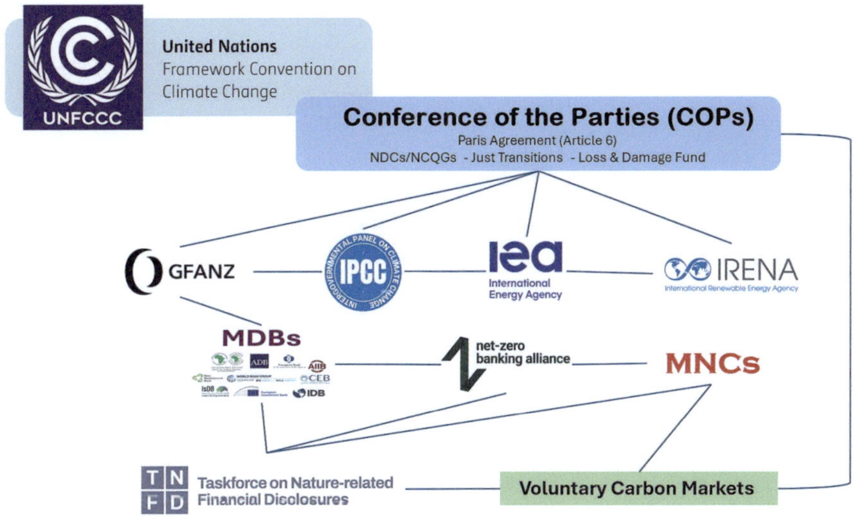

Fig. 7.4 The financial ecosystem for tackling climate change. Design by Layla Ali

to the possible use of carbon credits. COP-29 (2024) agreed under Articles 6.2 and 6.4 to open a door to credits from the voluntary carbon markets so long as regulations applied. Under the UN auspices come the Glasgow Financial Alliance for Net Zero (GFANZ) launched at COP-26 (2021), the Intergovernmental Panel on Climate Change (IPCC), the International Energy Authority (IEA), and the International Renewable Energy Agency (IRENA). Next come the Multilateral Development Banks (MDBs), the Net Zero Banking Alliance (NZBA), and the climate investments of the Multinational Companies (MDCs). Finally in Fig. 7.4 is the independent Taskforce on Nature-related Financial Disclosures (TNFD) supported by public funds only, and the Voluntary Carbon Markets (VCMs) which are either unregulated or self-regulated.

While following the money is a challenge, finding the money is the greater challenge. The model is predicated upon the hope (aspiration) that public sector capital will catalyse private sector capital into programmes to combat climate change and in support of adaptation projects. The context for this is given in Box 7.1.

Box 7.1 The UN Model as Financial Scaffolding

A worldwide response to climate change mitigation and adaptation requires a global financial system capable of mobilising the necessary funds, and building and supporting such a system needs a financial scaffolding to hold it up. Until 2025 that scaffolding has been the US dollar, when other currencies wobbled, the dollar remained strong, always in demand as the world's premier trading currency. The one key indicator that this remains the case is the US Treasury's bond market. Capital flight from that market will always be the signal that the scaffolding is no longer reliable.

The only alternative would be the one rejected by America at the Bretton Woods conference in July 1944, the creation of a package of world currencies. Today the only serious candidates for the package are the £ sterling, the European € Euro, the Japanese ¥ Yen, and the Chinese yuan (dollar/renminbi or peoples' currency). But the yuan is neither freely convertible nor anywhere near being a world trading currency. This is important because only countries with guaranteed strong currencies globally in demand as trading currencies are in a position to support the financial scaffolding of a global system of climate change mitigation and adaptation, which in reality to succeed has to be a new development programme centred around a green industrial transformation. In other words, a movement to create a package of world currencies needs to lead to a new level of global collaboration (not competition or merely cooperation) that *at minimum* prioritises financial transfers to support GHG mitigation in EMDEs.

The model of the UN up to 2025 has been based upon public funds being used to catalyse private funds. It has not succeeded. Rising average global temperatures are the clearest indicators of this, as GHGs continue to rise. The reactions of UN agencies directly involved have been to identify systematic failures as including (a) the non-scalability of many climate-related projects and programmes which undermine returns on private investments, and (b) the spreading of risk and uncertainty attached to these projects and programmes, which is giving rise to calls for multilateral banks to increase their share of the burden. Naturally, the lack of political will have also played its part. Part of the solution to (a) was seen by the TSVCM (2021) as the creation of a database platform for small-scale project suppliers to match with project financiers by providing as much price information and risk management assessments as possible to facilitate over-the-counter (OTC) spot and futures trading of credits [38]. The UN-Bloomberg Climate Data Steering Committee (CDSC) [39] in 2022 created the UN Net-Zero Data Public Utility (NZDPU) [40], to combine with the UNFCCC's Global Climate Action Portal [41]. Earlier the World Bank, Singapore Government and Carbon Action Data (CAD) Trust supported the International Emissions Trading Association (IETA) online database listing carbon credits by registry, by sector, and by country. Tracking data is important, but on its own it does not do the job.

Spreading the risk as in (b) has spawned the blended finance approach whereby public finances share or assume the risks involved in project finance. Related to this are partnerships and procurement contracts by governments and development agencies to partner with the private sector. All of this makes perfect sense, other than it is clearly not overcoming the principal hurdle, which is a combination of abatement projects that have uncertain rates of return for capital and the equally uncertain outcomes of R&D into new technologies for carbon reduction or removal. The answer to the former problem is to make carbon reduction and adaptation of a social and policy priority and accept that the social costs of climate change and global warming far outweigh the costs of mitigation. This is not a financial decision, but an economic one, and the real choice is a temporal one: which projects or programmes to prioritise because not everything can be tackled at once.

The answer to the second problem is less straightforward because the outcomes are less predictable. It makes good sense for governments to support the R&D of entrepreneurs who have credible ideas, whether through a national innovation bank or through tax incentives, but it is also the case that governments can sponsor their own R&D programmes: the postwar history of US Federal Government support for inventions and innovations, ranging from space to the Internet, and from medicines to computer technology, is often put into the shadow by contemporary focus upon the Big Tech companies who nevertheless benefit enormously from government grants. A National Academy of Science publication of 1995 is a good reminder [42].

Public and Private Financing of Carbon Abatement

On the public financial side, a large expansion of public debt is needed. Devotees of neo-liberal political-economy will react against this, but public debt within constraints (see below) is the necessary means by which economic resources can be shifted towards compelling social goals. This debt will pay for itself in two ways. First, it will accelerate the capacity and use of low-cost renewable energy on the supply side which will raise the profitability of state or privately owned manufacturing and services, including AI services. Further, despite the energy-hungry development of AI LLMs, the spread of AI *agents* is likely to be a major technological enhancement of carbon abatement as businesses (and households) are more easily able to calculate their GHG emissions and to manage their use of energy (save money) in more efficient ways [43]. Second, so long as the economic expansion is within the resources available in a country to avoid inflation and so avoid the pressure to raise interest rates, state spending on a green agenda can ensure full employment and a growth in tax revenues. Objections that, as resources get stretched, this could squeeze out private capital and therefore spark inflation can be overcome by re-directing private capital from resource-scare sectors into government bonds, sovereign carbon credits [44], or joint-enterprises. This will shift the public–private ratio in the mixed economy reflect the priority being given to combatting climate change.

This approach could be termed a planned economy, but not in the sense of a centrally controlled command economy where all aspects of economic and industrial life are closely managed or owned by the state. Rather, it is an approach that is more focused on specific policy priorities, in this case climate change mitigation, and in which the scarcity of resources, and therefore the need to use them efficiently, is just as fully recognised as it is within a planned enterprise. It also differs from a public–private partnerships (PPPs) approach that essentially outsource management, monitoring, operational decisions, and responsibilities to the private sector. A planned economy proposes active state monitoring and a shared operational responsibility for resource management and accountability as a way to prioritise environmentally friendly policies.

Such a shift is, of course, easier for developed than for developing countries, so ways to transfer technologies, knowledge, funding (economic), and investment (financial) need to be baked into any new system of global collaboration—see Box 7.1. COPs are not the mechanism to achieve this. They are too ponderous and often too transactional. At best they can play a complementary role involving representatives and viewpoints and learnings from all stakeholders. What is required is pressure upon governments in both developed and developing countries from their civil societies to work together to prioritise and tackle the public health issues, the social cost issues and the impoverishing household cost issues of the continued use of fossil fuels. These are universal issues, as are pollution and loss of biodiversity; they can unify civil societies across countries as they call for universal solutions. At the universal level, this is indeed a case of 'one size fits all'. There is no other approach that has this global appeal. Where this approach differs from current policy approaches is that it calls upon governments to do two things: (i) use a massive expansion in government debt to tackle climate change, shift the balance of the mixed economy, and abandon neo-liberalism; (ii) in the area of climate change, at the very least, create a global collaboration between governments as way to minimise capital flight and one that prioritises the funding of mitigation through a new collective financial system, based upon a basket of leading currencies, and devoted to this specific issue. In other words, funds for climate-related projects to EMDEs which are lacking the funds themselves should be separately designated or 'ring fenced'.

The reason for ring fencing is that, in terms of realpolitik, the creation of a new financial architecture for active global collaboration that manages to go beyond tacit cooperation would attract little international support if it ranged across contentious issues such as strategic trade, energy security, defence, immigration, and so on; hence, ring fencing could keep the focus narrowly upon a climate change policy consensus and the Paris Agreement, albeit for the immediate future without the participation of the USA.

Box 7.2 A New Global Approach

A new global approach could be achieved, for example, through a repurposing of existing multilateral banks or a specialist global bank focused exclusively on financing climate mitigation projects in EMDEs, prioritising the heaviest polluters, but also helping adaptation in the poorest of countries. One collective direction of travel for policy could be an internationally agreed semi-specialisation of climate-related trade, whereby different higher-income countries would be supported financially, possibly by a new global climate bank (which could be a separate unit of the World Bank) in their funding of climate-related projects in, and technology transfers to, EMDEs.

Examples might be Chinese exports of grid and battery infrastructure to EMDEs, of solar panels from Canada, China, and India to EMDEs, of windfarm turbines from Germany and other EU countries to EMDEs, of wave power technologies from Australia and South Korea to EMDEs, of small nuclear reactors from post-Putin Russia to EMDEs, of green technologies in agriculture, transportation, and building efficiency from Japan to EMDEs, of hydropower technology from Brazil, Scandinavian countries, Canada and the US to EMDEs, and so on. These 'exceptional' trades would be limited to EMDEs and would not impinge upon normal trading relations between the OECD countries, nor trades in non-climate-related projects with EMDEs. An internationally agreed policy of this nature would prioritise carbon reductions in the most polluting of countries which lack resources of their own to combat GHG emissions and the impact of climate change.

On the private financial side, there are four parallel tracks: carbon avoidance, carbon reduction, carbon removal, and carbon credits which can relate to any of those. The first three is where the focus has to be and where finances are most urgently needed, because they each relate to all parts of Scopes 1 (enterprises), 2 (energy sources), and 3 (supply chains). This is about the private sector cleaning up its act, with or without public carrots such as subsidies, tax breaks, or sticks, such as carbon taxes, withholding of business and planning licences, loss of procurement contracts. Investment in cleaner, greener production and distribution services lies at the heart of fighting climate change. While public finances and policies may be primarily responsible for making renewable energy available, the active involvement of the private sector in propagating and using it, in conjunction with materials that are either themselves green, or used in conjunction with carbon reduction and removal technologies, is the essential step to the Paris goals. Those goals should not be surrendered to circumstances. They along with the GHG associated with them are the only bellwether we have.

- **Carbon avoidance** is where enterprises adopt the use of processes using green energy (the easy part) and materials (the less easy part) that emit no GHGs.

Examples of such materials, even of recyclable materials, that emit no GHGs are not so easily to come by. For example, bamboo emits minimal GHG and can make green furniture but is unlikely to become a substitute for wood, wood-chip, or laminated products in shops like IKEA.

- **Carbon reduction**: The use of production methods and materials that minimise C02 emissions. This is more likely than total avoidance. An example would be green concrete which is composed of recycled materials such as slag, with claims it reduces C02 emissions by 30% [45] That still leaves a lot for removal.
- **Carbon removal**: There are a host of technologies vying for attention and investment, but most are in their infancy. The earliest of these technologies, Carbon Capture and Storage (CCS), was actually developed by the fossil-fuel industry to capture and separate C02 in its quest for natural gas in the 1930s, but now the sequestration of C02 is adopted for carbon removal and storage for its own sake. Critics argue that CCS gives oil and gas companies an excuse to continue drilling, but each CCS facility is also both expensive and expansive and needs to be custom-built for oil fields, so costs are difficult to reduce. A good account is given in the 2023 book *Climate Capitalism* [46]. The most widely used ways to remove C02 are conventional methods as part of land use, land-use change and forestry (LULUCF), responsible for 2GtC02 abatement in 2024. Other so-called novel methods, such as ocean alkalinity enhancement, direct air carbon capture and storage (DACCS), bioenergy with carbon capture and storage (BECCS), biochar and enhanced rock weathering, were responsible for no more than 1.3 million tonnes (0.0013 Gt CO_2) per year. According to the CDR.fyi newsletter, in 2025 'Biomass-based methods accounted for 99% of delivered tonnes. Biochar remains the only method delivering at commercial scale' [47]. DACCS, or sometimes just Direct Air Capture (DAC), is an example of just how difficult it is to scale these technologies. Depending upon the size of the apparatus, it costs between USD400 and USD1,000 per metric tonne of CO_2, although industry hopefuls are looking towards USD100 per metric tonne. There is also concern over its use of chemicals [48]. See Chap. 5 for a fuller discussion.
- **Carbon credits**: Very broadly, there are two sets of buyers (i) those who are emitting GHGs and are looking to offset them by purchasing credits that supposedly pay for GHG reductions elsewhere that would not otherwise have taken place (the 'additionality' requirement), so if the offset is tonne for tonne the result is deemed to be 'net zero'; (ii) investors who can benefit financially if the market (the price) for credits goes up, or who may be motivated by altruistic, philanthropic, or reputational reasons. These transactions can take place within (a) the mandatory (compliance) carbon markets (MCMs) where governments cap the emissions allowed by major polluters who have to buy credits if they fail to meet their caps, or can sell credits if they are well within their caps ('cap-and-trade' schemes); (b) a sovereign carbon credits market where the integrity of the credit is guaranteed (see above); and (c) the voluntary carbon markets (VCMs) which for the most part are unregulated and historically have a very poor integrity record which has held them back. A fundamental issue here is the principal-agent problem, when the project owner/manager hires the agent whose task is to represent the true nature

of the project to buyers/standards bodies. The principal has insider information that s/he may not share with the agent, or the agent may not share in a fair and transparent way (the 'moral hazard' problem) with the buyer, and no-one necessarily has a legal responsibility to measure, monitor, report, and validate (MMRV) the progress of the project. Poor quality or no-quality projects easily slip through the net. For this reason, many international companies protect their reputation by avoiding markets and brokers and invest directly in projects such as rainforests and wetland restorations, or in novel carbon capture projects such as BECCS. These issues are fully discussed in Chap. 5.

Areas of Carbon Urgency

The main polluters are well known: oil and gas, heavy industry, transport services, agriculture, but those most at risk include the oceans, which store at depth most of the world's carbon dioxide and methane [49], and the most vulnerable and most marginalised communities, among whom children and women from all cultures and countries are the majority—see Box 7.3.

The challenge for the oceans is to stop the various seabed ecosystems being drilled to death in search of precious minerals to support the inexorable march of electronic systems and devices, thereby also disrupting and releasing the ocean's capture of carbon [50]. In his 2023 book *The Blue Commons*, Guy Standing argues that the International Seabed Authority (ISA) created by the UN Convention of Law of the Sea (UNCLOS) has been ineffective in introducing global regulations, giving opportunity for several countries, such as Japan [51], and the USA [52] in competition with China, to plan the start of seafloor drilling and mining. A joint publication by the World Economic Forum (WEF) and the Friends of the Earth (FOE) in 2020 proposed various best practices for investments in the commercial use of oceans, but politically it is a very shy document [53].

Box 7.3 Climate Change and Gender

Given that women and girls are half the world's population, it *should* be expected that when the implications of climate change and global warming have researched the impact upon females, such as the implications for access to health, education, and jobs, especially in rural areas, should figure large in the data. They do not, and for all the well-known wrong reasons [54]. The paucity of good data, or data of any kind, is well noted in studies by the World Bank [55], the UNFCCC [56], and NASEM [57], among many others. A helpful review came in 2021, *Twenty Years of Gender Equality Research* [58], which surveyed over 15,000 peer-reviewed papers in the domain of business and economics.

This does not mean that the adverse effects of climate change on women and girls, notably in EMDEs, lack documentation. Examples from UN agencies include the International Finance Corporation (IFC) *Gender-Responsive Climate Governance and the Role of Women Leaders* (2024) [59] which as the title suggests has a focus upon the need for women taking on leadership roles around the so-called 'gender-climate nexus' to place gender firmly on the agenda, in effect an admission that gender has been an AOB ('any other business') item, and the UNFCCC submission to the 6th Report of the IPCC (2022) [56], where the focus is upon women *as agents of change* in the mobilisation of resources to combat climate change and global warming. The submission draws upon the LP-LEDS (long-term low-emission development strategies) submitted by EMDEs to the UNFCCC via a horrendously named Compliance Meeting Annually of the COP, or just CMA [60]. The Ministry of Foreign Affairs (Japan) conveniently offers a diagrammatic representation of the UN spaghetti nomenclature [61].

The UNFCCC also acknowledges that cultural factors play a major role in the way gender issues are managed locally, and 'considering women as a homogenous group must be avoided'. Indigenous women in remote areas are especially at risk. For example, when men move away from rural villages due to what the World Bank (see above) refers to as 'climate shock-induced outmigration', often women are left alone to care for the young, the fields and animals, family health, and are isolated and vulnerable to severe weather such as droughts or floods and the heightened risk of air-borne and water-borne diseases. And frequently they are without modern means of communications, either to receive weather alerts or to request help. Importantly, the World Bank report also cites evidence that where women are active participants, this 'results in more stringent climate change policies'.

Further, the World Bank commissions *Country Climate and Development Reports* (CCDRs), but notes of the 166 NDCs cited in the CCDRs, only 21 referenced gender and of these only '20% elaborated the language and 38% included information related to gender mainstreaming'. These are miserable percentages suggesting that until attitudes and awareness shift, insufficient effort will go into the collection and analysis of gender data without which gender-positive climate change policies will make little progress. Even among the developed OECD countries, only seven were reported to 'collect gender disaggregated data related to the environment (OECD, 2020)' [62]. The UN estimates it will take 22 years to close the data gap on women and their impact on sustainable development goals [63]. The UN-backed multi-partnership GEDA (Gender Environment Data Alliance) has been created to promote data related to the environment that is disaggregated by gender [64]. WEDO (Women's Environment and Development Organisation), a founding member of GEDA, in 2016 launched a Gender Climate Tracker App and website [65], which gathers data, news, and stories from across the globe.

A UK-sponsored report for ASEAN advocates *Integrating a Gender Lens in Voluntary Carbon Markets* [66], and an active source for mainstreaming gender issues and climate change is the Canada-based International Institute for Sustainable Development (IISD) which, for example, released a paper arguing for the issuing of gender-inclusive bonds [67]. But one example of an effort to offer equal representation to men and women in forestry management in Indonesia was ill-advised when a policy was decided of having one member per family on the management committee, effectively resulting in an exclusion of women from the decision-making [68]. Good intentions need good designs, and in this case a good design needed women to design it.

Après Trump, Le Déluge?

2025 was the year of the locust [69], or as some might see it, of the Antichrist [70]. Donald Trump, the 47th President of the USA, for the second time pulled America out of the UN Sustainable Development Goals (SDGs), out of the Paris Agreement, out of the UNFCCC and its climate change funding mechanism, GFANZ, out of its commitments to support EMDEs efforts to mitigate and adapt to climate change, and dismantled most of America's domestic mechanisms under the Inflation Reduction Act (IRA) for funding climate change mitigation and adjustment. He has also upended the stability of world trade with the threat of a tariff system that will inevitably raise the costs of supply chains that are vital to technologies to reduce GHG emissions, such as solar panels and wind farm turbines, and carbon capture and abatement technologies. During his first Presidency he appeared as a climate change denier, but emerged in his second Presidency as someone who sees soaring CO_2 and other GHG levels as simply the unavoidable (the price to be paid) consequence of further industrial growth in the USA which privileges fossil fuels.

As of early 2024, the Trump administration has systematically threatened or actually closed down government-funded research into climate change, data collection efforts, and websites that record that data [71]. Included is the intention to disband the National Weather Service and NOAA [72], and its funding for Princeton University's Cooperative Institute for Modelling the Earth System (CIMES) which provides important modelling for the IPCC. The Department of Commerce gave as its reason that CIMES 'promotes exaggerated and implausible climate threats' and contributes 'to a phenomenon known as "climate anxiety", which has increased significantly among America's youth' [73]. This is an unbelievably stupid statement. If, as the saying goes, there are none so blind as cannot see, everyone is blinded if there is nothing to see.

All the above follows the ultra-conservative Project 2025 [74] playbook [75]. Naturally, the fossil-fuel lobby is a significant backer of the President as well as a massive financial supporter of climate change disinformation across social media, and of a lobbying focus upon Republican Party members of the US Congress [76].

Après Le Déluge?

So what? Can the rest of the world continue the fight without the USA? The USA has been an integral part of the global financial model, exemplified by the workings of GFANZ and its funding institutions alongside public and private banks. That model has proved largely ineffectual in bridging the gap between the funds that are needed and those made available, and the withdrawal of the US is close to being fatal. However, and bearing in mind the old adage 'never let a good crisis go to waste', that means governments elsewhere will have to rethink their strategies based upon the fact that the US is not the most important polluter, nor is it the only country with sufficient resources to address the problem.

The principal sources of GHGs globally do not come from the USA, which, as of 2022, was responsible for around 11%. By far the most prolific was China (29%) with the other so-called BRICS countries (Brazil, Russia, India, and South Africa) adding a further 15.5% and the EU countries a further 11%—see Fig. 7.5. All other countries combined, each under 1%, are responsible for the remaining 35%. Logically, this means that the challenge of reducing GHG emissions could be solved without the direct participation of the USA. Of course, the big loss is the financial one.

It needs to be noted that up to date, the USA has been reducing its emissions faster than most. The JTC report from which Fig. 7.5 comes shows that along with the EU and Japan, the USA was one of only three to have a negative GHG compound average growth rate (CAGR) 1990–2022., although data from Climate Watch suggests US

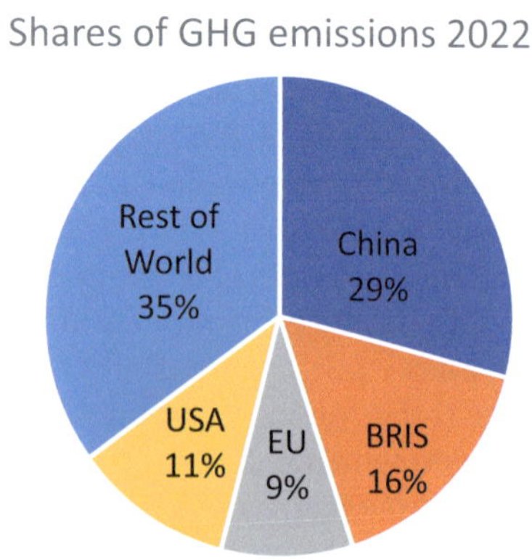

Fig. 7.5 Shares of GHG emissions 2022. *Source* Author, data from Joint Research Centre (JRC), the European Commission (2023) [77]. *Note* the pie chart shows RICS but show be designated as BRIS – to be changed

emissions actually rose until COVID-19 [78]. This is largely due to a structural shift in the US away from rust-belt polluting industries and especially the move away from coal, with the additional combined stimulation of investment incentives and subsidies under the Inflation Reduction Act (IRA). The IRA was really climate change action legislation hidden under a different guise to squeeze past Republicans in Congress under President Biden's regime.

The Polluters

Although the IRA is dismantled, it does highlight the significance of relevant legislation in the battle against GHGs, and something that can be repeated elsewhere in the world, as it has been. Especially important has been the capping of emissions on the most polluting of industries, which are shown in Fig. 7.6.

Electricity generation and usage in heating account for more than 50% of global GHG emissions, and all forms of transport (land, sea, and air) for more than 30%, although in the USA transport had overtaken power generation as the major polluter by 2017 [80]. Besides these priority areas, manufacturing and other industries together account for 25%. Lack of good data is a problem in fixing this problem. Data on reducing Scope 1 emissions by modifying industrial processes is not available unless specified in annual reports. This is a legal requirement in the UK and the EU for listed companies and companies of a certain size [81], but even these should be third party validated before they can be taken seriously. Some companies report to the Climate Disclosure Project [82], but it requires licences to access data on emissions reductions, and there seems to be no estimate of actual total tonnage of C02e avoided. One way forward for businesses is to use verifiable carbon 'insets' *within* an

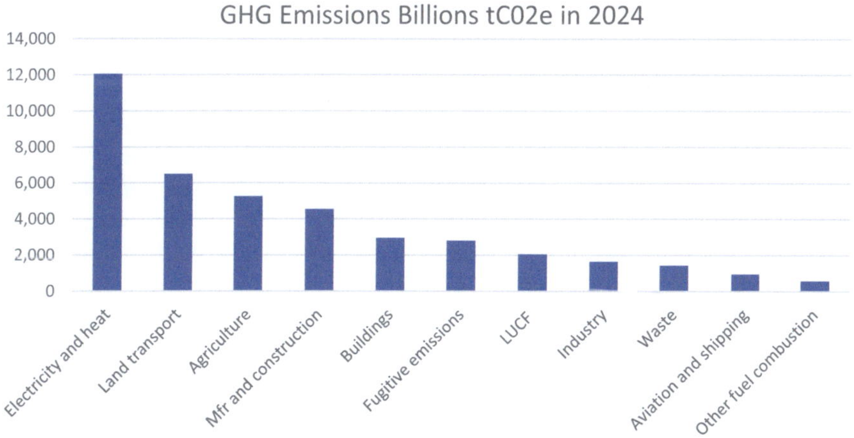

Fig. 7.6 GHG emissions (billions tonnes C02e) in 2024. *Source* Data from Our World in Data (2024) [79]

organisation so departments with an energy budget excess or deficit could trade with another within the same organisation, a solution supported by ex-Microsoft founder Bill Gates in his book *How to Avoid a Carbon Disaster* (2022).

Measuring, monitoring, and reporting supply chain Scope 3 emissions is the area of greatest weakness. There are guidelines issued [83], and examples, such as Harvard University's sustainability initiative that is tracking Scope 3 emissions across five sectors from which it buys services: construction, food, air travel and commuting, IT, and other purchased goods/services [84]. One report does suggest that upwards of 40% of public companies globally now do report Scope 3 estimates [85], but the practical difficulties of monitoring lengthy and complex supply chains inevitably raise doubts about the veracity of the reporting, especially where self-reporting may be the standard practice.

Ghost-In-The-Machine

A real ghost-in-the-machine is the financial sector, hidden within the commercial sector. Banks and financial institutions generally are not significant emitters of GHGs, but a study in 2024 by the Climate Policy Initiative (CPI) of over 600 financial institutions [86] showed they prioritized customers and investments in high-emitting sectors within their portfolios even 'where clear decarbonisation pathways exist', finding that direct finance to new fossil-fuel plants outstripped investments in renewables, reaching USD33.4 billion in 2023. Even central banks tend to work to a mandate that excludes the social costs of climate change, while providing support when the financial institutions that help generate those social costs encounter difficulties.

Looking at country emissions is one approach, identifying industrial sources is another, identifying enterprise or ownership of sources is yet another. An analysis from the World Bank in 2023 estimates 60% of global industrial emissions come from 157 large MNEs (multinational national enterprises), of which 10% comes from their direct activities (Scopes 1 and 2) and the remaining 50% from their supply chains (Scope 3); the energy sector alone accounting for 37% of the 60% [87]. Identifying industry sources of energy consumption is yet another approach. A Goldman Sachs report in 2024 suggested, on average, an AI ChatGPT query needs nearly 10 times as much electricity to process as a Google search, and that demand for data centre electricity would grow by 160% by 2030 [88], which is also the 'base case' for the IEA which warns that its 'lift-off' case is much higher while its 'high-efficiency' case is lower [89]. This is important because the structural shift away from smokestack industries such as steel and shipbuilding using twentieth century technologies is being replaced by a new set of energy-thirsty digital technologies. The only good news there is that this transition will *ultimately* see 100% renewable energy powering AI and data centres, and AI management agents will support more energy-efficient consumption. An unalloyed positive view on the potential for AI to assist in climate change mitigation across multiple industries is the roadmap from the Innovation for Cool Earth Forum (ICEF) [90].

What is Fit for Purpose?

The argument throughout this book, especially from Chap. 4 onwards, is that the evidence demonstrates that current public policies based upon the UN Model (see Box 7.1) are not working, meaning they are not providing adequate finance to support climate change mitigation, they are not catalysing sufficient funds from private capital markets, and they are having only limited success in substituting renewable energy for fossil fuels. On current UN-IRENA projections, fossil fuels will still constitute at least 10% of energy generation by 2050, and 'about USD7 trillion was spent on subsidizing the fossil-fuel industry in 2022, including through explicit subsidies, tax breaks, and health and environmental damages that were not priced into the cost of fossil fuels' [91]. COP-29 grappled with the aim of raising USD1.3 trillion per year globally from government sources; this compares with USD4.3 trillion required annually by 2030 if the 1.5C goal was to be reached by 2050 as estimated by the Climate Policy Initiative in 2021 [92]. All the evidence points to failure in terms of achieving the Paris goals, a failure despite the well-intentioned actions of the UN, and many of its agencies. As the saying goes, the path to hell is paved with good intentions, but what has gone wrong?

The fundamental problem is that a market system is not designed to handle externalities, despite the fact that the externalities are caused by human action, in other words, are endogenous to the capitalist system itself. Arguably, *any* system that privileges economic growth over the environment will face the same problem, and equally the Trump view (clearly shared by the fossil-fuel industry) that economic growth (GDP growth) implies GHG emissions and global warming as an *inevitable* consequence carries a lot of weight. The *inevitability* will be contested by advocates of green policies, and rightly so, but *only if* they imply a departure from a capitalist market economy, because the alternative has to involve economic and social priorities that directly clash with market values. A departure does not mean a complete transformation into a centralised command control economy because that offers an opposite problem, a single point of failure, and threatens to suppress an active role for civil society, a workers' state without workers' control. A transformation seen in historical and dialectical terms is always about a synthesis of a continuum and a change, the best of both. In the final analysis, as a review in 2024 of ESG policies by the Cambridge Institute for Sustainability Leadership argues: 'trends on climate and nature continue to head in the wrong direction. It is time to recognise that the market has failed to deliver' [93].

The Basis of Action Beyond Slogans

Clearly slogans are not enough, and real action rather than what the UN calls on its website 'carbon action' is required. At the personal level, one option is to support microfinance for local EMDE communities struggling to build resistance and adaptability to climate change [94]. Small manageable funds can make a big local difference, but while community-level action is possible and in many ways the easiest to motivate and put into action, linked community action is more effective and desirable. However, it will never be a substitute for the collective action of civil society at national and state level capable of enacting strong laws, and best of all linked-up state action at the international level, because effective action implies public finances will have to take precedence over private capital markets wherever catalysation is not working. Where this places strains upon national resources private capital either needs to be channeled in the bond market for passive savings, or into partnerships led by public sector funding, or nationalized where market failure is both obvious and damaging to the economy.

An Alternative and Transformative Perspective

This alternative perspective is almost certainly not viable unless there is a high degree of international collaboration around a green agenda and the socio-economic policies that go with it. Capital flight from any single pioneering country could undermine the entire project. Therefore, the withdrawal of the USA from the global agenda on fighting global warming, however temporary it may be, should be turned into an opportunity to broaden international alliances around a green agenda. Easier said than done is an understatement, and one fraught with political minefields. The biggest challenge (see Box 7.2) would be to bring China (India, Brazil, South Africa also, and Russia post-Putin) into a new global political economy using a bundle of trading currencies (including de facto the US dollar) principally or even *exclusively* to finance climate change mitigation and adaptation in all EMDEs. Creating a new International Green Bank would be the most decisive and effective way to achieve this. Besides supporting multilateral green bonds specifically for EMDEs, a Green Bank could also help finance the transfer of climate mitigation and adaptation technologies from its Member States to the Global South. Given that CCUS and CDR technologies are difficult to scale, this could provide a global market across the most polluting countries, where returns on investment would be highest. Trade and other issues would not necessarily be part of its agenda, *except for* technologies and R&D that specifically support renewable energy and GHG reductions, and cross-border carbon taxes.

Technologies, Climate Change, Health, and the Environment

As above, the obvious technologies would include solar, wind and wave power and grid infrastructure development, and 'novel' carbon removal technologies, alongside Earth measuring and monitoring technologies such as LIDAR (Light Detecting and

Ranging) [95] and CNNs (Convolutional Neural Networks) that provide algorithm-generated ground-level views from satellite images for reforestation projects [96]. Further, if possible, there would be collaboration around R&D, especially around nuclear fusion R&D and around carbon removal technologies where economies of scale will be crucial to their success.

International collaboration around combatting climate change could and should extend into the related area of health. Global surveillance of hazards and the spread of animal diseases go hand-in-hand with the satellite monitoring of fires and the spread of botanical diseases. COP-28 (2023) was the first ever to devote a day to climate-related health issues in a joint session between the WEF and the Oliver Wyman consulting group, where it was pointed out that climate change could result in an additional 14.5 million deaths, USD12.5 trillion in economic losses worldwide, and USD1.1 trillion in extra costs to healthcare systems by 2050. In a related paper, 55 million people annually face drought conditions, 40% of the world already face water scarcity according to the World Health Organization (WHO) with an estimated 700 million threatened with displacement due to drought by 2030 [97]. Climate change is not necessarily the only cause of a lack of water, as, for example, the lowering of water tables can and does result from indiscriminate urbanisation, but once the damage is done droughts prevent restoration. Higher temperatures and greater humidity in more regions of the world are also well known to heighten the risk of malaria-carrying mosquitoes and other pathogens.

Collaboration around health issues and wider environmental issues offers an opportunity to find a common thread of interest and commitment across nations and also across civil societies. Climate change is abstract, GHGs are material [98], but pollution, health, clean water, and the environment are everyday practical issues facing families and costing them real money.

Facing Realities

Most climate changes prior to the industrial revolution are attributed to very small variations in Earth's orbit that change the amount of solar energy our planet receives. The current warming trend is different because it is clearly the result of human activities since the mid-1800s, and is proceeding at a rate not seen over many recent millennia (NASA) [99].

The question that will inevitably come to dominate future generations is whether the current epoch, which is defined in geological terms as lasting several millions of years [100], is coming to end during an undetermined but historically short timescale, and about to give way to a geological epoch that is beyond the influence of human activity. The current epoch is called the Holocene (from the Greek words for 'whole' and 'new') or, alternatively, the Anthropocene, implying that since around 12,000 years or more ago human activities such as agriculture, urbanisation, deforestation, and pollution became the principal drivers of climate change [101].

It *probably* lies within the powers of humanity to halt and even reverse man-induced climate change and global warming. It *certainly* lies within the powers of humanity to halt and reverse levels of GHG pollution and of environmental degradation, and *possibly* to halt the loss of biodiversity [102]. Diversity of resources provides more ways to combat global warming. Diversity is the basis of sustainability, it spreads the risk. That is as true for human society as it is for every species on Earth.

References

1. National Geographic (2025) All About Climate https://education.nationalgeographic.org/resource/all-about-climate/
2. IPCC Reports https://www.ipcc.ch/reports/
3. Scientific American.com (29 April 2025) Trump Dismisses Scientists Writing Key Climate Report https://www.scientificamerican.com/article/trump-dismisses-scientists-writing-the-national-climate-assessment/
4. UNFCCC The Paris Agreement https://unfccc.int/process-and-meetings/the-paris-agreement
5. TNC (2023) Greening Sovereign Debt https://www.nature.org/content/dam/tnc/nature/en/documents/TNC_Greening_Sovereign_Debt_Factsheet_2023.pdf
6. ECB (2025) ECB adds indicator of nature loss in climate-related financial disclosures as portfolio emissions continue to decline https://www.ecb.europa.eu/press/pr/date/2025/html/ecb.pr250612~484d529368.en.html
7. Institute for Energy Economics and Finance Analysis (2023) Carbon Pricing: Governments Increasingly Make Polluters Pay for Climate Change https://ieefa.org/sites/default/files/2024-05/Carbon%20Pricing%20Govts%20Make%20Polluters%20Pay%20for%20Climate%20Change_May%202024.pdf
8. CPI (17 October 2024) Accelerating Sustainable Finance for Emerging Markets and Developing Economies https://www.climatepolicyinitiative.org/publication/accelerating-sustainable-finance-for-emerging-markets-and-developing-economies/
9. WWF (2025) Six ways loss of Arctic ice impacts everyone https://www.worldwildlife.org/pages/six-ways-loss-of-arctic-ice-impacts-everyone
10. Climate.Gov (August 2023) Climate Change: Global Sea Level https://www.climate.gov/news-features/understanding-climate/climate-change-global-sea-level#:~:text=Global%20mean%20sea%20level%20has,record%20(1993-present)
11. Coastadopt.Com (2017) Ocean acidification and its effects https://coastadapt.com.au/ocean-acidification-and-its-effects#:~:text=The%20oceans%20have%20absorbed%20between,are%20risks%20from%20ocean%20acidification
12. Livescience.com (October 2024) Key Atlantic current could collapse soon, 'impacting the entire world for centuries to come,' leading climate scientists warn https://www.livescience.com/planet-earth/rivers-oceans/key-atlantic-current-could-collapse-soon-impacting-the-entire-world-for-centuries-to-come-leading-climate-scientists-warn
13. News.UN.Org (June 2019) 24 billion tons of fertile land lost every year, warns UN chief on World Day to Combat Desertification https://news.un.org/en/story/2019/06/1040561#:~:text=What%20is%20desertification?,the%20productivity%20of%20the%20land
14. NASA.GOV (October 2024) Extreme Weather and Climate Change https://science.nasa.gov/climate-change/extreme-weather/#:~:text=As%20the%20climate%20changes%2C%20the,responding%20to%20extreme%20weather%20events

15. Weform.org (October 2023) Here's how extreme weather is affecting animal migration https://www.weforum.org/stories/2023/10/climate-crisis-impacting-animal-migration/#:~:text=Climate%20change%20and%20extreme%20weather,an%20impact%20on%20animal%20migration
16. WWF.Org (October 2024) Catastrophic 73% decline in the average size of global wildlife populations in just 50 years reveals a 'system in peril' https://www.worldwildlife.org/press-releases/catastrophic-73-decline-in-the-average-size-of-global-wildlife-populations-in-just-50-years-reveals-a-system-in-peril#:~:text=Some%20of%20the%20species%20populations,in%20the%20Sacramento%20River%20in
17. WHO.Int (February 2024) Biodiversity https://www.who.int/news-room/fact-sheets/detail/biodiversity#:~:text=Threats%20to%20biodiversity%20and%20health,benefits%20for%20health%20and%20medicine
18. UNHCR.Org (November 2023) Climate change and displacement: the myths and the facts https://www.unhcr.org/us/news/stories/climate-change-and-displacement-myths-and-facts
19. Copernicus.EU (2024) Global surface air temperature anomalies: 12 monthly averages https://climate.copernicus.eu/surface-air-temperature-september-2024
20. Climate.Copernicus.EU Surface air temperature for January 2025 https://climate.copernicus.eu/surface-air-temperature-january-2025#:~:text=January%202025%20was%20the%20warmest,define%20the%20pre-industrial%20level
21. GML.NOAA.GOV The Data: The Story Told from CO2 Samples https://gml.noaa.gov/education/isotopes/mixing.html#:~:text=Using%20the%20special%20isotopic%20fingerprints%20for%20each,a%20sample%20comes%20from%20each%20individual%20source.&text=So%2C%20we%20know%20that%20the%20ratio%20of,isotopic%20fingerprint%20of%20each%20of%20those%20sources
22. Wikipedia (28 February 2025) Kaya identity https://en.wikipedia.org/wiki/Kaya_identity
23. Our World in Data (2024) Kaya identity: drivers of CO_2 emissions World https://ourworldindata.org/grapher/kaya-identity-co2
24. Christoph Böhringer and Nicolas Rivers (September 2021) 'The energy efficiency rebound effect in general equilibrium' Journal of Environmental Economics and Management, V.109 https://www.sciencedirect.com/science/article/abs/pii/S0095069621000772#:~:text=Prior%20studies%20divide%20the%20rebound%20effect%20into,et%20al.%2C%202007%2C%20Gillingham%20et%20al.%2C%202016)
25. Peter Sainsbury (16 April 2025) The Kaya identity Carbon Risk, Substack https://carbonrisk.substack.com/about?utm_source=substack&utm_medium=email
26. IEA (2025) China https://www.iea.org/countries/china/efficiency-demand
27. IRENA.Org (June 2023) World Energy Transitions Outlook 2023: 1.5°C Pathway https://www.irena.org/Publications/2023/Jun/World-Energy-Transitions-Outlook-2023
28. BloombergNEF (August 2018) World Reaches 1,000GW of Wind and Solar, Keeps Going https://about.bnef.com/blog/world-reaches-1000gw-wind-solar-keeps-going/#:~:text=Investment,from%202018%20to%202022%20inclusive
29. E360.yale.ed (2019) Shifting U.S. to 100 Percent Renewables Would Cost $4.5 Trillion, Analysis Finds https://e360.yale.edu/digest/shifting-u-s-to-100-percent-renewables-would-cost-4-5-trillion-analysis-finds#:~:text=To%20achieve%20100%20percent%20renewable,price%20tag%20to%20$4%20trillion
30. BloombergNEF (February 2025) Global Cost of Renewables to Continue Falling in 2025 as China Extends Manufacturing Lead. https://about.bnef.com/insights/clean-energy/global-cost-of-renewables-to-continue-falling-in-2025-as-china-extends-manufacturing-lead-bloombergnef/
31. Politico (7 March 2025) Trump, Hegseth promise record $1 trillion Pentagon budget https://www.politico.com/news/2025/04/07/hegseth-trump-1-trillion-defense-budget-00007147
32. Dieter Helm (2017) Cost of Energy Review https://assets.publishing.service.gov.uk/media/5a749c26ed915d0e8e39997a/Cost_of_Energy_Review.pdf
33. CPI (November 2023) Global Landscape of Climate Finance 2023 https://www.climatepolicyinitiative.org/publication/global-landscape-of-climate-finance-2023/

34. WEF (June 2024) How to spot a nature crisis – and why we should use natural capital to combat it https://www.weforum.org/agenda/2024/06/3-traits-nature-crisis-natural-capital/
35. GBAC/IFC (2022) COP27: Implications for Financial Institutions https://ecagbac.org/cop27-implications-for-financial-institutions/
36. Media.odi.org (April 2019) Blended finance in the poorest countries: The need for a better approach https://media.odi.org/documents/12666.pdf
37. IIF (2020) TSVCM https://www.iif.com/tsvcm
38. TSVCM (2021) Taskforce on Scaling Voluntary Carbon Markets https://www.iif.com/Portals/1/Files/TSVCM_Report.pdf
39. CDP (2024) About Us https://www.cdp.net/en/info/about-us
40. CDP (2024) CDP and the Net-Zero Data Public Utility expand partnership to power the net-zero transition https://www.cdp.net/en/articles/media/cdp-and-the-net-zero-data-public-utility-expand-partnership-to-power-the-net-zero-transition
41. NZDPU About NZDPU https://nzdpu.com/about
42. NIH.Gov (1995) Supplement 1: The Evolution and Impact of Federal Government Support for R&D in Broad Outline https://www.ncbi.nlm.nih.gov/books/NBK45556/
43. IEA (2025) Energy and AI https://www.iea.org/reports/energy-and-ai
44. Carboncredits.com (17 October 2022) Sovereign Carbon Credits – The New Rival in the Carbon Market https://carboncredits.com/sovereign-carbon-credits-new-rival-carbon-market/#:~:text=The%20credits%20are%20generated%20both%20from%20nature%2Dbased,towards%20a%20country%27s%20Nationally%20Determined%20Contributions%20(NDCs)
45. The Constructor (2009–2021) What is Green Concrete? Its Applications and Advantages in Construction https://theconstructor.org/concrete/green-concrete/5566/
46. Akshat Rathi (2023) Climate Capitalism: Winning the Global Race to Zero Emissions Publisher: John Murray (Great Britain)
47. CDR.fyi (April 2025) 2025 Q1 Durable CDR Market Update https://www.cdr.fyi/blog/2025-q1-durable-cdr-market-update-back-at-basecamp
48. Kleinman Center for Energy Policy (October 2023) The Calculus on Direct Air Capture https://kleinmanenergy.upenn.edu/commentary/blog/the-calculus-on-direct-air-capture/#:~:text=Risks%20of%20DAC&text=While%20land%20intensity%20remains%20lower,disruption%20and%20negative%20community%20impacts
49. Oceanvisions.org (2025) Marine Carbon Dioxide Removal https://oceanvisions.org/ocean-based-carbon-dioxide-removal/
50. Fauna-Flora.org (2020) An assessment of the risks and impacts of seabed mining on marine ecosystems https://www.fauna-flora.org/wp-content/uploads/2023/05/FFI_2020_The-risks-impacts-deep-seabed-mining_Report.pdf
51. Mongabay (21 March 2024) Japan prepares to mine its deep seabed by decade's end https://news.mongabay.com/2024/03/japan-prepares-to-mine-its-deep-seabed-by-decades-end/
52. Mining-technology.com (14 April 2025) US aims to build deep-sea critical metals reserves to reduce Chinese reliance https://www.mining-technology.com/news/us-deep-sea-mining/?cf-view
53. Weforum & FOE (2020) The Ocean Finance Handbook: Increasing finance for a healthy ocean https://www3.weforum.org/docs/WEF_FOA_The_Ocean_Finance_Handbook_April_2020.pdf
54. John Ure (2024) Achieving the UN Sustainable Development Goals: Late of Too Late Chapter 4 https://bookstore.emerald.com/achieving-the-united-nations-sustainable-development-goals-pb-9781835494073.html
55. World Bank (January 2023) Placing Gender Equality at the Center of Climate Action https://documents1.worldbank.org/curated/en/099718102062367591/pdf/IDU08c737dd00f8580412b0aed90fce874ab09b0.pdf
56. UNFCCC (June 2022) Dimensions and examples of the gender-differentiated impacts of climate change, the role of women as agents of change and opportunities for women https://unfccc.int/sites/default/files/resource/sbi2022_07.pdf

57. NASEM (2025) Women's Empowerment, Population Dynamics and Socioeconomic Development https://nap.nationalacademies.org/download/27955
58. P. Berlingheri, et al. (2021) Twenty years of gender equality research: a scoping review based upon a new semantic indicator PLoS One https://journals.plos.org/plosone/article?id=10.1371/journal.pone.0256474#:~:text=Through%20this%20analysis%2C%20we%20build,of%20semantic%20importance%20and%20image
59. IFC (2024) Gender-Responsive Climate Governance and the Role of Women Leaders https://www.ifc.org/en/insights-reports/2024/gender-responsive-climate-governance-role-of-women-leaders
60. UNFCCC (2023) LT-LEDS Synthesis Report https://unfccc.int/lt-leds-synthesis-report
61. Ministry of Foreign Affairs (Japan) Structure of United Nations Climate Change Negotiations https://www.mofa.go.jp/ic/ch/page22e_000921.html
62. OECD (2020) Gender and Environmental Statistics. Exploring available data and developing new evidence https://www.oecd.org/environment/brochure-gender-and-environmental-statistics.pdf
63. Encarnacion et al. (2022) It will take 22 years to close SDG gender data gaps UN Women https://data.unwomen.org/features/it-will-take-22-years-close-sdg-gender-data-gaps
64. GEDA (2023) Addressing the Gender-Environment Data Gap https://genderenvironmentdata.org/
65. WEDO (2025) The Gender Climate Tracker Website https://genderclimatetracker.org/introduction
66. FCDO (2023) ASEAN Low Carbon Energy Programme: Integrating a Gender Lens in Voluntary Carbon Markets https://www.sddirect.org.uk/sites/default/files/2022-11/ASEAN%20Integrating%20gender%20into%20VCMs%20-%20Volume%20I%20-%20FINAL.pdf
67. IISD (January 2024) Mainstreaming Gender Equality and Social Inclusion in Nature-Based Solutions for Climate Change Adaptation https://www.iisd.org/publications/report/gender-equality-social-inclusion-nature-based-solutions
68. Theconversation.com (20 November 2024) Indonesian social forestry often excludes women from decisions, risking greater inequality https://theconversation.com/indonesian-social-forestry-often-excludes-women-from-decisions-risking-greater-inequality-238904?utm_source=linkedin&utm_medium=bylinelinkedinbutton
69. TheBible.com (2025) Joel 2:25–26 https://www.bible.com/bible/compare/JOL.2.25-26
70. Wikipedia (16 April 2025) Antichrist https://en.wikipedia.org/wiki/Antichrist#:~:text=In%20Christian%20eschatology%2C%20Antichrist%20refers,the%20Father%20and%20the%20Son.%22&text=The%20similar%20term%20pseudokhristos%20or,in%20the%20Book%20of%20Revelation
71. ProRepulica (24 April 2925) White House Proposal Could Gut Climate Modeling the World Depends On https://www.propublica.org/article/trump-noaa-budget-cuts-climate-change-modeling-princeton-gfdl
72. PBS News (24 September 2024) Fact-checking what Project 2025 says about the National Weather Service and NOAA https://www.pbs.org/newshour/politics/fact-checking-what-project-2025-says-about-the-national-weather-service-and-noaa
73. Towntopics.com (16 April 2025) Trump Administration Cuts Off Funding For PU Climate Research https://www.towntopics.com/2025/04/
74. Project2025.org (September 2024) Building for conservative victory through policy, personnel, and training https://www.project2025.org
75. Law.berkeley.edu (2025) A Guide to the Major Climate and Environmental Excerpts in the Project 2025 Report https://www.law.berkeley.edu/research/clee/research/other-research-initiatives/environmental-guide-to-project-2025-report/
76. Union of Concerned Scientists (10 October 2024) Fossil Fuel Accountability https://www.ucs.org/climate/accountability
77. JRC (2023) GHG emissions of all world countries - JRC/IEA 2023 Report (Crippa, M., et al) https://edgar.jrc.ec.europa.eu/report_2023#:~:text=Even%20in%202022%2C%20global%20GHG,F%2Dgases%20(2.6%25)

78. Climate Watch (2025) Historical GHG Emissions https://www.climatewatchdata.org/ghg-emissions?apcid=0065a83d31d8c2c5be4fb000&end_year=2022&start_year=1990&utm_campaign=climatewatch%20&utm_medium=email&utm_source=newsletter
79. Our World in Data (2024) Greenhouse Gas Emissions by Sector, World https://ourworldindata.org/grapher/ghg-emissions-by-sector?time=earliest..2020
80. e360.yale.edu (December 2017) Transportation Replaces Power in U.S. as Top Source of CO2 Emissions https://e360.yale.edu/digest/transportation-replaces-power-in-u-s-as-top-source-of-co2-emissions#:~:text=But%20according%20to%20new%20data,sector%20emits%201.8%20billions%20tons
81. Normative.io (July 2024) SECR: the UK carbon reporting legislation, explained https://normative.io/insight/secr-explained/
82. CDP (2025) Brand Hub https://www.cdp.net/en/brand-hub
83. EPA.Gov (2025) Scope 3 Inventory Guidance https://www.epa.gov/climateleadership/scope-3-inventory-guidance#:~:text=Description%20of%20Scope%203%20Emissions%20Scope%203,of%20both%20its%20upstream%20and%20downstream%20activities
84. Harvard University (2025) Harvard's Value Chain https://sustainable.harvard.edu/scope-3-emissions/
85. ESGtoday.com (April 2024) More than 40% of Public Companies Now Reporting on Scope 3 Emissions, but U.S. Lagging Far Behind: MSCI https://www.esgtoday.com/more-than-40-of-public-companies-now-reporting-on-scope-3-emissions-but-u-s-lagging-far-behind-msci/
86. Netzerofinancialtracker (2024) How are private finance institutions progressing on their Paris Agreement and Net Zero goals? https://netzerofinancetracker.climatepolicyinitiative.org/
87. World Bank (2023) The Effect of Multinational Enterprises on Climate Change Supply Chain Emissions, Green Technology Transfers, and Corporate Commitments https://openknowledge.worldbank.org/server/api/core/bitstreams/def956de-eac5-4ede-a366-e01331481c99/content
88. GoldmanSachs.Com (May 2024) AI is poised to drive 160% increase in data center power demand https://www.goldmansachs.com/insights/articles/AI-poised-to-drive-160-increase-in-power-demand
89. IEA (2024) Energy and AI: World Energy Outlook Special Report https://iea.blob.core.windows.net/assets/34eac603-ecf1-464f-b813-2ecceb8f81c2/EnergyandAI.pdf
90. ICEF (2024) Roa dmap https://www.icef.go.jp/roadmap/
91. UN.Org (2023) Carbon Action https://www.un.org/en/climatechange/raising-ambition/renewable-energy
92. CPI (2021) Global Landscape of Climate Finance 2021 https://www.climatepolicyinitiative.org/wp-content/uploads/2021/10/Full-report-Global-Landscape-of-Climate-Finance-2021.pdf
93. CSIL (2024) Survival of the Fittest: From ESG to Competitive Sustainability https://www.cisl.cam.ac.uk/files/from_esg_to_competitive_sustainability.pdf
94. Findevgateway.org (April 2022) What Do Microfinance Clients Need to Adapt to Climate Change? https://www.findevgateway.org/blog/2022/04/what-do-microfinance-clients-need-adapt-climate-change#:~:text=Microfinance%20can%20be%20a%20great,services%20focused%20on%20climate%20adaptation
95. National Ocean Service What is lidar? https://oceanservice.noaa.gov/facts/lidar.html
96. Geographyrealm (2018) Creating Ground-level Views from Satellite Imagery https://www.geographyrealm.com/creating-ground-level-views-satellite-imagery/
97. OliverWyman (2024) The Impact Of Climate Change On Human Health https://www.oliverwyman.com/our-expertise/insights/2024/jan/quantifying-climate-change-impact-on-human-health.html
98. The Nature Conservancy (2025) How much carbon was in the atmosphere when you were born? https://www.nature.org/en-us/get-involved/how-to-help/carbon-footprint-calculator/carbon-by-birth-year/
99. NASA (2024) Evidence https://science.nasa.gov/climate-change/evidence/#footnote_1

100. Energy Education (2018) Geologic time scale https://energyeducation.ca/encyclopedia/Geologic_time_scale#:~:text=Epochs%20tend%20to%20be%20a,(e.g.%20the%20Upper%20Triassic)
101. Natural History Museum What is the Anthropocene and why does it matter? https://www.nhm.ac.uk/discover/what-is-the-anthropocene.html#:~:text=For%20the%20last%2011%2C500%20years,had%20previously%20covered%20Earth%20disappeared
102. Andrew Seidl et al (2021) Nature, Ecology and Evolution Journal v.5, pp.530–539 (2021) 'The effectiveness of national biodiversity investments to protect the wealth of nature' https://www.nature.com/articles/s41559-020-01372-1#:~:text=The%20Biodiversity%20Finance%20Initiative%20provides%20a%20means%20for%20countries%20to

Index

A
A Just Transition, 179
Albedo, 6
Article 6 of the Paris Agreement, 98, 122
Atlantic Meridional Overturning Circulation (AMOC), 25
Atmosphere, 1–7, 9, 16, 18, 20

B
Battery power, 244
Bioenergy with Carbon Capture and Storage (BECCS), 145, 146

C
Carbon avoidance, 27, 91, 92, 96, 97, 108, 113, 137, 138, 143, 145, 146, 148, 152, 157, 160, 163, 165, 190, 195, 209, 210, 214, 220, 245, 249, 251, 252, 260, 261
Carbon Capture, Utilisation And Storage (CCUS), 27, 37, 38, 41, 43, 50
Carbon Dioxide (CO_2), 2–8, 12
Carbon Dioxide Removal (CDR), 27, 50
Carbon reduction, 27
China, 90, 96, 98, 103, 104, 114–123, 243, 244, 251, 253, 256, 260
Classical economics, 59, 61, 63
Climate and Health, 250, 253, 254, 259, 261
Climate and the Environment, 241, 244, 246, 254, 259, 261
Climate-Biodiversity-Environment, 207
Climate Finance Landscape, 180
Climate Mitigation and Adaptation, 182
Climate Policy Initiative (CPI), 245, 258
Climatology, 7, 10, 11, 14
Compliance or Mandatory Carbon Market (MCM), 91, 95–97, 101, 111, 122
Contract of Difference (CfD), 107
Corporate green finance, 188
Coupled Model Intercomparison Project (CMIP), 8, 19

D
Damage function, 64, 67–72, 77, 78, 81, 83, 84
Direct Air Carbon Capture and Storage (DACCS), 145
Discount rate, 57, 58, 68, 69, 72, 73, 83

E
Ecology economists, 73, 75, 76
Economics and nature, 62
Emerging Market and Developing Economies (EMDEs), 183–185, 189, 193, 198, 200, 218, 219, 222
Emission allowances vs. carbon credits/offsets, 135, 136
Emission Trading Schemes (ETS), 91, 96, 99, 104, 105, 110, 111, 114, 118, 119, 121–123
Endogenous models, 66
Energy budget, 239, 258
Energy efficiency and energy intensity, 43
Enhanced rock weathering, 145, 146
Environmental, Social, and Corporate Governance (ESG), 96, 97
Equilibrium Climate Sensitivity (ECS) models, 7, 15, 19

J. Ure, *Climate Change and Carbon Markets: Late or Too Late?*,
https://doi.org/10.1007/978-3-032-07619-9

European Union (EU), 89, 110, 112
Exogenous models, 64

F
Financial Taskforces, 195

G
Gender, 254, 255
Glasgow Financial Alliance for Net Zero (GFANZ), 198, 200, 201, 205, 206, 219
Global Climate Models (GCM), 7, 15, 16
Global Energy and Climate (GEC) models, 33
Global South, 179–181, 183, 190, 192, 193, 201, 219
Government policy options, 91
Green Bank, 193, 206
Green Finance, 177, 179–182, 186, 188, 189, 192, 194, 220, 222
Green Gap in Finances, 245
Green House Gases (GHG), 2
Green House Gas Protocol (GHG Protocol), 150, 155, 164

I
India, 90, 98, 103, 120, 122, 123, 251, 256, 260
Infrastructure, 244, 245, 251, 260
Institute and Faculty of Actuaries (IFoA), 15
Integrated Assessment Model (IAM), 64–66, 68–70, 72
Integrity Council for the Voluntary Carbon Market (ICVCM), 150, 153, 160, 161
Intergovernmental Panel on Climate Change (IPCC) – Pathways, 25, 49
Intergovernmental Panel on Climate Change (IPCC) - scenarios, 6
International Energy Authority (IEA), 26, 30–43, 45, 79, 80, 116, 117, 120
International Green Bank, 260
Internationally Transferred Mitigation Outcomes (ITMOs), 99, 102, 103, 122
International Renewable Energy Agency (IRENA) - Scenarios, 35, 36, 40–45
Isotopes, 1, 2, 16

K
Kaya Identity, 243
Kunming-Montreal Biodiversity Framework (GBF), 147

L
Land Use, Land-Use Change and Forestry (LULUCF), 145, 147

M
Martin Weitzman- economist, 68, 83
Measure, Monitor, Report and Verify (MMRV), 135, 139, 141, 144, 149, 151, 156, 162, 163, 165
Monitoring, Reporting and Assurance (MRA), 99
Multilateral Development Banks, 181, 185, 189

N
National Aeronautics and Space Administration (NASA), 5, 8, 11
Nationally Determined Contributions (NDCs), 90, 94, 97, 103, 122
National Oceanic and Atmospheric Administration (NOAA), 55
Natural capital, 147, 148
Nature-based Projects (NbP), 137, 140, 147, 148
Nature-based Solutions (NbS), 147
Neoclassical economics, 61, 63, 65, 68, 76, 83
Network of Central Banks and Supervisors for Greening the Financial System (NGFS), 177, 198–200
Net zero, 25–27, 31, 47, 49, 50
New Collective Quantified Goal (NCQG), 181, 184, 185
Nicolas Stern - economist, 58, 73
Nuclear power, 190, 191

P
Paris Agreement, 19, 49
Particle physics, 3, 7, 9
Primary and final energy consumption, 28

R
Radiation forcing, 6
Radiative Equilibrium, 6

Reducing Emissions from Deforestation and Forest Degradation plus sustainable management of forests and the conservation and enhancement of forest carbon stocks (REDD+ projects), 137, 140–144, 149, 156, 164
Registries and standards for carbon credits, 137
Renewable energy, 27, 30, 32–35, 41, 42, 44, 45
Risk and Uncertainty, 11

S
Scenarios, 7, 16, 17
Scepticism, 11, 15
Science, 1, 2, 7, 8, 10–14, 16
Scopes 1, 2 and 3, 28, 50
Singapore, 90, 91, 102, 121, 122
Steve Keen - economist, 68, 74, 76, 77, 84
Systems-Based Analysis (SBA), 73, 75

T
Technologies and climate, 240, 243, 249–252, 255, 258, 260
The Nature Conservancy (TNC), 241
The Taskforce on Scaling Voluntary Carbon Markets (TSVCM), 246, 248
Tipping points and disequilibria, 55, 58, 60, 62, 64, 78, 83
Transient Climate Response (TCR) models, 7

U
UN Conference of Parties (COP), 206
United Kingdom (UK), 90, 92, 97, 103, 104, 106–111, 113, 114, 122, 123
United Nations Environment Programme (UNEP), 179, 184, 193, 207, 211
United Nations Framework Convention on Climate Change (UNFCCC), 184, 185, 189, 207
United States of America (USA), 89, 90, 92, 94, 104, 105, 112, 118, 120, 122, 123

V
Voluntary Carbon Markets (VCM), 91, 96, 97, 101, 102, 107, 111, 114, 135, 136, 138–141, 143, 144, 150, 152, 153, 159–165

W
William Nordhaus - economist, 58, 64, 66, 68, 74, 76, 83
World Bank, 184, 189, 193, 194, 207, 214, 218
World Economic Forum (WEF), 201, 203, 204, 206, 210–213, 218
World Meteorological Organization (WMO), 7
World Weather Attribution (WWA), 17, 18

MIX
Papier aus verantwortungsvollen Quellen
Paper from responsible sources
FSC® C105338

If you have any concerns about our products,
you can contact us on
ProductSafety@springernature.com

In case Publisher is established outside the EU,
the EU authorized representative is:
Springer Nature Customer Service Center GmbH
Europaplatz 3, 69115 Heidelberg, Germany

Printed by Libri Plureos GmbH
in Hamburg, Germany